ANATOMY OF SEED PLANTS

NEW YORK · LONDON   JOHN WILEY & SONS, INC.

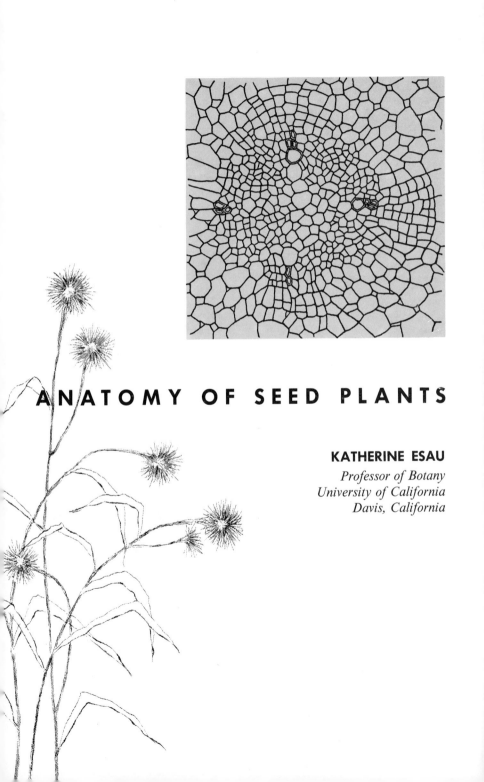

ANATOMY OF SEED PLANTS

KATHERINE ESAU

Professor of Botany
University of California
Davis, California

PREFACE

THE WRITING OF A SECOND BOOK IN PLANT ANATOMY BY THE SAME author needs, perhaps, a brief comment. The idea of the present book began to develop as soon as my first book, *Plant anatomy,* John Wiley and Sons, 1953, appeared. Even during the writing of that book the publishers expressed a preference for a relatively short text designed for a one-semester course. My idea, however, was to write a comprehensive treatise in which I could develop the concepts fully and include much detail on ontogeny of plant structure. Very considerately, my publishers raised no objections and cooperated to the fullest extent. Appreciating their attitude, I offered to write a short text later. After the first book underwent classroom tests and when reviews by teachers and investigators became available, the idea to write a new, shorter text became increasingly more appealing to me. Furthermore, the intensive research in plant anatomy of the last decade made me wish to bring the subject matter up to date.

The book may best be characterized by comparing it with my first book. The new book and the old are essentially alike in their basic approach: both follow the "classic" method of discussing the cells and tissues first, then the plant parts composed of these units. The new book, however, contains fewer developmental details, the concepts are defined with less of the historical background, and the terms are not projected back to their Greek and Latin origins. On the other hand, the present book contains a glossary, which was prepared with especial regard to the latest developments of concepts and terminologies. The bibliographies at the ends of chapters are short; moreover, they consist mostly of references published since the appearance of *Plant anatomy*. This limitation is pointed out in the first chapter, with the thought that the larger book may be consulted for the older and the classic works in plant anatomy. This treatment seemed to be the best solution of the problem of dealing with the vast increase of botanical literature during the last decade.

In the organization of the present book an effort was made to achieve brevity not by deleting topics but by combining closely related subject matter. Thus, the protoplast and the cell wall are reviewed not in separate chapters as in the old book but jointly with parenchyma, collenchyma, and sclerenchyma. The vascular cambium is described after the xylem tissue—a sequence that facilitates the explanation of the arrangement of cells in the cambium. Finally, the apical meristems are taken up in the chapters on root and stem instead of separately. Thus the activity of these meristems may be studied in close relation to the plant parts they produce. This approach is especially suitable for introducing the modern researches on apical meristems, which emphasize the causal aspects of the specific organization of plant organs. Two other features in the arrangement of material should be mentioned. First, the chapters dealing with embryo development are placed before the discussion of the individual tissues and organs. These chapters are used for introducing the concepts of growth, differentiation, and organization of the plant. Second, the root is treated before the stem because it is least difficult to introduce the concepts pertaining to the division into vascular and nonvascular regions in the plant by referring to the relatively simple root structure.

Since this book is intended primarily for students who have had a relatively limited experience in the study of plants, a few words about the significance of plant anatomy among the plant sciences should be helpful. Whether we deal with plants as horticulturists, agronomists, foresters, plant pathologists, or ecologists, we must know what the plant is like and how it functions. We gain this knowledge by studying the structure of the plant, its development, and its various activities. It would be ideal to pursue the study of the plant from all these aspects jointly, but we find it more efficient to concentrate on one aspect at a time. Therefore we divide the study of organisms into separate areas: first the two broad ones, morphology and physiology; then each of these two into more circumscribed areas, such as cytology, anatomy, taxonomy, ecology, and others. Obviously the various areas are not sharply separated; moreover, the study of one area invariably raises questions that are properly answerable only by reference to some other area.

In this interrelationship of sciences, anatomy plays a major role. A realistic interpretation of plant function by the physiologist must rest on a thorough knowledge of the structure of cells and tissues associated with that function. Notable examples of functions, the understanding of which has been materially enhanced by studies of structures of the parts concerned, are photosynthesis, movement of water, translocation of food, and absorption by roots. Knowledge of plant anatomy is also indispensable for the advancement of research in plant pathology. The effect of a parasite cannot be fully understood unless the normal structure of the invaded plant part is known. Furthermore, the warding off of the effects of the parasite, or even the resistance of the parasite itself, may be revealed by structural changes or structural peculiarities of the host. The explanation of the successes or failures of many horticultural practices, such as grafting, pruning, vegetative propagation, and the associated phenomena of callus formation, wound healing, regeneration, and development of adventitious roots and buds, are more meaningful if the structural features underlying these phenomena are properly understood. Fruitful results are obtained by the ecologist when he relates the behavior of plants in different environments with structural peculiarities of plants growing in those environments. It seems significant that one of the most active fields in modern plant research—the study of development of form and organization—is often carried out on the basis of constant correlation between biochemical and structural changes in the experimental plant. By this correlation a much more complete picture of development is obtained than would result from considering the biochemical changes separately from those in the number, sizes, and structure of cells. Finally, plant anatomy is interesting for its own sake. It is a gratifying experience to follow the ontogenetic and evolutionary development of structural features and gain the realization of the high degree of complexity and the remarkable orderliness in the organization of the plant.

ACKNOWLEDGMENTS

Various persons mentioned in the legends for the illustrations kindly provided material of one kind or another and made possible to include so many new illustrations in the book. Professor John E. Sass was particularly generous in lending numerous negatives and slides. Mr. Paul L. Conant loaned the slides used for the key for identification of woods. Mrs. Margery P. Mann ably assisted with the finishing of the photomicrographs.

I also wish to acknowledge the helpful reviews of the manuscript by Professor Vernon I. Cheadle, Professor Charles LaMotte, and Mr. Charles H. Lamoureux. Dr. James J. Dunning has lightened the task of preparing the glossary by making the initial collection of words and definitions. I am very grateful to Mrs. Mary M. Brinton who typed the manuscript with devotion and exceptional accuracy.

KATHERINE ESAU

December 1959

CONTENTS

1. INTRODUCTION

THIS BOOK DEALS WITH THE INTERNAL STRUCTURE OF EXTANT SEED plants. Angiosperms are emphasized, but some of the features of the vegetative parts of gymnosperms also are reviewed. For the angiosperms the anatomy of the flower, fruit, and seed is described in the concluding chapters.

The seed plant has a highly evolved body that bears evidences of structural and functional specialization expressed in the differentiation of this body, externally, into organs and, internally, into various categories of cells, tissues, and tissue systems. Three vegetative organs, the root, the stem, and the leaf, are commonly recognized; and the flower is interpreted as an assemblage of organs, some of which are reproductive (stamens and carpels), others sterile (sepals and petals). With regard to the internal structure, one stresses the distinctive features of cells and tissues and establishes types on the basis of these distinctions.

This particularizing of the plant body and the associated categorizing of its parts are logical and convenient approaches to the study of the plant because they bring into focus the structural and functional specialization of its parts, but they must not be emphasized to the degree that they might obscure the essential unity of the plant. This unity is clearly perceived if the plant is studied developmentally, a method that reveals the gradual emergence of the organs and tissues from a relatively undifferentiated body of the young embryo. A similar change, from less differentiated to more differentiated, from less particularized to more particularized, has occurred in the evolution of the seed plants, so that one commonly conceives of the root, the stem, the leaf, and the floral organs as parts phylogenetically interrelated and of the distinctive cells and tissues as derivatives of unspecialized cells of the type now called parenchyma cells. Even a static view of the parts of an adult plant reveals their unity and interdependence; the same tissue systems are common to all of them.

Thus, the separation of the plant into organs can be made only approximately. It is impossible, for example, to draw a clear demarkation between shoot and root and between stem and leaf, and the flower in many ways resembles the vegetative shoot. The internal structures, similarly, are not sharply delimited, and the categories of cells and tissues show much intergrading.

INTERNAL ORGANIZATION OF THE PLANT BODY

The plant body consists of morphologically recognizable units, the *cells,* each enclosed in its own cell wall and united with other cells by means of a cementing intercellular substance. Within this united mass certain groupings of cells are distinct from others structurally or functionally or both. These groupings are referred to as *tissues.* The structural variations of tissues are based on differences in the component cells and their type of attachment to each other. Some tissues are structurally relatively simple in that they consist of one cell type; others, containing more than one cell type, are more or less complex.

The arrangement of tissues in the plant as a whole and in its major organs reveals a definite structural and functional organization. Tissues concerned with conduction of food and water—the vascular tissues—form a coherent system extending continuously through each organ and the entire plant. These tissues connect places of water intake and food synthesis with regions of growth, development, and storage. The nonvascular tissues are similarly continuous, and their arrangements are indicative of specific interrelations (such as between storage and vascular tissues) and of specialized functions (such as support or storage.) To emphasize the organization of tissues into large entities, revealing the basic unity of the plant body, the expression *tissue system* has been adopted.

As was implied at the beginning of this chapter, the classification of cells and tissues is a somewhat arbitrary matter because of the frequent occurrence of intermediate forms between categories of cells and tissues. Nevertheless, for purposes of an orderly description of plant structure the establishment of categories is necessary. Moreover, if the classifications develop from broad comparative studies, in which the variability and the intergrading of characters are clearly revealed and properly interpreted, they not only will be descriptively useful but will also reflect the natural relation of the entities classified.

In agreement with Sachs (1875), in this book the principal tissues of a vascular plant are grouped on the basis of topographic continuity into three tissue systems, the *dermal,* the *vascular,* and the *fundamental* (or *ground*) system. The dermal system includes the *epidermis,* that is, the primary outer protective covering of the plant body, and the *periderm,* the protective tissue that supplants the epidermis, mainly in plant parts that undergo secondary increase in thickness. The vascular system contains two kinds of conducting tissues, the *phloem* (food conduction) and the *xylem* (water conduction).

The fundamental system includes tissues that, in a sense, form the ground substance of the plant but at the same time show various degrees of specialization. The main ground tissues are *parenchyma* in all its varieties, *collenchyma,* the thick-walled supporting tissue related to parenchyma, and *sclerenchyma,* the main supporting tissue with thick, hard, often lignified walls.

Within the plant body the various tissues are distributed in characteristic patterns depending on plant part or plant group or both. Basically the patterns are alike in that the vascular tissue is imbedded in the ground tissue and the dermal tissue forms the outer covering. The principal variations in patterns depend on the relative distribution of the vascular and the ground tissues. In a dicotyledon, for example, the vascular tissue of the stem forms a hollow cylinder, with some ground tissue enclosed by the cylinder (*pith*) and some located between the vascular and the dermal tissues (*cortex*). In the leaf, the vascular tissue forms an anastomosing system imbedded in the ground tissue, here differentiated as mesophyll. In the root, the vascular cylinder may enclose no pith, but a cortex is present.

The cells and tissues of the plant are derived from the zygote (sometimes the egg itself) through the intermediate stages represented by the embryo. The embryonic stage, however, is not completely abandoned after the embryo develops into the adult plant. Plants have the unique property of open growth resulting from the presence of embryonic tissue zones, the *meristems,* in which the addition of new cells continues while other plant parts reach maturity. Meristems at apices of roots and shoots, the *apical meristems,* produce cells whose derivatives differentiate into new parts of root and shoot. This growth is called *primary* and the resultant plant body, the primary body. In many plants the stems and roots are increased in thickness by addition of vascular tissues to the primary body. This thickening growth is produced by the *vascular cambium* and is called *secondary* growth. Commonly secondary growth also involves the formation of periderm by the meristem *phellogen.* The vascular cambium and the phellogen are referred to as *lateral meristems* because of their position parallel with the sides of stem and root.

SUMMARY OF TYPES OF CELLS AND TISSUES

The literature on plant anatomy published since the issue of the first book on the subject by the present author (Esau, 1953) does not require a substantial revision of the summary of types of cells and tissues given in that book. The same summary, slightly revised, is given below.

Epidermis. Epidermal cells form a continuous layer on the surface of the plant body in the primary state. They show various special characteristics related to their superficial position. The main mass of epidermal cells, the epidermal cells proper, vary in shape but are often tabular. Other epidermal cells are guard cells of the stomata and various trichomes, including root hairs. The epidermis may contain secretory and sclerenchyma cells. The principal distinctive features of the epidermal cells of the aerial parts of the plant are the cuticle on the outer wall and the cutinization of the outer and of some or all of the other walls. The epidermis gives mechanical protection and is concerned with restriction of transpiration and with aeration. In the stems and roots having secondary growth, the epidermis is commonly replaced by the periderm.

Periderm. The periderm comprises cork tissue or *phellem,* cork cambium or *phellogen,* and *phelloderm.* The phellogen occurs near the surface of axial organs having secondary growth and is itself secondary in origin. It arises in the epidermis, the cortex, the phloem, or the pericycle and produces phellem toward the outside, phelloderm toward the inside. Phelloderm may be small in amount or absent. The cork cells are commonly tabular in form, are compactly arranged, lack protoplasts at maturity, and have suberized walls. The phelloderm usually consists of parenchyma cells.

Parenchyma. Parenchyma cells form continuous tissues in the cortex of stem and root and in the leaf mesophyll. They also occur as vertical strands and rays in the vascular tissues. They are primary in origin in the cortex, the pith, and the leaf and primary or secondary in the vascular tissues. Parenchyma cells are characteristically living cells, capable of growth and division. The cells vary in shape, are often polyhedral, but may be stellate or much elongated. Their walls are often primary, but secondary walls are not uncommon. Parenchyma is concerned with photosynthesis, storage of various materials, wound healing, and origin of adventitious structures. Parenchyma cells may be specialized as secretory or excretory structures.

Collenchyma. Collenchyma cells occur in strands or continuous cylinders near the surface of the cortex in stems and petioles and along the veins of foliage leaves. It is uncommon in roots. Collenchyma is a living tissue closely related to parenchyma; in fact, it is usually regarded as a form of parenchyma specialized as supporting tissue in young organs. The shape of cells varies from short prismatic to much elongated. The most distinctive feature is the uneven thickening of primary walls.

Sclerenchyma. Sclerenchyma cells may form continuous masses, or they may occur in small groups or individually among other cells. They may develop in any or all parts of the plant body, primary or secondary. They are strengthening elements of mature plant parts. Sclerenchyma cells have thick, secondary, often lignified walls and may lack protoplasts at maturity. Two forms of cells are distinguished, sclereids and fibers. The sclereids vary in shape from polyhedral to elongated and may be much branched. Fibers are generally long slender cells.

Xylem. Xylem cells form a structurally and functionally complex tissue which, in association with the phloem, is continuous throughout the plant body. It is concerned with water conduction, storage, and support. The xylem may be primary or secondary in origin. The principal water-conducting cells are the tracheids and the vessel members. The vessel members are joined end to end into vessels. Storage occurs in the parenchyma cells, which are arranged in vertical files and, in the secondary xylem, also as rays. Mechanical cells are fibers and sclereids.

Phloem. Phloem cells form a complex tissue. The phloem tissue occurs throughout the plant body, together with the xylem, and may be primary or secondary in origin. It is concerned with conduction and storage of food and with support. The principal conducting cells are the sieve cells and sieve-tube members, both enucleate at maturity. Sieve-tube members are joined end to end into sieve tubes and are associated with companion cells, which are special parenchyma cells. Other phloem parenchyma cells occur in vertical files. Secondary phloem also contains parenchyma in the form of rays. Supporting cells are fibers and sclereids.

Secretory structures. Secretory cells—cells producing a variety of secretions—do not form clearly delimited tissues but occur within other tissues, primary and secondary, as single cells or as groups or series of cells, and also in more or less definitely organized formations on the surface of the plant. The principal secretory structures on plant surfaces are glandular epidermal cells and hairs and various glands, such as, floral and extrafloral nectaries, certain hydathodes, and digestive glands. The glands are usually differentiated into secretory cells on their surfaces and nonsecretory cells supporting the secretory. Internal secretory structures are secretory cells, intercellular cavities or canals lined with secretory cells (resin ducts, oil ducts), and secretory cavities resulting from disintegration of secretory cells (oil cavities). Laticifers may be placed among the internal secretory structures. They are either single cells (nonarticulated laticifers), usually much branched, or series of cells united through partial dissolution of walls (articulated laticifers). Laticifers contain a fluid called latex, which may be rich in rubber. They are commonly multinucleate.

GENERAL REFERENCES

The bibliographic citations appearing at the ends of chapters 2 to 22 were selected largely from the most recent literature, but the extended lists of references given in Esau's *Plant Anatomy* (1953) were also used for the presentation and the interpretation of the subject matter. This introductory chapter gives a selected list of books in plant anatomy and some in plant morphology. Most of the books pertain to seed plants, but some works dealing with the structure of lower vascular plants are also included.

Aleksandrov, V. G. *Anatomiĩa rasteniĩ.* [Anatomy of plants.] Moskva, Sovetskaĩa Nauka. 1954.

Bailey, I. W. *Contributions to plant anatomy.* Waltham, Massachusetts, Chronica Botanica Company. 1954.

Biebl, R., and H. Germ. *Praktikum der Pflanzenanatomie.* Wien, Springer. 1950.

Boureau, E. *Anatomie végétale.* 3 vols. Paris, Presses Universitaires de France. 1954, 1956, 1957.

Chamberlain, C. J. *Gymnosperms, structure and evolution.* Chicago, University of Chicago Press. 1935.

De Bary, A. *Comparative anatomy of the vegetative organs of the phanerogams and ferns.* (English translation by F. O. Bower and D. H. Scott.) Oxford, Clarendon Press. 1884.

Deysson, G. *Éléments d'anatomie des plantes vasculaires.* Paris, Sedes. 1954.

Eames, A. J. *Morphology of vascular plants. Lower groups.* New York, McGraw-Hill Book Company. 1936.

Eames, A. J., and L. H. MacDaniels. *An introduction to plant anatomy.* 2nd ed. New York, McGraw-Hill Book Company. 1947.

Esau, K. *Plant anatomy.* New York, John Wiley and Sons. 1953.

Foster, A. S. *Practical plant anatomy.* 2nd ed. Princeton, D. Van Nostrand Company. 1949.

Foster, A. S., and E. M. Gifford, Jr. *Comparative morphology of vascular plants.* San Francisco, W. H. Freeman and Company. 1959.

Haberlandt, G. *Physiological plant anatomy.* London, Macmillan and Company. 1914.

Hasman, M. *Bitki anatomisi.* [Plant anatomy.] Istanbul, Matbaasi. 1955.

Hayward, H. E. *The structure of economic plants.* New York, The Macmillan Company. 1938.

Hector, J. M. *Introduction to the botany of field crops.* 2 vols. Johannesburg, South Africa, Central News Agency Ltd. 1938.

Jackson, B. D. *A glossary of botanic terms.* 4th ed. New York, Hafter Publishing Co. 1953.

Jane, F. W. *The structure of wood.* New York, The Macmillan Company. 1956.

Jeffrey, E. C. *The anatomy of woody plants.* Chicago, University of Chicago Press. 1917.

Küster, E. *Pathologische Pflanzenanatomie.* 3rd ed. Jena, Gustav Fischer. 1925.

Linsbauer, K. *Handbuch der Pflanzenanatomie.* Vol. 1 and Following. Berlin, Gebrüder Borntraeger. 1922–1943.

Mansfield, W. *Histology of medicinal plants.* New York, John Wiley and Sons. 1916.

Metcalfe, C. R., and L. Chalk. *Anatomy of the dicotyledons.* 2 vols. Oxford, Clarendon Press. 1950.

Molish, H. *Anatomie der Pflanze.* 6th ed. Revised by K. Höfler. Jena, Gustav Fischer. 1954.

Popham, R. A. *Developmental plant anatomy.* Columbus, Ohio, Long's College Book Company. 1952.

Rauh, W. *Morphologie der Nutzpflanzen.* Heidelberg, Quelle und Meyer. 1950.

Record, S. J. *Identification of the timbers of temperate North America.* New York, John Wiley and Sons. 1934.

Reuter, L. Protoplasmatische Pflanzenanatomie. In: *Protoplasmatologia.* Vol. XI. pp. 1–113. Wien, Springer. 1955.

Sachs, J. *Textbook of botany.* Oxford, Clarendon Press. 1875.

Smith, G. M. *Cryptogamic botany.* Vol. 2: *Bryophytes and Pteridophytes.* New York, McGraw-Hill Book Company. 1938.

Solereder, H. *Systematic anatomy of the dicotyledons.* Oxford, Clarendon Press. 1908.

Solereder, H., and F. J. Meyer. *Systematische Anatomie der Monokotyledonen.* Berlin, Gebrüder Borntraeger. No. 1, 1933; No. 3, 1928; No. 4, 1929; No. 6, 1930.

Stover, E. L. *An introduction to the anatomy of seed plants.* Boston, D. C. Heath and Company. 1951.

Troll, W. *Praktische Einführung in die Pflanzenmorphologie.* Part 1: *Der vegetative Aufbau.* Part 2: *Die blühende Pflanze.* Jena, Gustav Fisher. 1954 and 1957.

2. THE EMBRYO

A STUDY OF EMBRYOGENY, OR FORMATION OF EMBRYO, REVEALS THE origin of vegetative parts of the plant and the inception of tissue organization. Embryogeny, therefore, can serve as a useful introduction to a study of the structure of adult plants. The formation of the embryo, however, is by no means a simple topic to present. Embryonic development varies considerably in different groups of plants, and many disagreements still exist regarding the significance of the early partitioning in the young embryo and the interpretation of certain parts of the mature embryo. Furthermore, whereas a voluminous literature deals with the early stages of embryonic development, only scanty information is available on the late stages, during which the tissue systems and the apical meristems attain their final embryonic organization.

Originally the term embryo was used in botany to designate the young sporophyte within a seed, but later it became more elastic and came to connote any plant in its initial stages of development (cf. Wardlaw, 1955). In this book the discussion is limited to the angiosperm embryo.

DICOTYLEDON EMBRYO

Parts of embryo

The basic parts of a mature dicotyledon embryo are the embryo axis and the two first foliar structures, the cotyledons. Since the axis occurs below (*hypo*) the cotyledons, it is referred to, at least in part, as the hypocotyl (fig. 2.1, *E*). The qualification "in part" is necessary because at its lower end the embryo axis bears the incipient root of the new plant. This entity is often only an apical meristem of the root covered by the rootcap, but sometimes the lower end of the axis develops certain root characteristics before the seed germinates. Such an embryonic root is referred to as the radicle. Since it is not always obvious whether a radicle or only an apical meristem of the root is present, the embryo axis may be called hypocotyl-root axis.

In the mature embryo, some meristematic tissue remains reserved at the top of the axis between the two cotyledons. This tissue is the apical meristem of the future shoot, or the shoot apex (fig. 2.1, *E*). Sometimes a small shoot with one or more leaf primordia develops from this meristem before the embryo matures. The resulting bud is called the epicotyl (*epi*, above), a term sometimes used interchangeably with plumule.

Origin and development of parts

The embryo develops within the ovule, usually from the fertilized egg, or zygote. Although the early growth of the embryo appears to follow a simple plan, in most dicotyledons the first division of the zygote already reveals a determined course of development: of the first two cells formed, the one that is nearer the micropyle (the proximal cell) becomes the lower part of the embryo, and the other (the distal cell) the upper part of the embryo. In other words, the embryo shows polarity; it has a root pole and a shoot pole. In fact, the cytologic appearance of the egg—the presence of a large vacuole in its proximal end and dense cytoplasm and nucleus in its distal end— suggests that polarity may be established before fertilization.

Typically the first division is transverse or more or less inclined with regard to the long axis of the zygote. Further transverse divisions may follow, or vertical divisions also may appear in some parts of the embryo. Whatever the combination of divisions, their sequence is orderly. Generally the embryo at first assumes a cylindrical (filamentous) or club-like form (fig. 2.1 *A*). Soon thereafter the distal part of the embryo becomes the locus of most active cell division. As a result, the distal part enlarges into a nearly spherical structure (fig. 2.1, *B*). With this change the initial distinction between the embryo body and its stalk, or suspensor, is introduced. Before this stage is reached the embryo is often called the proembryo.

In the stages that follow, the embryo undergoes a change in symmetry. The spherical body with radial symmetry develops into a

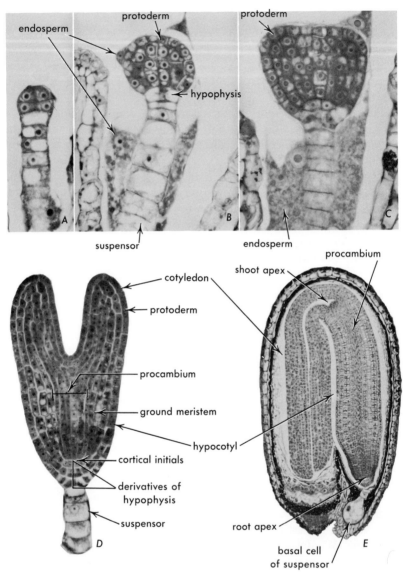

Fig. 2.1. *Capsella* embryos in different stages of development. In *A*, arrangement of cells in the uppermost two tiers indicates beginning of development of embryo body as distinct from suspensor. In *B*, embryo body is globular and has a protoderm; the upper suspensor cell is a hypophysis—a cell participating in embryo formation. In *C*, bilateral symmetry has been established just before emergence of cotyledons. *D*, embryo with all basic tissue regions blocked out. *E*, almost mature embryo. (*A–D*, ×435; *E*, ×64. *A, D*, from A. S. Foster and E. M. Gifford, Jr., *Comparative Morphology of Vascular Plants*. San Francisco, W. H. Freeman and Company, 1959. *B, C, E*, courtesy of E. M. Gifford, Jr.)

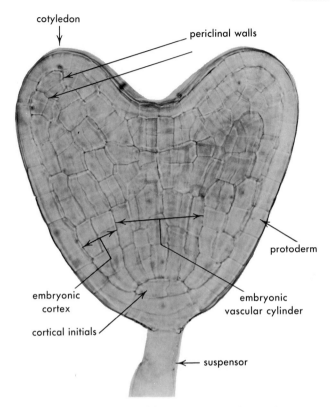

cotyledon

periclinal walls

protoderm

embryonic
cortex

cortical initials

embryonic
vascular cylinder

suspensor

Fig. 2.2. Embryo of *Diplotaxis tenuifolia* at emergence of cotyledons. The embryo was treated with enzymes to remove the cytoplasm and thus clearly to reveal the arrangement of cells. (From Von Stosch, *Ztschr. für Mikros. und mikros. Tech.*, 62, 1955.)

distally flattened structure with bilateral symmetry. The flattening immediately precedes the initiation of the two cotyledons (fig. 2.1, *C*). These first foliar structures are initiated by localized divisions right and left at the ends of the longer diameter in the upper part of the flattened body (fig. 2.2). The cotyledons first protrude slightly (figs. 2.2 and 2.1, *D*) and then by further divisions and by enlargement of cells develop into leaf-like structures (fig. 2.1, *E*). The axis below the cotyledons differentiates a root meristem or a radicle at its lower end and becomes the hypocotyl-root axis.

As was pointed out, the divisions in the developing embryo follow an orderly sequence, and the sequences have specific characteristics in the different groups of plants. To a limited extent the details of embryonic development have been used in taxonomy and phylogeny (cf. Lebègue, 1952; Maheshwari, 1950; Souèges and Crété, 1952;

Takhtadzhîan, 1954). In studies of the physiological aspects of embryogeny, investigators also inquire into the factors determining the orderliness of that process (cf. Wardlaw, 1955). Some phases of early growth of the embryo have been elucidated by culturing embryos in vitro (Rappaport, 1954).

Beginning of tissue organization

The development of the cotyledons is accompanied or preceded by changes in the internal structure that initiate the organization of tissue systems (Meyer, 1958). The future epidermis, the *protoderm* (often called *dermatogen*), is formed by periclinal divisions close to the surface (fig. 2.1, *B, C*). When the cotyledons develop, the protoderm is propagated on the surface of the extending cotyledons by anticlinal divisions (fig. 2.1, *D*). An accelerated vacuolation of certain parts of the embryo indicates the development of the *ground meristem* (fig. 2.1, *D*), the precursor of the ground tissue. The remaining less-vacuolated tissue, in both the hypocotyl-root axis and the cotyledons, constitutes the meristem of the future primary vascular system, the *procambium*. Through subsequent longitudinal divisions and elongation the procambial cells assume a narrow elongated shape. The form of the vascular meristem as a whole varies in embryos of different plants, but it is consistently that of a single orderly system continuous between the cotyledons and the hypocotyl-root axis. The vascular system of the young seedling, which develops from the embryo, is an enlarged and differentiated replica of the embryonic procambial system.

Apical meristems

The growth of the future plant from the embryo is made possible by the organization of the apical meristems of shoot and root. These meristems appear at the two opposite poles of the embryo axis, the proximal or root pole and the distal or shoot pole (fig. 2.1, *E*).

The apical meristem of the shoot may be regarded as a small residuum of embryonic tissue located between the cotyledons. In adult plants, the vegetative apical meristem has a characteristic cellular organization (chapter 16). The apical meristem in the embryo may or may not acquire such organization before the seed germinates; it also may or may not initiate the shoot, or epicotyl, while the embryo is still within the seed.

The apical meristem of the root is usually organized by a characteristic sequence of cell divisions in one or more of the lower tiers of cells of the embryo. Sometimes the most distal suspensor cell, which is contiguous with the enlarged body of the embryo proper, is chiefly concerned with the formation of the root meristem (and the rootcap) and is called the hypophysis (fig. 2.1, *B*). The apical meristem of the embryonic root may or may not assume the same cellular organization as that of the growing root (chapter 14), but it has a rootcap. In some dicotyledons, adventitious root primordia differentiate in the embryo hypocotyl (Steffen, 1952).

MONOCOTYLEDON EMBRYO

In the early stages of development, the proembryo stages, the embryos of dicotyledons and monocotyledons follow similar sequences of cell division, and both become cylindrical or club-shaped bodies (fig. 2.3, *A*). The difference in development becomes evident when the formation of the cotyledon begins. In the absence of a second cotyledon the monocotyledon embryo does not become two-lobed at the distal end (fig. 2.3, *B–D*). The single cotyledon, moreover, dominates the development to such an extent that it often appears as though it were a direct continuation of the axis of the embryo (figs. 2.3, *E*, and 2.4, *B*). In other words, the cotyledon seemingly occupies a terminal position, whereas the epicotyl meristem—that is, the apical meristem of the shoot—is found at the side of the cotyledon. The ontogenetic relation between the cotyledon and the apical meristem of the shoot is a subject of much debate because it bears upon the morphologic nature of the monocotyledonous plant. Some workers, for example, interpret the cotyledon as truly terminal and regard the monocotyledonous plant as a sympodium of lateral shoots, each producing a terminal leaf and a new lateral shoot meristem (Souèges, 1954). Others maintain that the terminal position of the cotyledon is only apparent and results from a displacement of the apical meristem by the vigorously growing cotyledon. The proponents of this view support their contention by showing that the degree of terminality is variable. In some monocotyledons the cotyledon and the apical meristem arise side by side from the distal part of the embryo (Baude, 1956; Haccius, 1952), and the extreme terminal and the truly lateral positions of this meristem intergrade through transitional types of positions. The concept that the single cotyledon in the monocotyledons is not truly terminal is further strengthened by the occurrence in dicotyledons of species normally having only one cotyledon and an embryogeny resembling that of the monocotyledons (Haccius, 1954).

Embryo development in onion (Liliaceae) and in grasses (Gramineae) illustrates embryogeny in the monocotyledons.

Onion embryo

Early divisions lead to formation of a club-shaped embryo (fig. 2.3, *A*). Later the embryo becomes an almost spherical body on a thin suspensor (fig. 2.3, *B, C*). The cotyledon develops upward from the nearly spherical body (fig. 2.3, *D, E*). A slight depression, or notch, on one side of the embryo—the site of the future apical meristem —indicates that the upwardly growing part is not a simple prolongation of the embryo axis. Nevertheless, because of its vigorous growth the cotyledon appears to be terminal and the depressed site of the apical meristem lateral.

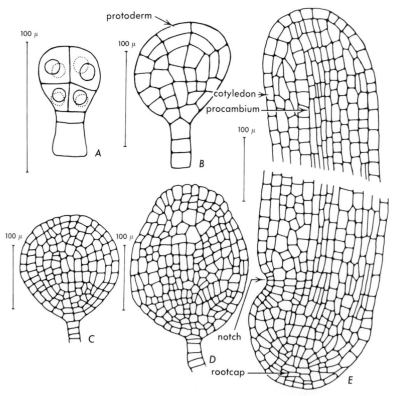

Fig. 2.3. Embryo of *Allium Cepa* (onion) in several stages of development. *A*, embryo body is distinct from suspensor. *B*, protoderm has been initiated in distal tiers. *C*, embryo body is still spherical. *D*, elongation at the distal end—initiation of cotyledon— has begun. *E*, embryo still immature but with tissue regions blocked out; also the notch —future site of apical meristem of shoot—is present.

The depression is shallow at first (fig. 2.3, *E*) but increases in depth as the tissue on its margin grows out. This marginal growth is a sheath-like extension from the cotyledon (fig. 2.4, *A*). The apical meristem arises as a small mound of embryonic cells at the bottom of the depression and initiates the first foliage leaf primordium (fig. 2.4, *A*). When the seed germinates, the first leaf emerges from the enclosure through a slit above the sheath. The apical meristem of the root and the rootcap become organized at the base of the short hypocotyl. (A detailed study of the root-initiation stage of embryogeny is found in Guttenberg et al., 1954.)

The mature embryo has a protoderm, a somewhat-vacuolated ground meristem, and a less vacuolated procambium. The latter extends from the root meristem to the base of the cotyledon where it widens out and forms a short branch directed toward the epicotyl apex and a long branch extending through the cotyledon (fig. 2.4, *B*).

Grass embryo

Attention was called to the difficulties of interpreting the morphology of the monocotyledonous embryo. The development and structure of the grass embryo are so complex that this embryo may be said to raise more morphological problems than any other plant embryo. In the following discussion only one of the interpretations of embryo parts is used. (For examples of various views see Pankow and Guttenberg, 1957, Roth, 1955, 1957, and Wardlaw, 1955.)

A grass embryo reaches a relatively high degree of differentiation. When seen within the mature caryopsis, the embryo is appressed to the endosperm by a massive cotyledon, the scutellum (fig. 2.5). In a section through the median longitudinal plane of the scutellum the embryo axis appears laterally attached to the scutellum (fig. 2.5, *B*). The lower part of the axis is a primordial root (radicle) which bears an apical meristem and a rootcap at the lower end. The root and its rootcap are enclosed in a coleorhiza which, in the young embryo, is continuous with the suspensor.

Above the radicle is the cotyledonary node (thus there is no hypocotyl distinct from the radicle), then follows the epicotyl with several foliar primordia. The outermost of these is the coleoptile which in most grasses is a hollow cone with a small pore near its apex. (In *Streptochaeta*, a primitive grass, the coleoptile is leaf-like and is open on one side; Reeder, 1953.) The coleoptile encloses several ordinary leaf primordia and the apical meristem of the shoot. The part of the axis between the scutellar node and the coleoptile is an internode which is often called the mesocotyl. In some Gramineae a small out-

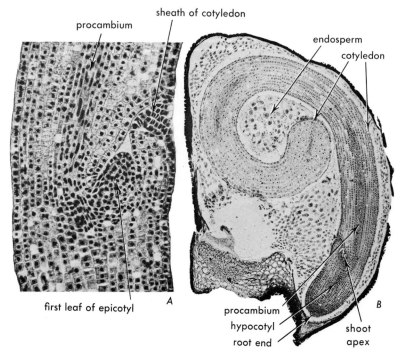

Fig. 2.4. Mature embryo of *Allium Cepa* (onion). *A*, median part of embryo with epicotyl (sectioned through its first leaf) and cotyledon enclosing it. *B*, mature embryo within seed in longitudinal section. (*A*, ×100; *B*, ×25. *B*, from Esau, *Plant Anatomy*, John Wiley and Sons, 1953.)

growth, the epiblast, is present opposite the scutellum. It is sometimes interpreted as a rudimentary second cotyledon. A complex procambial system develops in the embryo. It occurs in the radicle, the scutellar node, the scutellum, the coleoptile, and some of the other foliar primordia.

The developmental details are best known for the embryos of *Zea mays*, maize (e.g., Abbe and Stein, 1954; Kiesselbach, 1949), and of *Triticum*, wheat (e.g., Roth, 1957). About five days after pollination the maize embryo becomes club-shaped (fig. 2.6, *A*). The upper enlarged part gives rise to the main body of the embryo; the lower part is the suspensor. A ten-day-old embryo is elongated and thickened on one side because of the growth of the scutellum (fig. 2.6, *B*). Opposite the scutellum on the embryo axis is the epicotyl apex. It becomes a small rounded prominence surrounded by a circular welt of tissue, the incipient coleoptile. As the coleoptile develops further (fig. 2.6, *C*), leaf primordia are initiated, and the growth of the epicotyl is reoriented from a lateral to a vertical direction (fig. 2.6, *D*). The scutellum enlarges and grows around the groove in the endosperm in which the embryo axis is located (fig. 2.5, *A*) and eventually covers the scutellar node. In contrast to the maize embryo the wheat embryo possesses an epiblast. It develops rather late in embryogeny, after the scutellum has elongated considerably and the coleoptile has partly enclosed the apical meristem.

At the lower end of the embryo axis, above the suspensor, the radicle and rootcap are organized. The radicle is at first united with the coleorhiza tissue but becomes separated from it as the embryo matures (fig. 2.5, *B*). The method of development of the radicle raises the interesting question of identity of the coleorhiza and the radicle. If the coleorhiza is not part of the suspensor but of the hypocotyl (Roth, 1957) or if it is a suppressed primary root (Pankow and Guttenberg, 1957), then the radicle originates endogenously (that is, in-

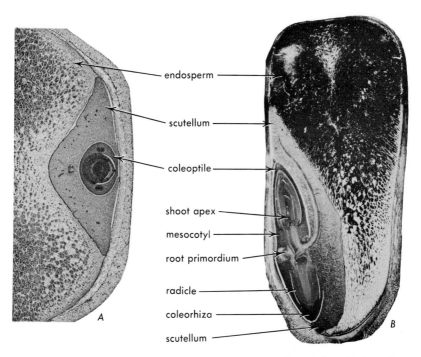

endosperm

scutellum

coleoptile

shoot apex

mesocotyl

root primordium

radicle

coleorhiza

scutellum

A

B

Fig. 2.5. *Zea mays* (corn) embryo within the caryopsis, collected about 30 days after pollination. *A,* transverse section, with part of the caryopsis removed. *B,* longitudinal section. (*A,* ×12; *B,* ×14. Courtesy of J. E. Sass. *A,* from J. E. Sass, *Botanical Microtechnique,* 3rd ed. Iowa State College Press, 1958.)

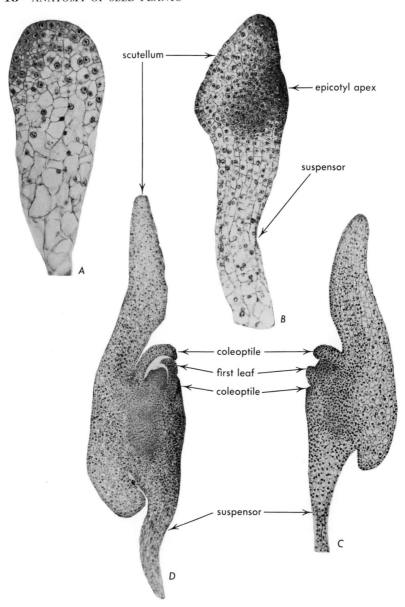

Fig. 2.6. *Zea mays* (corn) embryo in several stages of development. The material was collected the following numbers of days after pollination: *A,* 5; *B,* 10; *D, C,* 15. (*A,* ×230; *B,* ×130; *C,* ×96; *D,* ×48. Courtesy of J. E. Sass. *D,* from J. E. Sass, *Botanical Microtechnique,* 3rd ed. The Iowa State College Press, 1958.)

ternally) like a lateral branch root or an adventitious root (chapters 14 and 15). In monocotyledons lacking a coleorhiza and in most dicotyledons the radicle is generally regarded as having an exogenous (that is, external) origin. Above the scutellar node additional roots are initiated (fig. 2.5, *B*). These are called seminal adventitious roots. After germination, during the growth stage known as tillering, further adventitious roots develop at the nodes of the main and lateral shoots.

REFERENCES

Abbe, E. C., and O. L. Stein. The growth of the shoot apex in maize: embryogeny. *Amer. Jour. Bot.* 41:285–293. 1954.

Baude, E. Die Embryoentwicklung von *Stratiotes aloides* L. *Planta* 46:649–671. 1956.

Guttenberg, H. von, H.-R. Heydel, and H. Pankow. Embryologische Studien an Monocotyledonen. II. Die Entwicklung des Embryos von *Allium giganteum* Rgl. *Flora* 141:476–500. 1954.

Haccius, B. Die Embryoentwicklung bei *Ottelia alismoides* und das Problem des terminalen Monokotylen-Keimblattes. *Planta* 40:433–460. 1952.

Haccius, B. Embryologische und histogenetische Studien an "monokotylen Dikotylen." I. *Claytonia virginica* L. *Österr. Bot. Ztschr.* 101:285–303. 1954.

Kiesselbach, T. A. The structure and reproduction of corn. Univ. Nebraska Coll. Agr., *Agric. Exp. Sta. Res. Bul.* 161. 1949.

Lebègue, A. Recherches embryogéniques sur quelques Dicotylédones Dialypétales. *Ann. des Sci. Nat., Bot. Ser.* 11. 13:1–160. 1952.

Maheshwari, P. *An introduction to the embryology of angiosperms.* New York, McGraw-Hill Book Company. 1950.

Meyer, C. F. Cell patterns in early embryogeny of the McIntosh apple. *Amer. Jour. Bot.* 45:341–349. 1958.

Pankow, H., and H. V. Guttenberg. Vergleichende Studien über die Entwicklung monokotyler Embryonen und Keimpflanzen. In:*Bot. Studien.* No. 7:1–39. 1957.

Rappaport, J. In vitro culture of plant embryos and factors controlling their growth. *Bot. Rev.* 20:201–225. 1954.

Reeder, J. R. The embryo of *Streptochaeta* and its bearing on the homology of the coleoptile. *Amer. Jour. Bot.* 40:77–80. 1953.

Roth, I. Zur morphologischen Deutung des Grasembryos und verwandter Embryotypen. *Flora* 142:564–600. 1955.

Roth, I. Histogenese und Entwicklungsgeschichte des *Triticum*-Embryos. *Flora* 144:163–212. 1957.

Souèges, R. L'origine du cône végétatif de la tige et la question de la "terminalité" du cotylédon des Monocotylédones. *Ann. des Sci. Nat., Bot. Ser.* 11. 15:1–20. 1954.

Souèges, R., and P. Créte. Les acquisitions les plus récentes de l'embryogénie des Angiospermes (1947–1951). *Année Biol. Ser.* 3. 28:9–45. 1952.

Steffen, K. Die Embryoentwicklung von *Impatiens glanduligera* Lindl. *Flora* 139:394–461. 1952.

Takhtadzhîân, A. L. Voprosy evolîûzionnoĭ morfologii rasteniĭ. [Problems of evolutionary morphology of plants.] Leningrad, University Press. 1954.

Wardlaw, C. W. *Embryogenesis in plants.* New York, John Wiley and Sons. 1955.

3. FROM THE EMBRYO
TO THE ADULT PLANT

AFTER THE SEED GERMINATES, THE APICAL MERISTEM OF THE SHOOT forms in regular sequence leaves and nodes and internodes, that is, increments of shoot growth (fig. 3.1). Apical meristems in the axils of leaves may produce axillary shoots (fig. 3.2, *A*). These, in turn, may have axillary shoots. As a result of such activity of the apical meristems the plant bears a system of branches on the main stem. The apical meristem of the root located at the base of the hypocotyl—or of the radicle, as the case may be—forms the taproot (or the primary root; fig. 3.1). In many plants the taproot produces branch roots (or secondary roots; fig. 3.1, *E*) from new apical meristems originating deep in the taproot (endogenous origin). The branch roots produce further branches in their turn. Thus, a much branched root system results. In some plants, especially commonly in monocotyledons, the root system of the adult plant develops from adventitious roots arising on the stem.

The growth outlined above comprises the vegetative stage in the life of a seed plant. At an appropriate time, determined in part by an endogenous rhythm of growth (Bünning, 1953) and in part by environmental conditions, especially light (Parker and Borthwick, 1950) and temperature (Thompson, 1953), the vegetative apical meristem of the shoot is changed into a reproductive apical meristem, that is, in angiosperms, into a floral apical meristem, which produces a flower or an inflorescence. The vegetative stage in the life cycle of the plant is thus succeeded by the reproductive stage.

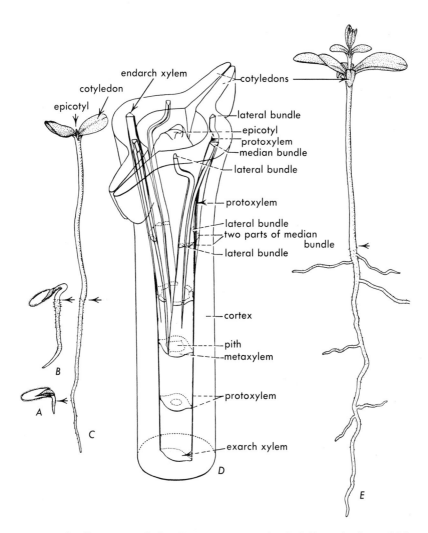

Fig. 3.1. Seedling structure in flax (*Linum usitatissimum*). *A–C, E,* germinating seed (*A*) and three stages of seedling development showing: growth of taproot (below arrowhead) and appearance of branch roots; elongation of hypocotyl (above arrowhead); unfolding of cotyledons and development of epicotyl. *D,* xylem system of the transition region through which root and cotyledons are connected. Phloem would be on outer periphery of xylem. (*A–C, E,* drawn by Alva D. Grant; *D,* from Crooks, *Bot. Gaz.,* 1933.)

MERISTEMS AND ORIGIN OF TISSUES

The formation of new cells, tissues, and organs through the activity of the apical meristems involves division of cells. Certain cells of the meristems repeat divisions in such a way that one product of a division becomes a new body cell whereas the other remains in the meristem. In other words, certain cells in the meristem have, in a sense, a dual role: they perpetuate themselves and form new body cells. The self-perpetuating cells may be referred to as the initials, their products by division, the derivatives. The concept of initials and derivatives should, however, include the qualification that the initials are not inherently different from their derivatives and may become supplanted by their derivatives. The concept of initials and derivatives is taken up from various aspects in connection with the descriptions of the apical meristems of the root (chapter 14) and the shoot (chapter 16). Suffice it to point out here that, according to a common view, certain cells in the meristems act as initials mainly because they occupy the proper position for such an activity and that the apical meristems of roots and shoots of the higher vascular plants (gymnosperms and angiosperms) contain groups of initials.

The initials and their immediate derivatives compose the apical meristems (fig. 3.2). These derivatives usually divide also and produce one or more generations of cells before the cytologic changes, denoting differentiation of specific types of cells and tissues, occur near the tip of root or shoot. Furthermore, divisions continue at levels where such changes are already discernible. In other words, growth, in the sense of cell division, is not limited to the very tip of root or shoot but extends to levels considerably removed from the area usually called the apical meristem. In fact, the divisions some distance from the apex are more abundant than at the apex (cf. Buvat, 1952). In the shoot, a more intensive meristematic activity is observed at levels where new leaves are initiated than at the tip; and during the elongation of the stem, cell division extends several internodes below the apical meristem.

In meristematic activity cell division is combined with an enlargement of the new products of division. Generally, from the younger to the older meristematic tissue the degree of cell enlargement increases (fig. 3.2) and eventually becomes the main factor in the increase in width and length of the particular region of root or shoot. The no longer dividing cells—they may be still enlarging, however—gradually differentiate into the specific cells characteristic of the region of shoot or root where these cells occur. Thus, the various phenomena of growth and differentiation overlap in the same cell; furthermore, at

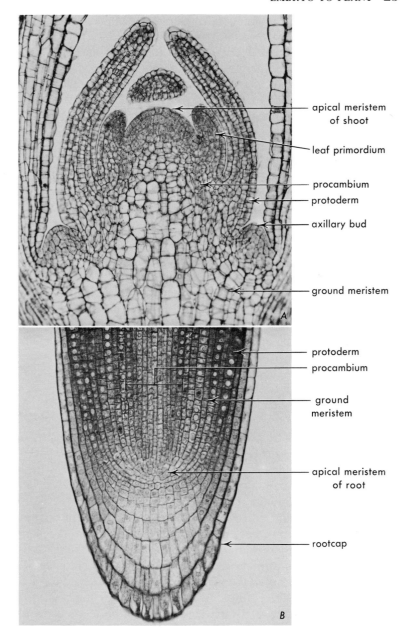

apical meristem
of shoot

leaf primordium

procambium
protoderm

axillary bud

ground meristem

protoderm
procambium

ground
meristem

apical meristem
of root

rootcap

Fig. 3.2. Shoot tip (*A*) and root tip (*B*) of seedling of flax (*Linum usitatissimum*) in longitudinal sections. Both illustrate apical meristems and derivative meristematic tissues. *A*, primordia of leaves and of axillary buds are present. *B*, rootcap covers apical meristem. (*A*, ×184; *B*, ×200. *A*, from J. E. Sass, *Botanical Microtechnique*, 3rd ed., The Iowa State College Press, 1958.)

the same level of shoot or root, different regions may be in different stages of growth and differentiation.

In view of the gradual change from the apical meristems to the adult tissues and the intergrading of the phenomena of cell division, cell enlargement, and cell differentiation, one cannot restrict the term meristem to the apex of shoot or root. The shoot and root parts where the future tissues and organs are already partly determined but where cell division and cell enlargement are still in progress are also meristematic. If a distinction between the apical meristem and the subjacent levels is desired, one may speak of the apical meristems and the meristematic tissues below them, and one may use the terms root tip or shoot tip in a somewhat broad sense to include the apical meristem and the subjacent meristematic tissues.

DIFFERENTIATION AND SPECIALIZATION

The progressive change from the structurally relatively simple meristematic tissue to the complex and variable tissues and combinations of tissues in the adult plant body is referred to as differentiation. The change from the undifferentiated meristematic state to the differentiated adult state involves the chemical constitution of cells as well as their morphologic characteristics (cf. Commoner and Zucker, 1953), and it may be analyzed in terms of single cells, a tissue, a tissue system, an organ, or the plant as a whole (Bonner, 1952). Differentiation may be looked upon as a process, first, of becoming different from the meristematic precursors and, second, of becoming different from the neighboring cells or tissues. The second aspect implies that similar meristematic cells may pass through different steps in their development into mature cells, an event that brings about a diversity of structure in an initially relatively homogeneous tissue.

When we compare cells that have completed their differentiation we recognize that some become much more distinct from the meristematic cells than others and that the higher degree of change is associated with a more pronounced specialization with reference to the role that the cells play in the plant body. A high degree of specialization is attained, for example, by the water-conducting cells in the xylem, cells that have relatively thick walls and no living contents at maturity, and by the food-conducting sieve elements in the phloem, cells that lack nuclei at maturity. A less profound change occurs during the differentiation of a photosynthetic parenchyma cell in the mesophyll of a leaf. Such a cell may assume a noticeably different shape than its meristematic precursor, but its wall thickens only moderately and its protoplast remains complete. The most distinguishing feature in the development of such a cell is the acquisition of numerous chloroplasts. A photosynthetic cell is also concerned with various activities other than photosynthesis, and, what is particularly important, it may be stimulated to resume meristematic activity. Wounding of a leaf, for example, may induce division of mesophyll cells and subsequent formation of suberized protective tissue (cf. Bloch, 1952). Formation of callus tissue along cut surfaces is another example of resumption of cell division by parenchyma, and such parenchyma may be many years old (Barker, 1953). Fully differentiated but living cells may resume meristematic activity spontaneously, as well. The formation of periderm in stems, for example, results from such a resumption of meristematic activity. The development of adventitious shoots and roots by reactivation of division in differentiated parenchyma cells also my be spontaneous. The foregoing examples clearly point out the variability in the degree of differentiation and specialization among the cells of the same plant body. One can say in this connection that some cells are more determined than others if their potentialities for further growth are compared.

The presence of meristematic potentialities in many types of fully developed but living cells makes it difficult to think of these cells as mature. Nevertheless, it is often convenient to speak of cells as being mature when they complete their differentiation. Thus, mature cells may be nonliving, or they may have active protoplasts capable of resuming meristematic activity.

EXPRESSIONS OF ORGANIZATION IN THE PLANT BODY

A significant aspect of differentiation is that cells with certain characteristics appear in definite positions in the plant body, in other words, that differentiation results in the appearance of tissue patterns in the plant. Much of the modern research in developmental morphology and physiology is concerned with the possible causes of the development of the complex but orderly internal patterns in plants. This research may consist of developmental studies of cells, tissues, embryos, parts of plants, or entire plants; of studies of abnormal growth (Bloch, 1954); of studies of cells, tissues, or larger parts severed from plants and cultivated in vitro (Gautheret, 1953; Steward et al., 1958); of studies of responses of plants to various stimuli or surgical treatments (cf. Wardlaw, 1952); or of studies of the effect of genes upon development (Hansen, 1957). All this work leads to one conclusion: the plant is an organized entity in which the development follows a definite course that imparts to the plant its characteristic structure.

One of the expressions of organization is polarity (cf. Bloch, 1943), a property of the plant mentioned in the discussion of the embryonic development. The embryo was characterized as an axiate structure with a root pole and a shoot pole. Polarity, which may be detected as soon as the embryo begins to develop, continues to be one of the dominant conditions in the differentiation of the plant body (Bünning, 1953). Its effect may be manifested by the changes in character of structures from one end of the axis to the other or by the physiological distinctions expressed in the differential behavior of cells or larger units. The phenomenon of polarity may be strikingly demonstrated experimentally. In the regeneration of plants from cells freely suspended in a culture medium, organized growth begins with formation of vascularized nodules. As soon as such a nodule produces a root, polarity becomes evident: opposite the root a shoot pole is organized, and the nodule develops into a plantlet resembling a normal seedling (Steward et al., 1958).

It is instructive to analyze the vascular structure of a normal seedling in terms of polarity. As was shown in chapter 2, the embryo initiates the organization of the future plant in the arrangement of its partly differentiated meristematic tissues: the protoderm, the procambium, and the ground meristem. This organization becomes more clearly defined in the young seedling after an increment of root has been added and the hypocotyl has elongated. An examination of the internal structure of a dicotyledonous seedling shows that the root nature of the root pole of the axis is expressed in the differentiation of the type of vascular cylinder characteristic of roots of dicotyledons. The shoot end of the plant, on the other hand, shows a close relation, typical of higher vascular plants, between the vascular system of the axis (hypocotyl in the young seedling) and the leaves (cotyledons in the young seedling) borne on the axis: the vascular tissue of the hypocotyl appears to branch above into strands that can be followed into the cotyledonary blades (fig. 3.1, *D*). These strands in the hypocotyl may be referred to as cotyledonary traces.

Between the two levels, those of the shoot and the root, a connection exists between the cylindrical vascular system of the root and the system of strands of the upper hypocotyl (fig. 3.1, *D*). If one follows this connection level by level, beginning for example with the root, one gains the impression that the root structure gradually changes into shoot structure. The compact cylinder of the root is replaced by a less compact structure higher up. If no pith is present in the root, a pith may be evident higher. Still higher the vascular tissue appears to separate into two or more units, the cotyledonary traces.

In addition to these simple differences in the general form of the vascular system between one level and another, a complex difference involving the direction of differentiation of the xylem elements is commonly present. The first mature xylem elements (protoxylem) in the root occur in peripheral positions in the vascular cylinder. The subsequently maturing elements (metaxylem) appear successively closer to the center. In other words, as seen in transection the direction of maturation of the vascular elements is centripetal. The xylem showing this order of differentiation is called exarch xylem. In the cotyledonary bundles (or first in the epicotyl in some plants) the order of xylem maturation is reversed. The earliest mature xylem elements are located farthest from the periphery, and the subsequent xylem elements mature in the centrifugal direction. The xylem showing a centrifugal order of maturation is called endarch. The connection between the exarch xylem of the root and the endarch xylem of the shoot occurs through a part of the vascular system in which the relative position of the early and late xylem elements is intermediate between those in the shoot and the root.

The region of the seedling where the root system and shoot system are connected and where the structural details change from level to level in relation to the differences between the two systems is referred to as the transition region (fig. 3.1, *D*). The foregoing discussion dealt with a relatively simple type of transition region in a dicotyledon. Many dicotyledons have more complex root and shoot connections, and in the monocotyledons the presence of only one cotyledon is associated with an asymmetrical structure of the transition region (cf. Boureau, 1954; Hayward, 1938). In the gymnosperms the frequent presence of more than two cotyledons contributes to the complexity of the transition region (Boureau, 1954).

The change in character of the histologic pattern in the successive levels of the transition region is gradual and, therefore, suggests that graded influences from the root pole and the shoot pole are responsible for the development of the particular pattern. Workers detect evidences of graded influences in the differentiation of many other patterns in the plant body, even to the extent that these influences are detectable in single cells, and speak in this connection of gradients of differentiation (Prat, 1951).

The phenomena that have been defined as polarity and gradients of differentiation are thought to explain, at least in part, the occurrence of structural patterns in plants. One important aspect of the development of patterns is that the course of differentiation of a cell is determined by its position within the general pattern. Stated in another way, the cell is subject to a positional restraint during its development. This restraint may be modified to a degree by environmental influences, diseases, or other injurious agencies and also by removing small parts of the plant and subjecting them to various treatments in tissue cultures (Gautheret, 1953). It seems to be difficult, however, to redirect the general polarity in a specialized plant as a whole or in its parts. In fact, work with tissue cultures suggests that even small fragments of tissue may retain the polarity that was originally established by the position of this fragment in the plant body (cf. Bloch, 1943, and Wardlaw, 1952). On the other hand, separation of the tissue into single cells may lead to the organization of a new plant with its own polarity system (Steward et al., 1958).

PRIMARY AND SECONDARY GROWTH

The development of the adult plant from the seedling results from the activity of the apical meristems and the growth and differentiation of their derivatives. The stage of development that ends with the maturation of the more or less direct derivatives of the apical meristems is referred to as primary growth. A complete plant body with roots, stems, leaves, flowers, fruits, and seeds and with its dermal system (epidermis), the ground-tissue system, and the vascular system is produced by primary growth. Small dicotyledonous annuals and most of the monocotyledons complete their life cycle by primary growth. The majority of the dicotyledons and gymnosperms, however, show a secondary stage of growth, resulting from the activity of a special vascular meristem, the vascular cambium. This meristem increases the amount of vascular tissues and causes thereby the thickening of the axis (stem and root). The formation of the protective tissue, periderm, which replaces the epidermis, is also regarded as part of the secondary growth. The secondary addition of vascular tissues and protective covering makes possible the development of large, much branched plant bodies, such as trees.

REFERENCES

Barker, W. G. Proliferative capacity of the medullary sheath region in the stem of *Tilia americana. Amer. Jour. Bot.* 40:773–778. 1953.

Bloch, R. Polarity in plants. *Bot. Rev.* 9:261–310. 1943.

Bloch, R. Wound healing in higher plants. II. *Bot. Rev.* 18:655–679. 1952.

Bloch, R. Abnormal plant growth. *Brookhaven Symposia in Biology* No. 6: 41–54. 1954.

Bonner, J. T. *Morphogenesis. An essay on development.* New Jersey, Princeton University Press. 1952.

Boureau, É. *Anatomie végétale.* Vol. I. Paris, Presses Universitaires de France. 1954.

Bünning, E. *Entwicklungs- und Bewegungsphysiologie der Pflanze.* 3rd ed. Berlin, Springer. 1953.

Buvat, R. Structure, évolution et fonctionnement du méristème apical de quelques Dicotylédones. *Ann. des Sci. Nat., Bot. Ser.* 11. 13:202–300. 1952.

Commoner, B., and M. L. Zucker. Cellular differentiation: an experimental approach. In: *Growth and Differentiation in Plants.* W. E. Loomis, ed. pp. 339–392. Ames, Iowa State College Press. 1953.

Gautheret, R. Recherches anatomiques sur la culture des tissus de rhizomes de topinambour et d'hybrides de soleil et de topinambour. *Rev. Gén. de Bot.* 60:129–218. 1953.

Hansen, A. The expression of the gene dwarf-1 during the development of the seedling shoot in maize. *Amer. Jour. Bot.* 44:381–390. 1957.

Hayward, H. E. *The structure of economic plants.* New York, The Macmillan Company. 1938.

Parker, M. W., and H. A. Borthwick. Influence of light on plant growth. *Annu. Rev. Plant Physiol.* 1:43–58. 1950.

Prat, H. Histo-physiological gradients in plant organogenesis. II. *Bot. Rev.* 17:693–746. 1951.

Steward, F. C., M. O. Mapes, and K. Mears. Growth and organized development of cultured cells. II. Organization in cultures grown from freely suspended cells. *Amer. Jour. Bot.* 45:705–708. 1958.

Thompson, H. C. Vernalization of growing plants. In: *Growth and Differentiation of Plants.* W. E. Loomis, ed. pp. 179–196. Ames, Iowa State College Press. 1953.

Wardlaw, C. W. *Phylogeny and morphogenesis.* London, Macmillan and Company. 1952.

4. PARENCHYMA

THE PARENCHYMA TISSUE IS THE MAIN REPRESENTATIVE OF THE GROUND tissue and may be found in all plant organs as a continuous tissue, as, for example, in cortex and pith of stems (fig. 4.1 *A, B*), cortex of roots, ground tissue of petioles (fig. 4.2), and mesophyll of leaves (chapter 18). Parenchyma cells are also components of certain complex tissues, notably the vascular tissues. The various metabolic activities of the plant are carried on in the protoplasts of parenchyma cells. The presence of normal nucleate protoplasts in parenchyma cells indicates that the cells are relatively unspecialized. Indeed, the parenchyma tissue is frequently characterized as being potentially meristematic. The phenomena of wound healing, regeneration, formation of adventitious roots and shoots, and union of grafts are made possible through a resumption of meristematic activity by parenchyma cells.

As a potential meristem the parenchyma tissue is not far removed, developmentally, from the meristematic tissue and may, therefore, be regarded as developmentally primitive. This characterization does not imply, however, that the cells are simple in their internal organization. On the contrary, they are more complex than some of the highly specialized cells because they possess living protoplasts. Thus, morphologically most parenchyma cells may be described as simple, but physiologically they must be termed complex.

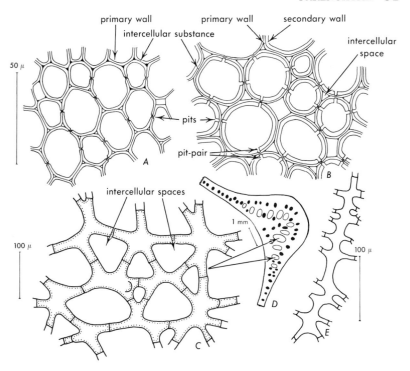

Fig. 4.1. Shape and wall structure of parenchyma cells. (Cell contents are omitted; only symbolically indicated in *C*.) *A, B,* parenchyma from the stem pith of birch (*Betula*). In younger stem (*A*) the cells have only primary walls; in older (*B*), secondary walls occur also. *C, D,* parenchyma of the aerenchyma type (*C*), which occurs in lacunae of petioles and midribs (*D*) of *Canna* leaves. The cells have many "arms." *E,* long "armed" cell from the mesophyll of a disc flower of *Gaillardia.*

SHAPE OF CELLS

Parenchyma cells vary in form, but typically the ground-tissue parenchyma consists of cells that are not much longer than wide and may be nearly isodiametric. However, parenchyma cells may also be considerably elongated or variously lobed (fig. 4.1, *C–E*). Even if the parenchyma cells are approximately isodiametric, they are not spherical but have many facets along which they are in contact with neighboring cells. In relatively homogeneous parenchyma, such as is found in the pith and cortex of stems, the number of facets approaches fourteen. The occurrence of smaller and larger cells in the same tissue, the development of intercellular spaces, and the change of cells from nearly isodiametric to some other shape are factors associated with changes in the average number of facets per cell. (For variations in number of facets see table in Matzke and Duffy, 1956.)

CELL WALL

The cell walls in the active vegetative ground parenchyma, including the mesophyll of leaves, are relatively thin (figs. 4.2 and 4.3), and their principal carbohydrate components are cellulose, hemicelluloses, and pectic substances (but see Bishop et al., 1958). Like all walls of plants those of parenchyma cells are cemented to walls of adjacent cells by an intercellular substance (fig. 4.1, *A*) composed chiefly of pectic compounds. The thin cementing lamella is commonly called the *middle lamella*. The recognition of the cementing substance has clarified a fundamental feature of cell walls, namely, that each cell has its own wall and that the partitions seen betweeen two adjacent cells consist of a cementing substance and two walls belonging to two cells. The double nature of the partitions between cells becomes clearly revealed during the formation of the so-called schizogenous intercellular spaces (figs. 4.1, *B,* and 4.2) when the two walls separate along the middle lamella.

The cell wall arises during cell division at the stage in mitosis known as the telophase (fig. 4.4, *A*). At early telophase the two

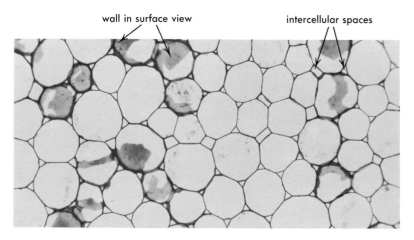

wall in surface view   intercellular spaces

Fig. 4.2. A thin-walled type of parenchyma, with regularly shaped cells and schizogenous intercellular spaces, from petiole of celery (*Apium*). (×250.)

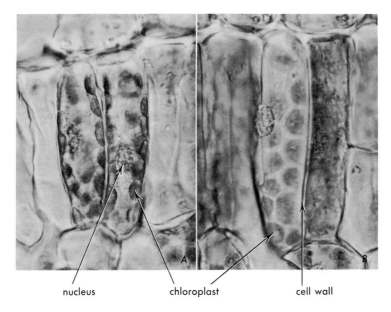

nucleus   chloroplast   cell wall

Fig. 4.3. Parenchyma cells with chloroplasts in mesophyll of privet (*Ligustrum californicum*). Both from living untreated tissue. *A*, chloroplasts in side view. *B*, chloroplasts in surface view. (Both, ×740.)

daughter nuclei are connected to each other by a spindle-shaped system of fibrils referred to as the phragmoplast. In the equatorial plane of the phragmoplast one perceives a plate of differentially staining material (fig. 4.4, *A, B;* face view in fig. 4.4, *C*). The exact nature of this *cell plate,* when it is first discernible, is not known, but is may possibly contain the pectic materials that later constitute the middle lamella. Each of the protoplasts of the two new cells lays down, next to the middle lamella, a wall containing cellulose, hemicelluloses, and pectic substances. New wall layers appear also all around the protoplast on the inside of the original parent-cell wall in continuity with the wall at cell plate (fig. 4.4, *D*). As the new cells increase in volume the parent wall is stretched (fig. 4.4, *D*) and is finally ruptured (fig. 4.4, *E*). Sometimes more than two cells may accumulate within one parent wall before the latter disappears as a distinct entity.

In meristematic tissues, the walls originate in cells that are still growing, and, therefore, the newly formed walls must increase in surface area as the cell enlarges. This extension does not result in a thinning down of the new wall because more wall material is added to the expanding wall. The growth of young walls is a complex phe-

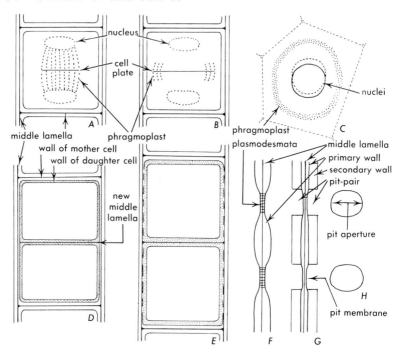

Fig. 4.4. Formation of wall during cell division (*A–E*) and structure of walls (*F–H*). *A,* formation of cell plate in the equatorial plane of phragmoplast at telophase. *B, C,* phragmoplast now appears along the margin of the circular cell plate (in side view in *B;* in surface view in *C*). *D,* cell division is completed and each sister cell has formed its own primary wall (stippled). *E,* sister cells have enlarged, their primary walls have thickened, and the mother cell wall has been torn along the vertical sides of sister cells. *F,* wall composed of middle lamella and two primary wall layers of two adjacent cells. The thin areas in wall are pits (primordial pits) traversed by plasmodesmata. *G,* wall composed of middle lamella, two primary wall layers, and two secondary wall layers. Secondary wall is interrupted in pit regions. *H,* outlines of pits as they appear in surface view.

nomenon and is accomplished through an intimate association between the cytoplasm and the wall (Scott et al., 1956; Wardrop, 1956).

Whereas the wall is relatively thin in early stages of growth of the cell, in later stages it usually increases in thickness, sometimes considerably. A cell may have no other wall than that which is deposited during the growth of the cell. This wall was originally defined by the wood anatomists as the *primary wall*. Parenchyma cells in the cortex, pith, and leaf mesophyll usually have such primary walls. Sometimes the primary walls become very thick, as, for example, in the endosperm of certain seeds (fig. 4.5, *B*) and in the collenchyma tissue (chapter 5). After the cell completes its growth it may deposit an additional wall thickening. Such thickening constitutes the *secondary wall* (Committee on Nomenclature, 1933).* Fibers and sclereids (chapter 6) are good examples of cells with secondary walls. However, parenchyma cells such as those of the xylem and pith (fig. 4.1, *B*) often have secondary walls also. Moreover, the walls of parenchyma cells having secondary thickenings are commonly impregnated with lignin, that is, they are lignified.

The walls of parenchyma cells have thin areas that constitute the *pits* (fig. 4.1, *A, B*). (The pits in the primary walls are called *primary pit-fields* or *primordial pits* by the wood anatomists; cf. Committee on Nomenclature, 1957.) In primary walls the pits appear as depressions of various depths depending on the thickness of the wall (fig. 4.1, *A,* and 4.4, *F*). When secondary walls are formed, sharp interruptions occur in these walls over the pits in the primary walls so that, in the secondary wall, the pit is usually a clearly circumscribed canal or cavity (fig. 4.1, *B*, and 4.4, *G, H*). Commonly a pit in the wall of one cell has a counterpart in the adjacent cell. The middle lamella and two thin layers of primary wall located at the bottom of the pit constitute the *pit membrane* separating the pits of the two adjacent cells. The two opposite pits, together with the pit membrane, constitute a *pit-pair* (fig. 4.1, *B*).

According to the prevalent concept, the protoplasts of parenchyma cells are in communication with each other by means of fine cytoplasmic strands, the *plasmodesmata* (singular *plasmodesma*), that penetrate the walls (cf. Meeuse, 1957). Plasmodesmata may be aggregated in the pits (fig. 4.4, *F*), or they may be scattered throughout a given wall (fig. 4.5). In electron micrographs they have been recognized as strands continuous from protoplast to protoplast (e.g., Strugger, 1957).

The schizogenous intercellular spaces mentioned earlier are of common occurrence in parenchyma. The mesophyll of a leaf or the cortex of a stem or a root illustrate the typical lacunose condition of parenchyma. In some plants, for example those of aquatic or marshy habitats, the intercellular spaces may be especially large and may involve tearing of cell walls in their formation. Parenchyma with very large intercellular spaces is termed aerenchyma (fig. 4.1, *C*).

---

* The classification of walls into primary and secondary given above is generally not utilized by students of fine wall structure employing electron microscopy. They tend to restrict the term primary wall to an early part of the primary wall as defined above. (See also chapter 6.)

intercellular space   lumen of cell

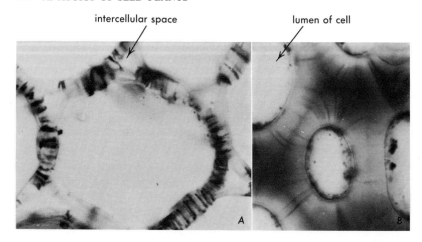

Fig. 4.5. Walls with plasmodesmata, which are not restricted to pits as in fig. 4.4, *F*. *A*, parenchyma cell from leaf of *Primula puberulenta*. Plasmodesmata present in all walls, including some of those next to intercellular spaces. *B*, cell from endosperm of persimmon (*Diospyros*) with thick walls traversed by plasmodesmata. (*A*, ×600; *B*, ×1200. *A*, from Lambertz, *Planta*, 1954.)

CONTENTS

The contents of parenchyma cells vary largely in relation to the metabolic activities of these cells. Because the activities are diverse throughout the plant, a great variety of materials is found in parenchyma cells in different parts of the plant body. An enumeration of the contents of parenchyma cells would be a listing of components of protoplasts of living cells.

For a discussion of the detailed characteristics of the components of living protoplasts, especially those of the nucleus, the cytoplasm with its membranes, the vacuoles, the plastids, the mitochondria, and other small particulates, the reader is referred to books and reviews in cytology (e.g., Sharp, 1943; Steffen, 1955*b*). In this chapter, only some of the common components of protoplasts are described, chiefly for the purpose of illustrating variations among parenchyma cells.

Plastids

Many parenchyma tissues contain chloroplasts, but as a specialized photosynthetic tissue the leaf mesophyll has particularly large numbers of chloroplasts (fig. 4.3). Chloroplasts are generally absent in the underground parts, but in the aerial parts they are not necessarily restricted to layers near the surface—that is, layers closest to light—for they may be found in parenchyma cells of the vascular tissues and even in the pith (Zavalishina, 1951). Parenchyma containing chloroplasts is often called chlorenchyma.

The chloroplasts are receiving much attention from investigators employing electron microscopy and numerous papers report on the fine structure of these bodies (cf. Granick, 1955). The most generally accepted interpretation is that the chloroplasts of the higher plants contain chlorophyll-bearing units, the grana, which are embedded in a colorless protein matrix, or stroma, all enclosed in a membrane. Each granum consists of several discs stacked one above the other so that the granum has the shape of a short cylinder (fig. 4.6). Fine membranes connect one granum to another. During photosynthesis chloroplasts may accumulate starch. This carbohydrate originates within the stroma, and as the grains enlarge they appear to push the grana apart (Mühlethaler, 1955).

The chloroplasts develop from relatively simple proplastids present in the meristematic cells. According to some investigators (e.g., Strugger, 1954), each proplastid contains one granum, the primary granum, that gives rise to the other grana and to the membranes

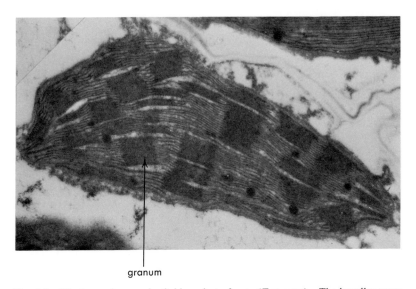

granum

Fig. 4.6. Electron micrograph of chloroplast of corn (*Zea mays*). The lamellar grana (denser regions) are connected with one another by membranes (less dense regions). (×28,000. From Hodge, McLean, and Mercer, *Jour. Biophys. and Biochem. Cytol.*, 1955.)

interconnecting the grana; others (e.g., Heitz and Maly, 1953) recognize no primary granum and suggest that the grana differentiate from the stroma.

Plastids that have no color are the leucoplasts. Upon exposure to light leucoplasts may develop into chloroplasts. A close relation between chloroplasts and leucoplasts is also suggested by the report that grana occur in the leucoplasts in underground parts of certain plants (Bartels, 1955). Leucoplasts frequently contain starch. In starch-storing organs the leucoplasts forming starch are referred to as amyloplasts.

Plastids containing coloring substances other than chlorophyll are the chromoplasts. (Sometimes the chloroplast is classified as a green chromoplast.) The best known chromoplasts are those containing carotenoid pigments and having yellow, orange, or red color. They are found in petals of many flowers (fig. 4.7, *A*) and in various fruits (fig. 4.7, *B*). Grana have been reported in chromoplasts of some plants (Bartels, 1955). The pigmented bodies of the carrot root (fig. 4.7, *D*) and tomato fruit (fig. 4.7, *C*), so commonly used to illustrate the chromoplasts, are of considerable interest because in fully developed state they have the shape of filaments, discs, ribbons, spirals, or polygonal plates; many resemble crystals. Some workers, in fact, interpret these pigmented bodies as crystallized pigment; others, however, regard them as special forms of plastids (Straus, 1953). In young carrot and tomato cells the chromoplasts have the form of ordinary plastids.

Mitochondria

The mitochondria (or chondriosomes) are small cytoplasmic bodies ranging between those that are near the limit of resolution by the light microscope and those several microns in length. They are difficult to separate from proplastids and other small bodies such as, for example, the lipoidal granules. They may be spherical or rod shaped and are composed mainly of protein and lipid. Electron micrographs reveal a double membrane around a mitochondrion and membrane folds in the interior (Steffen, 1955*b*). Mitochondria are considered to be highly significant physiologically, apparently associated with respiratory and accumulatory mechanisms (Robertson, 1957).

Ergastic substances

The cytoplasm, nucleus, plastids, and mitochondria are cytoplasmic components of the protoplast. The nonprotoplasmic components are

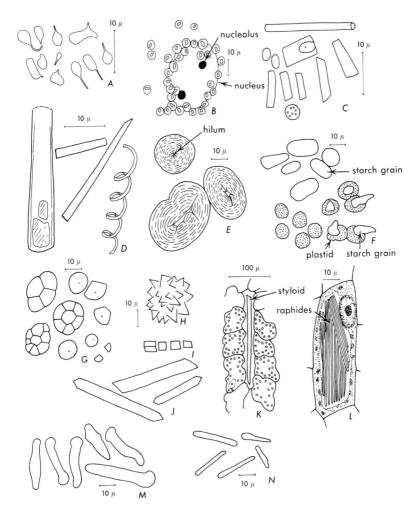

Fig. 4.7. Plastids and ergastic inclusions of protoplasts. *A*, chromoplasts from disc flowers of *Gaillardia*. *B*, nucleus (with two nucleoli) and chromoplasts from pericarp of red pepper (*Capsicum*). The bodies in the plastids are deep orange-red in fresh material. *C*, pigment bodies from pericarp of tomato (*Lycopersicon*). *D*, pigment bodies from the root of carrot (*Daucus*). *E*, starch grains from seed of bean (*Phaseolus*). *F*, starch grains and plastids from rhizome of *Iris*. The plastids are leucoplasts (or elaioplasts) that form starch and oil. *G, H*, simple and multiple starch grains (*G*) and a druse crystal (*H*) from root of sweet potato (*Ipomoea*). *I, J*, crystals from secondary phloem of *Juglans* (*I*) and *Pinus* (*J*). *K*, styloid crystal in an elongated cell among mesophyll cells of *Iris*. *L*, cell with raphid crystals from root tip of *Vitis*. *M, N*, starch grains from laticifers of two different species of *Euphorbia*.

commonly classified as ergastic substances. There is a close relation and interchange between the protoplasmic and the nonprotoplasmic components, but the division is nevertheless convenient for descriptive purposes. The ergastic substances are reserve or waste products that result from cellular activities. Although ergastic substances occur in the fluid of vacuoles, because of the physiologic importance of vacuoles they are usually treated separately.

One of the most common ergastic substances is starch, which develops in the form of grains in the plastids (fig. 4.7, *F*). Next to cellulose, starch is the most abundant carbohydrate in the plant world (Radley, 1954). As was mentioned earlier, storage starch is formed by leucoplasts called amyloplasts. An amyloplast may contain one or more starch grains. The numerous starch grains originating together may form a compound starch grain (fig. 4.7, *G*). It has also been reported that starch may be formed directly in the cytoplasm in the endosperm of cereals (chapter 22).

Starch grains vary in shape (fig. 4.7, *E–G, M, N*) and commonly show layering (fig. 4.7, *E*) centered around a point, the hilum, that may be in the center of the grain or to one side. Splits that often radiate from the hilum appear to result from dehydration of the grains. Some ascribe the layering to an alternation of starch-rich and water-rich layers (e.g., Hess, 1955), others to an alternation of two carbohydrates, amylose and amylopectin (e.g., Mühlethaler, 1955). The amylose is more highly soluble in water than the amylopectin, and when the grain is placed in water the differential swelling of the two substances brings out the layering.

Storage starch occurs in parenchyma of the cortex and pith, in parenchyma cells in the vascular tissues, and in parenchyma of fleshy leaves (bulb scales), rhizomes, tubers, fruits, cotyledons, and the endosperm of seeds.

Another common group of ergastic substances are the tannins. Tannins represent a heterogeneous group of substances that are widely distributed in the plant body. In some of their forms the tannins are very conspicuous in sectioned material. They appear as coarsely or finely granular masses or as bodies of various sizes colored yellow, red, or brown.

No tissues lack tannins entirely; they occur even in meristematic cells. They are abundant in leaves of many plants, in vascular tissues, in the periderm, in unripe fruits, in seed coats, and in pathologic growths. They occur in the cytoplasm and the vacuole and may impregnate the walls. They may be present in many cells of a given tissue or in isolated cells scattered through the tissue (tannin idioblasts). They may be located in much enlarged cells, called tannin sacs.

Some other ergastic substances should be mentioned, at least briefly. Among these are the various crystals (fig. 4.7, *H–L*) that occur as single or aggregated types and are mostly composed of calcium oxalate. Solid protein may be present as ergastic substance (Steffen, 1955*a*). It may occur in the form of amorphous aleuron grains, found in the endosperm and embryo of many seeds, and in the form of protein crystalloids, so called because unlike the mineral crystals they imbibe water and swell. Crystalloids occur within some aleuron grains and in nuclei.

Fats and related substances occur as ergastic substances. They probably occur in every living cell in at least small amounts and may be found in plastids or cytoplasm (Scott, 1955). Fats are frequent reserve materials in seeds, spores, embryos, and meristematic cells. They may occur as solid bodies or as droplets. Various parts of the plant may serve as economic sources of fats. The flowering plants, particularly the dicotyledons, are the main producers of commercially used plant fats, whereas the gymnosperms and the lower vascular plants are relatively unimportant as practical sources (Eckey, 1954).

REFERENCES

Bartels, F. Cytologische Studien an Leukoplasten unterirdischer Pflanzenorgane. *Planta* 45:426–454. 1955.

Bishop, C. T., S. T. Bayley, and G. Setterfield. Chemical constitution of the primary cell walls of *Avena coleoptiles*. *Plant Physiol.* 33:283–289. 1958.

Committee on Nomenclature. International Association of Wood Anatomists. *Glossary of terms used in describing woods. Trop. Woods* 1933(36):1–12. 1933. 1957(107):1–36. 1957

Eckey, E. W. *Vegetable fats and oils.* ACS Monograph Series. New York, Reinhold. 1954.

Granick, S. Plastid structure, development and inheritance. In: *Handbuch der Pflanzenphysiologie* 1:507–564. 1955.

Heitz, E., and R. Maly. Zur Frage der Herkunft der Grana. *Ztschr. für Naturforsch.* 8*b*:243–249. 1953.

Hess, C. Über die Rhytmik der Schichtenbildung beim Stärkekorn. *Ztschr. für Bot.* 43:181–204. 1955.

Matzke, E. B., and R. M. Duffy. Progressive three-dimensional shape changes of dividing cells within the apical meristem of *Anacharis densa*. *Amer. Jour. Bot.* 43:205–225. 1956.

Meeuse, A. D. J. Plasmodesmata (vegetable kingdom). In: *Protoplasmatologia.* Vol. II A lc, pp. 1–43 Wien, Springer. 1957.

Mühlethaler, K. Untersuchungen über den Bau der Stärkekörner. *Ztschr. für Wiss. Mikros. und Mikros. Technik* 62:394–400. 1955.

Radley, J. A. *Starch and its derivatives.* Vol. 1. 3rd ed. New York, John Wiley and Sons. 1954.

Robertson, R. N.   Electrolytes in plant tissue.   *Endeavour* 16:193–198.   1957.

Scott, F. M.   The distribution and physical appearance of fats in living cells—introductory survey.   *Amer. Jour. Bot.* 42:475–480.   1955.

Scott, F. M., K. C. Hamner, E. Baker, and E. Bowler.   Electron microscope studies of cell wall growth in the onion root.   *Amer. Jour. Bot.* 43:313–324.   1956.

Sharp, L. W.   *Fundamentals of cytology.*   New York, McGraw-Hill Book Company. 1943.

Steffen, K.   Einschlüsse.   In: *Handbuch der Pflanzenphysiologie* 1:401–412.   1955*a.*

Steffen, K.   Chondriosomen und Mikrosomen (Sphärosomen).   In: *Handbuch der Pflanzenphysiologie* 1:574–613.   1955*b.*

Straus, W.   Chromoplasts—development of crystalline forms, structure, state of the pigments.   *Bot. Rev.* 19:147–186.   1953.

Strugger, S.   Die Proplastiden in den jungen Blättern von *Agapanthus umbellatus* L'Hérit.   *Protoplasma* 43:120–173.   1954.

Strugger, S.   Der elektronenmikroskopische Nachweis von Plasmodesmen mit Hilfe der Uranylimprägnierung von Wurzelmeristemen.   *Protoplasma* 48: 231–236.   1957.

Wardrop, A. B.   The nature of surface growth in plant cells.   *Austral. Jour. Bot.* 4:193–199.   1956.

Zavalishina, S. F.   Khloroplasty v tkaniakh steli u pokrytosemianykh rastenii.   [Chloroplasts in stele tissues in angiosperms.]   *Akad. Nauk SSSR Dok.* 78:137–139.   1951.

5. COLLENCHYMA

Collenchyma and sclerenchyma are thick-walled tissues that are considered to be specialized with reference to support; that is, they are supporting, or mechanical, tissues. They differ from one another mainly in wall structure and condition of the protoplast. Collenchyma has relatively soft pliable primary walls; sclerenchyma has relatively hard rigid secondary walls. Collenchyma cells retain active protoplasts at maturity and are capable of further growth and division (fig. 5.1, C). The protoplasts of sclerenchyma cells, if they persist at all, appear to be rather inactive; at least they are not known to be able to resume meristematic activity.

Collenchyma and parenchyma are similar in having active protoplasts, and both commonly contain chloroplasts. The difference between the two tissues lies chiefly in the thicker walls and greater length of collenchyma cells. Where the two tissues are in contact, however, they intergrade both in cell length and in wall thickness.

CELL WALL

The structure of cell walls in collenchyma is the most distinctive characteristic of this tissue. The walls are thick and glistening in fresh sections, and often the thickening is unevenly distributed. They contain, in addition to cellulose, pectic and certain other wall substances (Preston and Duckworth, 1946; Roelofsen and Kreger, 1951) but no lignin. Since the pectic substances are hydrophilic, collenchyma walls are rich in water. This feature can be demonstrated by treating fresh sections of collenchyma with alcohol. The dehydrating action of alcohol causes a noticeable contraction of collenchyma walls. Because intensive dehydration is involved in the paraffin method used for the preparation of permanent slides, collenchyma walls are usually considerably thinner in such slides than in fresh sections or sections processed without much dehydration.

The distribution of the wall thickening in collenchyma shows several patterns. If the wall is unevenly thickened it attains its greatest thickness either in the corners of the cell or on two opposite walls, the inner and the outer tangential walls. The collenchyma with wall thickenings localized in the corners is commonly called angular collenchyma, the one with the thickenings on the tangential walls, lamellar or plate collenchyma. As the wall ages its pattern may change through deposition of additional wall layers. Thus the initially angular pattern, for example, may be obscured (fig. 5.1, A), as the outline of the cell lumen becomes circular in transverse sections (Duchaigne, 1955).

Collenchyma may or may not contain intercellular spaces. If spaces are present in the angular type of collenchyma, the thickened walls occur next to the intercellular spaces. Collenchyma with such characteristic distribution of wall thickening is sometimes classified as a special type, the lacunar collenchyma (fig. 5.1, B).

The pectic substances play an important role in the formation of intercellular spaces—the schizogenous intercellular spaces mentioned in chapter 4. The weakening of the middle lamella (probably through enzyme action) permits the adjacent primary walls to separate from each other. When the collenchyma tissue develops no intercellular spaces, the corners where several cells meet show an especially thick accumulation of pectic substances. This appearance seems to result from an accumulation of intercellular material in the potential intercellular spaces. The rate of such accumulation apparently varies, for intercellular spaces may form in early stages of development, only to be closed later by pectic substances. Where the intercellular spaces are large the pectic material may fail to fill them. In such instances crests or wart-like accumulations of pectic material protrude into the intercellular spaces (Carlquist, 1956; Duchaigne, 1955).

The middle lamella may appear differently stained from the rest of the wall. Then the characteristically large accumulations of intercellular substance in the corners where several cells meet are discernible. However, the middle lamella is often closely united with the earliest

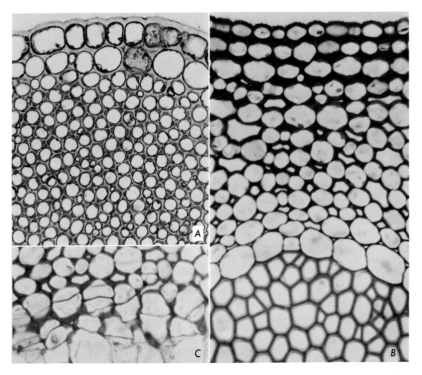

Fig. 5.1. Collenchyma tissue in cross sections. *A*, petiole of celery (*Apium graveolens*). Collenchyma is of the angular type. *B*, stem of *Ambrosia*. Collenchyma of the lacunar type, that is, with intercellular spaces and wall thickening next to these. Unevenly thickened walls of collenchyma sharply contrast with evenly thickened walls of sclerenchyma (below in *B*). *C*, collenchyma of *Ambrosia* with cells that underwent divisions near a surface injury. (All, ×320. *B*, *C*, slide courtesy of N. H. Boke.)

layers of the primary walls; as a result, the middle lamella as such may be indistinguishable. Thus, the differentially stained layer that may be seen in the median position in the collenchyma wall (fig. 5.1, *A*) is not necessarily the middle lamella alone. For collenchyma, as well as for other tissues, the term *compound middle lamella* is used to designate the three-ply structure composed of the middle lamella (intercellular substance) and the earliest primary wall layers from two adjacent cells.

Collenchyma walls exemplify thick primary walls. The primary wall was defined in chapter 4 as the wall that is formed while the cell is still increasing in size. The thickening of collenchyma walls is deposited while the cell is growing. In other words, the cell wall increases simultaneously in surface area and in thickness. Because this thickening is so great, wall growth in collenchyma is a striking and complicated phenomenon that has not yet been completely clarified (Magin, 1956; Majumdar and Preston, 1941; Wardrop, 1955).

The wall thickening of collenchyma is deposited in successive fibrillar layers of submicroscopic dimensions (Beer and Setterfield, 1958). As viewed through a light microscope, collenchyma walls also may show layering, possibly related to aggregations of submicroscopic layers into thicker ones. The layering may also be brought about by an alternation of layers of different physical and chemical structure. For example, layers rich in pectic substances may alternate with those that have proportionally less pectic material. Pits are often present in collenchyma walls, especially in those that are rather uniform in thickness (Duchaigne, 1955).

DISTRIBUTION IN THE PLANT

In discussing the distribution of collenchyma one should distinguish between the thick-walled tissue that arises independently of the vascular tissues and occurs in the peripheral regions of stem and leaf, that is, the collenchyma in the strict sense, and the thick-walled parenchyma associated with vascular tissues (Duchaigne, 1955). This parenchyma, which occurs in the peripheral part of the phloem (outer part of vascular bundle), or the peripheral part of the xylem (inner part of the vascular bundle), or completely surrounds the vascular bundle, consists of long cells with thick primary walls. The wall thickening may resemble that of collenchyma, especially the type with uniformly thickened walls. It is often called collenchyma, but because of its association with the vascular tissues it has a history of development somewhat different from that of the independent collenchyma. The fine structure of walls in the two kinds of tissues may also be different. The elongated cells with thick primary walls associated with the vascular bundles may be referred to as collenchymatous parenchyma cells if their resemblance to collenchyma cells must be stressed. This designation may be applied to parenchyma resembling collenchyma in any location in the plant. The present discussion deals only with the independent peripheral collenchyma.

The peripheral position of collenchyma is highly characteristic. The tissue is located either directly beneath the epidermis (fig. 5.1, *B*) or one or a few layers removed from it (fig. 5.1, *A*). In stems, collenchyma frequently forms a continuous layer around the circum-

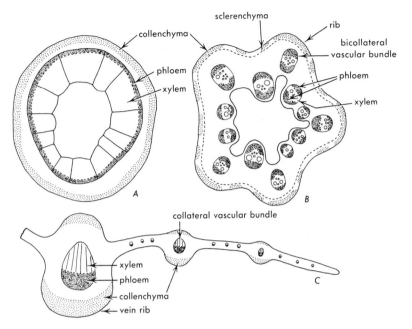

Fig. 5.2. Diagrams illustrating distribution of collenchyma (stippled) in stems of a woody species (*A, Tilia*), a herbaceous vine (*B, Cucurbita*) and a leaf (*C*). All cross sections.

ference of the axis (fig. 5.2, *A*). Sometimes it occurs in strands, usually conspicuous along ridges (ribs, fig. 5.2, *B*) that are found in many herbaceous stems and in those woody stems that have not yet undergone secondary growth. In the petioles the distribution of collenchyma shows patterns similar to those encountered in the stems. In the leaf blade it occurs in the ribs accompanying the larger vascular bundles (veins), sometimes on both sides of the rib (fig. 5.2, *C*), sometimes on one side only, usually the lower. Roots rarely have collenchyma.

STRUCTURE IN RELATION TO FUNCTION

Collenchyma appears to be particularly adapted for support of growing leaves and stems. Its walls begin to thicken early during the development of the shoot, but this thickening is plastic and capable of extension. It therefore does not hinder the elongation of stem and leaf. In a more advanced state of development collenchyma continues to be a supporting tissue in plant parts (many leaves, some herbaceous stems) that do not develop much sclerenchyma. In connection with the discussion of the supporting role of collenchyma it is of interest that in developing plant parts subjected to mechanical stresses (by exposure to wind, attachment of weights to inclined shoots) the wall thickening in collenchyma begins earlier and becomes more massive than in plants not subjected to such stresses (Razdorskiĭ, 1955; Venning, 1949).

Mature collenchyma is a strong flexible tissue consisting of long overlapping cells (they may reach 2 millimeters in length according to Duchaigne, 1955), with thick nonlignified walls. In their tensile strength collenchyma cells favorably compare with fibers. In old plant parts collenchyma may harden, or it may change into sclerenchyma by a deposition of secondary lignified walls. If it does not undergo these changes, its role as a supporting tissue may become less important because of the development of sclerenchyma in the deeper parts of the stem or petiole. Moreover, in stems with secondary growth, the xylem becomes the chief supporting tissue because of the predominance of cells with lignified secondary walls and the abundance of long overlapping cells in this tissue.

REFERENCES

Beer, M., and G. Setterfield. Fine structure in thickened primary walls of collenchyma cells of celery petioles. *Amer. Jour. Bot.* 45:571–580. 1958.

Carlquist, S. On the occurrence of intercellular pectic warts in Compositae. *Amer. Jour. Bot.* 43:425–429. 1956.

Duchaigne, A. Les divers types de collenchymes chez les Dicotylédones: leur ontogénie et leur lignification. *Ann. des Sci. Nat., Bot. Ser.* 11. 16:455–479. 1955.

Magin, T. L'ontogénie du collenchyme chez *Lamium album* L. *Rev. de Cytol. et de Biol. Vég.* 17:219–258. 1956.

Majumdar, G. P., and R. D. Preston. The fine structure of collenchyma cells in *Heracleum Sphondylium* L. *Roy. Soc. London, Proc. Ser. B.* 130:201–217. 1941.

Preston, R. D., and R. B. Duckworth. The fine structure of the walls of collenchyma in *Petasites vulgaris* L. *Leeds Phil. Lit. Soc., Proc.* 4:343–351. 1946.

Razdorskiĭ, V. F. *Arkhitektonika rastenii.* [Architectonics of plants.] Moskva, Sovetskaĭa Nauka. 1955.

Roelofsen, P. A., and D. R. Kreger. The submicroscopic structure of pectin in collenchyma walls. *Jour. Exp. Bot.* 2:332–343. 1951.

Venning, F. D. Stimulation by wind motion of collenchyma formation in celery petioles. *Bot. Gaz.* 110:511–514. 1949.

Wardrop, A. B. The mechanism of surface growth in parenchyma of *Avena* coleoptiles. *Austral. Jour. Bot.* 3:137–148. 1955.

6. SCLERENCHYMA

THE BASIC STRUCTURAL FEATURES OF SCLERENCHYMA WERE INTRODUCED in chapter 5 by comparing sclerenchyma with collenchyma, the two principal supporting tissues. As was pointed out there, sclerenchyma cells have secondary walls that are deposited over the primary after the latter completes its extension growth. Secondary walls are present also in water-conducting cells of the xylem and frequently in the parenchyma cells of that tissue, as well. Moreover, parenchyma cells in various other tissue regions may become sclerified. Thus, secondary walls are not unique in sclerenchyma cells, and, therefore, the delimitation between typical sclerenchyma cells and sclerified parenchyma on the one hand and water-conducting cells on the other is not sharp. Classifying the intergrading forms among these various cells is an example of the frequent difficulty of establishing categories within biological materials.

In this chapter sclerenchyma is discussed with reference to those mechanical, or supporting, cells that chiefly lend hardness or rigidity to tissues. Sclerenchyma cells are usually divided into two categories, the sclereids and the fibers. These two classes of cells again are not sharply separated, but in general the fiber is a long slender cell, many times longer than wide (fig. 6.1 and chapters 8 and 11), whereas sclereids vary in form from an approximately isodiametric to a considerably elongated one, and some kinds of sclereids are much branched.

Sclerenchyma cells may or may not retain their protoplasts at maturity. This variability adds to the difficulty of distinguishing between sclerenchyma cells and sclerified parenchyma cells.

CELL WALL

Since sclerenchyma cells have thick secondary walls they are convenient for the study of the structural features of cell walls. The initial detailed information on the microscopic and submicroscopic structure of plant cell walls was gained from investigations on thick secondary walls of various fibers and of tracheids in the xylem. The study of the less accessible primary wall was undertaken after the techniques became more refined through the work on secondary walls.

When a secondary wall is present the following wall layers are recognized (fig. 6.1): (*a*) middle lamella, or intercellular substance, with the pectic compounds constituting the basic component; (*b*) primary wall, with cellulose as the main component, accompanied by other noncellulosic substances, such as hemicelluloses and pectic compounds; (*c*) secondary wall, with cellulose as the main component, accompanied by various noncellulosic substances but usually lacking true pectic compounds. Lignin may or may not be present. When lignification occurs, it begins in the middle lamella then proceeds to the primary wall, and finally reaches the secondary wall.

In cells having secondary walls, the primary wall is commonly very thin. Moreover, the two primary walls and the middle lamella are firmly united into one structure, the compound middle lamella (chapter 5). Sections of sclerenchyma not specifically prepared to reveal the middle lamella usually show no clear distinction between the primary walls and the middle lamella. A further complication arises when the first layers of the secondary wall become firmly attached to the primary wall. Then the first part of the secondary wall enters into the composition of the compound middle lamella.

As the main component of the cell walls, cellulose forms the framework of the wall, whereas the other substances incrust or inlay this framework (Preston, 1952). The incrusting substances can be removed without loss of form of the cell wall, leaving the cellulose framework intact. Electron micrographs show that this framework is composed of a system of fine fibrils, or microfibrils (fig. 6.2). The spaces among the microfibrils are occupied by the incrusting substances in the intact cell wall.

By methods of X-ray crystallography and polarization optics the cellulose has been shown to be a crystalline substance, meaning that

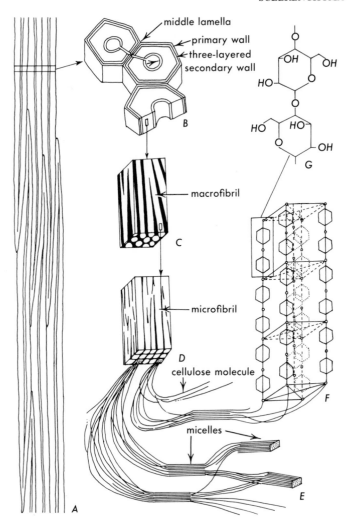

middle lamella
primary wall
three-layered
secondary wall

B

macrofibril

C

microfibril

D
cellulose molecule

G

F

micelles

E

A

Fig. 6.1. Detailed structure of cell walls. *A*, strand of fiber cells. *B*, cross section of fiber cells showing gross layering: a layer of primary wall and three layers of secondary wall. *C*, fragment from middle layer of secondary wall showing macrofibrils (white) of cellulose and interfibrillar spaces (black), which are filled with noncellulosic materials. *D*, fragment of a macrofibril showing microfibrils (white), which may be seen in electron micrographs (fig. 6.2). The spaces among microfibrils (black) are also filled with noncellulosic materials. *E*, structure of microfibrils: chain-like molecules of cellulose, which in some parts of microfibrils are orderly arranged. These parts are the micelles. *F*, fragment of a micelle showing parts of chain-like cellulose molecules arranged in a space lattice. *G*, two glucose residues connected by an oxygen atom—a fragment of a cellulose molecule.

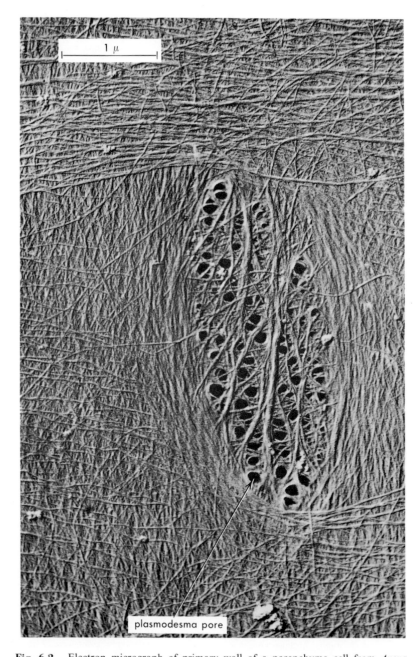

Fig. 6.2. Electron micrograph of primary wall of a parenchyma cell from *Avena* coleoptile. The longitudinal axis of the cell was in the direction of the 1μ scale. The parallel-oriented microfibrils occurred in one of the angles of the cell. Elsewhere the microfibrils are less definitely oriented. The plasmodesmata pores are grouped in an oval area—a primordial pit. (×26,000. From Böhmer, *Planta,* 1958.)

its molecules are arranged in a strictly regular way. A simple demonstration of the crystalline state of cellulose can be made by the use of sheets of polaroid, one placed between the source of light and the microscope, the other between the ocular and the eye. The light passing through a polaroid vibrates in one plane only; it is plane polarized. If the two polaroids are at right angles to each other (that is, if they are "crossed") with regard to their planes of polarization of light, the field in a polarizing microscope remains dark. If a crystalline substance, such as cellulose, intervenes between the two polaroids in crossed position, some light reaches the eye. Because of its crystalline nature and the associated property of double refraction, the cellulose changes the plane of polarization of the light that reaches the polaroid above the ocular and thus permits some light to enter the eye. In certain positions—the extinction positions—the birefringence (double refraction) is not revealed.

Substances affecting the light as cellulose does are classified as optically anisotropic. Substances not affecting the light are optically isotropic. The pectic substances are generally regarded to be isotropic and in conformity with this view the middle lamella usually appears dark when viewed through a polarizing microscope with crossed polaroids. The primary and secondary walls, on the contrary, are anisotropic and doubly refractive because of their high content of cellulose (fig. 6.3).

To explain the crystalline nature of cellulose it is necessary to

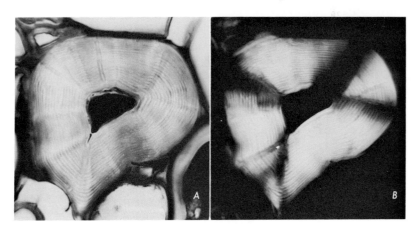

Fig. 6.3. Sclereid from root cortex of fir (*Abies*), as seen with nonpolarized (*A*) and polarized (*B*) light. Because of crystalline nature of cellulose cell wall shows double refraction and appears bright in polarized light (*B*). Wall shows concentric lamellation. (Both, ×890.)

consider the molecular structure of this carbohydrate (fig. 6.1). The cellulose molecule is a chain-like series of glucose residues $C_6H_{10}O_5$ united with each other through the oxygen atoms. These chain-like molecules, which vary in length, are combined into bundles. In some parts of these bundles—in the micelles—the chains are arranged very orderly into three-dimensional lattices with regular distances between the chains and also, of course, between pairs of glucose residues. This regular three-dimensional distribution of glucose residues and of the component atoms is the basis of the crystalline properties of cellulose.

The bundles of cellulose molecules just described are aggregated into larger units, the microfibrils, which are revealed by the electron microscope (fig. 6.2). In the primary wall the microfibrils are much interwoven and, in its earlier layers, are oriented transversely to the long axis of the cell. In contrast, the secondary wall shows approximately parallel orientation of the microfibrils, and the microfibrils are inclined to the long axis of the cell. The difference between the primary and secondary walls, as seen with the electron microscope, is not as sharp as the above statements might indicate. As was mentioned previously, the primary wall is formed chiefly during the enlargement of the cell. Consequently the wall itself undergoes surface enlargement. This growth affects the original orientation of microfibrils; they become more and more disoriented. The subsequently formed layers of primary wall show an increasingly greater orderliness of orientation of the microfibrils, and, finally, the secondary wall has the most orderly alignment of these units (cf. Böhmer, 1958; Bosshard, 1952; Wardrop and Cronshaw, 1958). The recognition of the change in the orientation of microfibrils in the successive primary wall layers should eventually lead to an agreement between the wood anatomists and the electron microscopists (cf. footnote in chapter 4) with regard to the use of the terms primary wall and secondary wall. (See, for example, Belford et al., 1958.)

The microfibrils usually are aggregated into coarser fibrils, the macrofibrils (Bailey, 1958), that are often visible with the light microscope. Many other patterns may be visible in thick secondary walls through a light microscope. Concentric layering, for example, is very common (fig. 6.3). It is related to the concentric deposition of the successive layers of wall material and to chemical and physical variations among the layers. Many fibers and tracheids, moreover, show a division of the wall into three zones (fig. 6.1) differentiated from each other by the angle of inclination of the microfibrils.

The most conspicuous markings on the secondary walls are the pits.

As was pointed out in chapter 4, in the secondary wall pits are interruptions in the continuity of the walls and usually occur over the primordial pits in the primary walls.   If the pit cavity formed by the interruption in the secondary wall is of the same diameter through the whole depth of the wall or if it widens or narrows gradually toward the interior of the cell, the pit is called *simple* (fig. 4.4, *G*).   If, however, the secondary wall overarches the pit cavity so that the latter is suddenly narrowed down toward the interior of the cell, the pit is called *bordered*.   Bordered pits are most common in the xylem elements and are considered in detail in chapter 8.

SCLEREIDS

The sclereids are widely distributed in the plant body and vary much in shape.   These cells usually have thick secondary walls, strongly lignified, and are provided with numerous, commonly simple, pits.   Sclereids have been categorized on the basis of form, but because of intergrading of the various forms it is sometimes difficult to assign sclereids to a specific category (cf. Arzee, 1953*a;* Rao, 1951).

The distribution of sclereids among other cells is of especial interest with regard to problems of differentiation in plants.   Sclereids may occur in more or less extensive layers or clusters, but frequently they appear isolated among other types of cells from which they may differ sharply by their thick walls and often bizarre shapes.   As isolated cells they are classified as idioblasts.   Their differentiation as idioblasts poses many still unresolved questions regarding causal relations in the development of tissue patterns in plants (cf. Foster, 1956).

Sclereids occur in the epidermis, the ground tissue, and the vascular tissues.   In the following paragraphs the sclereids are described by examples of sclereids in the different parts of the plant body, excluding those sclereids that occur in the vascular tissues.

Sclereids in stems.   A continuous cylinder of sclereids occurs on the periphery of the vascular region in the stem of *Hoya carnosa* and groups of sclereids in the pith of the same plant and in that of *Podocarpus*.   These sclereids have moderately thick walls and numerous pits (fig. 6.4, *C, D*).   In shape and size they resemble the adjacent parenchyma cells.   This resemblance is often taken as an indication that such sclereids develop from parenchyma cells, in other words, that they are by origin sclerified parenchyma cells.   Their sclerification, however, has advanced so far that they may be grouped with the sclereids rather than parenchyma cells.   This simple type of sclereid resembling a parenchyma cell in shape is called a stone cell or brachysclereid.

A contrasting type of sclereid, a much branched astrosclereid, is found in the cortex of *Trochodendron* stem (fig. 6.4, *J*). Somewhat less profusely branched sclereids occur in the cortex of the douglas fir, *Pseudotsuga taxifolia*.

Sclereids in leaves. Leaves are an especially rich source of sclereids with regard to variety of form. In the mesophyll, two main distributional patterns of sclereids are recognized: the diffuse, with sclereids dispersed in the leaf tissue (e.g., *Trochodendron*, *Osmanthus*, *Olea*, *Pseudotsuga*), and the terminal, with the sclereids confined to the ends of the small veins (e.g., certain members of Polygalaceae, Capparidaceae, Rutaceae, and others; cf. Foster, 1955, 1956). In some protective foliar structures like the clove scales of garlic (*Allium sativum*) the sclereids form part of or the entire epidermis (fig. 6.4, *K*).

Sclereids with definite branches or only with spicules occur in the ground tissue of *Camellia* petiole (fig. 6.4, *H, I*) and in the mesophyll of *Trochodendron* leaf. The mesophyll of *Osmanthus* and *Hakea* contains columnar sclereids, ramified at each end (fig. 6.4, *G*). *Monstera deliciosa, Nymphaea* (Water Lily), and *Nuphar* (Yellow Pond Lily) have sclereids that resemble branched plant hairs and are called trichosclereids (chapter 19). The branches of these sclereids extend into large intercellular spaces, or air chambers, characteristic of the leaves of these species. Branched sclereids may be found in leaves of conifers such as *Pseudotsuga taxifolia*.

The sclereids of the olive (*Olea europaea*) leaf are of considerable interest because of their great length (fig. 6.4, *L, M*). They average 1 millimeter in length (Arzee, 1953*a*) and may appropriately be called fiber-like or filiform sclereids. They originate in both palisade and spongy parenchyma and permeate the mesophyll in the form of a dense mat (Arzee, 1953*b*).

Sclereids in fruits. Sclereids occupy various positions in fruits. *Pyrus* (pear) and *Cydonia* (quince) have single or clustered stone cells, or brachysclereids, scattered in the fruit flesh (fig. 6.4, *A, B*). In the formation of clusters in the pear, cell divisions occur concentrically around some earlier formed sclereids, and the new cells also become sclereids (Sterling, 1954). The radiating pattern of parenchyma cells around the mature clusters of sclereids is related to this mode of development. The sclereids of pear and quince often show the so-called branched, or ramiform, pits resulting from a fusion of one or more cavities during the increase in thickness of the wall.

The apple (*Malus*) furnishes another example of sclereids in the fruit. The cartilaginous endocarp enclosing the seeds consists of

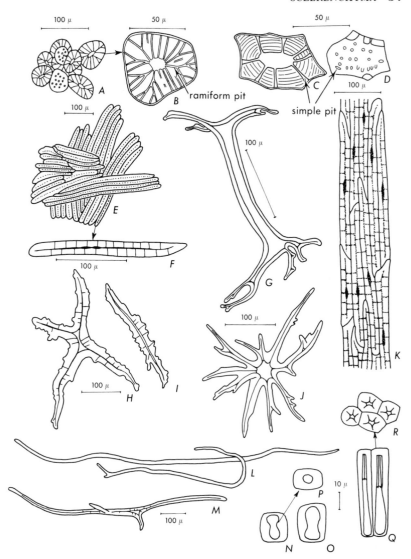

Fig. 6.4. Sclereids. *A, B,* stone cells from fruit flesh of pear (*Pyrus*). *C, D,* sclereids from stem cortex of wax plant (*Hoya*), in sectional (*C*) and surface (*D*) views. *E, F,* sclereids from endocarp of fruit of apple (*Malus*). *G,* columnar sclereid with ramified ends; from palisade mesophyll of *Hakea*. *H, I,* sclereids from petiole of *Camellia*. *J,* astrosclereid from stem cortex of *Trochodendron*. *K,* layer of sclereids from epidermis of clove scale of garlic (*Allium sativum*). *L, M,* filiform sclereids from leaf mesophyll of olive (*Olea*). *N–P,* sclereids from subepidermal layer of seed coat of bean (*Phaseolus*), "hourglass cells"; seen from the side (*N, O*) and from above (*P*). *Q, R,* epidermal sclereids, macrosclereids, with fluted wall thickenings; from seed coat of bean (*Phaseolus*) seen from the side (*Q*) and from above (*R*).

obliquely oriented layers of elongated sclereids (fig. 6.4, *E, F*). Sclereids also compose the hard shells of nut-like fruits and the stony endocarp of stone fruits.

Sclereids in seeds. The hardening of seed coats during ripening of the seeds often results from a development of secondary walls in the epidermis and in the layer or layers beneath the epidermis. The leguminous seeds furnish a good example of such sclerification. In seeds of bean (*Phaseolus*), pea (*Pisum*), and soybean (*Glycine*), columnar sclereids, macrosclereids (fig. 6.4, *R, Q*), compose the epidermis and prismatic sclereids (fig. 6.4, *N–P*) or bone-shaped osteosclereids (chapter 22) occur beneath the epidermis. The hard seed coat of the coconut (*Cocos nucifera*) contains sclereids with numerous ramiform pits.

FIBERS

Like the sclereids the fibers may be found in various parts of the plant. In the dicotyledons, fibers are particularly common in the vascular tissues. They are the phloem fibers and the xylem, or wood, fibers. In the monocotyledons, fibers may completely enclose each vascular bundle like a sheath, form a strand on one or both sides of a vascular bundle ("bundle caps"), or form strands or layers that appear to be independent of the vascular tissues. The fibers of the xylem are considered in chapter 8. The present chapter is concerned only with the fibers located outside the xylem, the extraxylary fibers, which include the phloem fibers of the dicotyledons and the fibers of the monocotyledons, whether associated with the vascular bundles or not.

The fibers are long cells, with more or less thick secondary walls, and usually occur in strands (fig. 6.5). These strands constitute the "fibers" of commerce. The process of retting used in the extraction of fibers from the plant consists of a separation of the fiber bundles from the associated nonfibrous cells. Within a strand the fibers overlap (fig. 6.1), a feature that imparts strength to the fiber bundles. In contrast to collenchyma walls, the fiber walls are not highly hydrated. They are, therefore, harder than the collenchyma walls and are elastic rather than plastic. The fibers serve as supporting elements in plant parts that are no longer elongating. The degree of lignification varies, and pits are relatively scarce.

Phloem fibers occur in many stems. The flax (*Linum usitatissimum*) stem has only one band of fibers, several cell layers in depth, located on the outer periphery of the vascular cylinder (fig. 6.5). These fibers

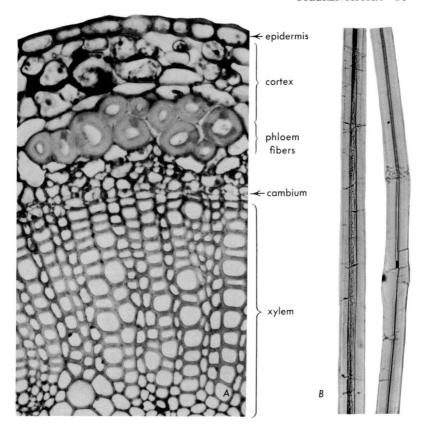

← epidermis

cortex

phloem fibers

← cambium

xylem

A

B

Fig. 6.5. Flax fibers. *A*, fibers in cross section of stem of flax (*Linum usitatissimum*). *B*, fragments of isolated fibers. (*A*, ×280; *B*, ×236; *B*, from C. H. Carpenter and L. Leney, *91 Papermaking Fibers*, Tech. Publ. 74, College of Forestry at Syracuse, 1952.)

originate in the earliest part of the primary phloem but mature as fibers after this part of the phloem ceases to function in conduction. The flax fibers are, therefore, primary-phloem fibers. In stems of *Sambucus* (Elderberry), *Tilia* (Basswood), *Liriodendron* (Tulip Tree), *Vitis* (Grapevine), *Robinia pseudoacacia* (Black Locust) and many others, fibers occur on the periphery of the phloem (primary-phloem fibers) and also within the secondary phloem (secondary-phloem fibers). Conifers may have secondary-phloem fibers (e.g., *Sequoia*, *Thuja*).

Stems of some dicotyledons have primary fibers on the periphery of the vascular cylinder that do not originate as part of the phloem tissue but outside of it. These fibers are referred to in this book as perivascular fibers. The term pericycle is often used with reference to these fibers, as well as to the primary-phloem fibers. (See chapter 16 for an evaluation of the term pericycle.)

Economic fibers

The phloem fibers of the dicotyledons constitute the bast fibers of commerce (Harris, 1954). Some of the well-known sources and usages of bast fibers are hemp (*Cannabis sativa*), cordage; jute (*Corchorus capsularis*), cordage, coarse textiles; kenaf (*Hibiscus cannabinus*), coarse textiles; flax (*Linum usitatissimum*), textiles (e.g., linen), thread; and ramie (*Boehmeria nivea*), textiles. The phloem fibers are classified as soft fibers because, whether lignified or not, they are relatively soft and flexible.

The fibers of the monocotyledons—usually called leaf fibers (Harris, 1954) because they are obtained from leaves—are classified as hard fibers. They have strongly lignified walls and are hard and stiff. Examples of sources and uses of leaf fibers are abaca or Manila hemp (*Musa textilis*), cordage; bowstring hemp (*Sansevieria*, entire genus), cordage; henequen (*Agave*, various species), cordage, coarse textiles; New Zealand hemp (*Phormium tenax*), cordage; pineapple fiber (*Ananas comosus*), textiles; and sisal (*Agave sisalana*), cordage.

The length of individual fiber cells varies considerably in the different species. Examples of ranges of lengths in millimeters may be cited from Harris' (1954) handbook. Bast fibers: jute, 0.8–6.0; hemp, 5–55; flax, 9–70; ramie, 50–250. Leaf fibers: sisal, 0.8–8.0; bowstring hemp, 1–7; abaca, 2–12; New Zealand hemp, 2–15.

In commerce the term fiber is often applied to materials that include, in botanical sense, other types of cells beside fibers and also to structures that are not fibers at all. In fact, the leaf fibers of the monocotyledons commonly include vascular elements. Cotton fibers are epidermal hairs of seeds of *Gossypium* (chapter 7); raffia constitutes leaf segments of *Raphia* palm; rattan, stems of *Calamus* palm.

DEVELOPMENT OF SCLEREIDS AND FIBERS

The development of branched and long sclereids and of the usually long fibers involves remarkable intercellular adjustments that suggest the existence in these cells of a degree of independence from the positional influences stressed in chapter 3. The primordium of a branched sclereid may not differ in appearance from neighboring parenchyma cells. Later, however, instead of enlarging uniformly it develops processes which elongate into branches. During this elongation, the branches not only invade the intercellular spaces but also force their way between walls of other cells. Presumably a softening of the middle lamella occurs between these cells by a still unknown process. (As may be remembered, a similar softening is thought to be involved in the formation of schizogenous intercellular spaces.) Thus, the sclereid establishes new contacts during its growth and is able to attain a much larger size than its neighbors despite the fact that it grows within a continuous tissue of cells which also are increasing in size.

The growth of cells by intrusion between walls of other cells is called intrusive growth, and it is contrasted with the coordinated growth that involves no separation of walls. Thus, a group of similar cells in a homogeneous parenchyma tissue grows by coordinated growth, with the pairs of conjoined primary walls presumably expanding at the same rate without a breakage of connection along the middle lamella. Coordinated growth does not prevent some cells from becoming longer than others. If a cell ceases to divide while its neighbors continue to do so the nondividing cell will become longer than its neighbors without affecting the intercellular relation of walls. In the growth of a sclereid, coordinated growth of the main body of cells is combined with intrusive growth in the elongating parts of its branches.

A fiber also shows a combination of coordinated and intrusive growth. A young fiber primordium increases in length without changing cellular contacts while the adjacent parenchyma cells are actively dividing. Somewhat later, the fiber primordium attains additional length by intrusive growth carried out at both its apices. During its elongation the fiber may show repeated nuclear division without the formation of new walls. Thus, the fiber protoplast may become multinucleate.

The interesting phenomenon of apical intrusive growth has been studied in detail with regard to flax fibers (Schoch-Bodmer and Huber, 1951). By measuring young and old internodes and the fibers contained in these internodes the authors calculated that by coordinated growth alone the fibers could become 1 to 1.8 centimeters long. Actually they found fibers ranging in length between 0.8 and 7.5 centimeters. Thus, lengths over 1.8 centimeters must have been attained by apical intrusive growth. The authors found microscopic evidence of such growth. In transverse sections taken about 2 millimeters below the shoot apex the appearance of small cells among rather wide fiber primordia indicated the intrusion of the narrow tips of growing fibers. Longitudinal sections taken at the same levels showed that, in contrast to the main body of the young fiber, its tips had very thin walls, contained dense cytoplasm with chloroplasts, and were not plasmolyzable. When during the intrusive growth the tip is obstructed in its progress by some combination of cells, the tip curves or forks. Thus, the occurrence of bent and forked ends in fibers and sclereids is further evidence of intrusive growth.

The great lengths attained by fibers and some sclereids introduces a certain complexity into the phenomenon of secondary thickening of the walls in these cells. As was mentioned previously, the secondary wall commonly develops over the primary after the latter ceases to expand. In sclereids and fibers showing protracted growth, the older part of the cell ceases to grow while the apices continue to elongate. The older (that is, median) part of the cell then begins to form secondary wall layers before the growth at the tips is completed. From the median part of the cell the secondary thickening progresses toward the tips and is completed after the tips cease to grow.

REFERENCES

Arzee, T. Morphology and ontogeny of foliar sclereids in *Olea europaea*. I. Distribution and structure. *Amer. Jour. Bot.* 40:680–687. 1953*a*. II. Ontogeny. *Amer. Jour. Bot.* 40:745–752. 1953*b*.

Bailey, I. W. The structure of tracheids in relation to the movement of liquids, suspensions, and undissolved gases. In: *The Physiology of Forest Trees.* pp. 71–82. K. V. Thimann, ed. New York, Ronald Press Company. 1958.

Belford, D. S., A. Myers, and R. D. Preston. Spatial and temporal variation of microfibrillar organization in plant cell walls. *Nature* 181:1251–1253. 1958.

Böhmer, H. Untersuchungen über das Wachstum und den Feinbau der Zellwände in der *Avena*-Koleoptile. *Planta* 50:461–497. 1958.

Bosshard, H. H. Elektronenmikroskopische Untersuchungen im Holz von *Fraxinus excelsior* L. *Schweiz. Bot. Gesell. Ber.* 62:482–508. 1952.

Foster, A. S. Structure and ontogeny of terminal sclereids in *Boronia serrulata*. *Amer. Jour. Bot.* 42:551–560. 1955.

Foster, A. S. Plant idioblasts: remarkable examples of cell specialization. *Protoplasma* 46:184–193. 1956.

Harris, M., ed. *Handbook of textile fibers.* Washington, Harris Research Laboratories. 1954.

Preston, R. D. *The molecular architecture of plant cell walls.* New York, John Wiley and Sons. 1952.

Rao, T. A. Studies on foliar sclereids. A preliminary survey. *Indian Bot. Soc. Jour.* 30:28–39. 1951.

Schoch-Bodmer, H., and P. Huber. Das Spitzenwachstum der Bastfasern bei *Linum usitatissimum* and *Linum perenne*. *Schweiz. Bot. Gesell. Ber.* 61:377–404. 1951.

Sterling, C. Sclereid development and the texture of Bartlett pears. *Food Research* 19:433–443. 1954.

Wardrop, A. B., and J. Cronshaw. Changes in cell wall organization resulting from surface growth in parenchyma of oat coleoptiles. *Austral. Jour. Bot.* 6:89–95. 1958.

$7.$ EPIDERMIS

THE EPIDERMIS IS A SYSTEM OF CELLS, VARIABLE IN STRUCTURE AND function, that together constitute the covering of the plant body in its primary state. Many of the structural features of the epidermis can be related to the roles this tissue plays as the layer of cells in contact with the external environment of the plant. The presence of the wax-like material, cutin, within the outer wall and on its surface (*cuticle*) restricts transpiration. Stomata are concerned with gaseous exchange. Because of the compact arrangement of cells and the presence of the relatively tough cuticle, the epidermis offers mechanical support. The epidermis of young roots is specialized with regard to absorption in having thin walls with a thin cuticle (Scott et al., 1958) and bearing root hairs. The epidermis may have other accessory functions that are associated with various special structural features.

The epidermis is usually one layer of cells in thickness. In some plants, however, the protoderm of leaves divides parallel with the surface (periclinally), and the derivatives of these divisions divide again so that a tissue of several ontogenetically related layers is produced. Such a tissue is referred to as *multiple epidermis* (fig. 7.1). (The velamen of roots—see chapter 14—is also a multiple epidermis.) The outermost layer of a multiple epidermis in a leaf assumes epidermal characteristics, whereas those beneath commonly develop into a tissue with few or no chloroplasts, which therefore is distinct from the underlying mesophyll. One of the functions ascribed to such a tissue is storage of water. In some plants subepidermal layers resemble those of a multiple epidermis but are derived from the ground tissue. Mul-

multiple epidermis   young lithocyst

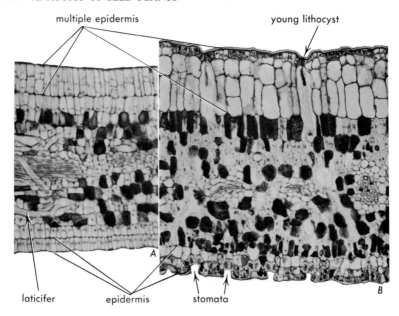

laticifer   epidermis   stomata   B

Fig. 7.1. Cross sections of leaf of *Ficus elastica* showing development of multiple epidermis. In upper epidermis of *A,* many cells have divided periclinally; in *B,* this epidermis is three to four cells in depth. Lithocysts do not divide. Lower epidermis also becomes multiple. (*A,* ×190; *B,* ×180.)

tiple epidermis, therefore, can be identified as such only by developmental studies.

The epidermis may last through the life of a given plant part, or it may be later replaced by another protective tissue, the periderm (chapter 12).

COMPOSITION OF EPIDERMIS

The epidermis consists of relatively unspecialized cells composing the groundmass of the tissue and the more specialized cells dispersed through this mass (fig. 7.2). The ground cells of the epidermis vary in depth but are often tabular in shape. In elongated plant parts such as stems, petioles, vein ribs of leaves, and leaves of most monocotyledons, the epidermal cells are elongated parallel with the long axis of the part (fig. 7.3). In leaves, petals, ovaries, and ovules the epidermal cells may have wavy vertical (anticlinal) walls.

The epidermal cells have living protoplasts and may store various products of metabolism. They contain plastids, but the plastids do not usually develop much, if any, chlorophyll. Small, typically granular chloroplasts, however, may be present (e.g., spinach; Kaja, 1954). Starch occurs in the epidermal plastids (Weber et al., 1955).

Among the more highly specialized epidermal cells are the guard cells of the stomata, which are discussed in detail farther below. Various other epidermal cells with special structural features or cell contents occur in many plants. In the Gramineae, for example, small cells filled with silica (silica cells) and cells with suberized walls (cork cells) occur among the ordinary epidermal cells (fig. 7.3, *B*). Gramineae and other monocotyledons often have rows of enlarged epidermal cells, the bulliform cells (chapter 19).

Plants of various groups may contain fiber-like cells in the epidermis. In the Gramineae the epidermal fibers may become over 300 microns in length. Cells containing tannins, oils, crystals, and other materials are often scattered as idioblasts in the epidermis. Sometimes consid-

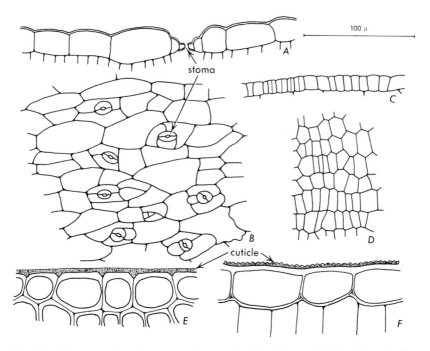

Fig. 7.2. Structure of epidermis. *A–D,* outer (*A, B*) and inner (*C, D*) epidermis of *Capsella* pericarp in sectional (*A, C*) and surface (*B, D*) views. *E,* epidermis from *Sambucus* stem. Cuticle forms ridges inserted between cells. *F,* epidermis of *Helleborus* leaf with cuticle, which has a wavy surface.

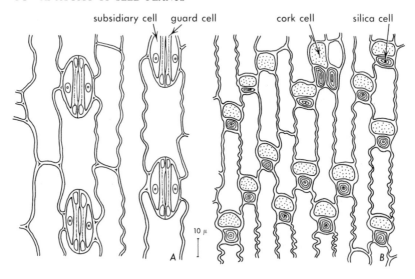

Fig. 7.3. Epidermis of a grass—sugar cane (*Saccharum*)—in surface views. *A,* lower epidermis of leaf with stomata. *B,* epidermis of stem with cork cells and silica cells.

erable expanses of the epidermis consist of specialized cells such as sclereids (chapter 6) or secretory cells (chapter 13). Epidermal appendages or trichomes, such as various plant hairs, occur in plants in many forms and sizes.

CELL WALL

The epidermal walls vary in thickness in different plants and in different parts of the same plant. In the thin-walled epidermis the outer wall is frequently the thickest. Epidermis with exceedingly thick walls is found in the leaves of conifers (chapter 19). In these leaves the wall thickening, which is probably secondary in nature, may almost obliterate the lumen of the cells and be lignified. Pits generally occur in the anticlinal and the inner periclinal (next to the mesophyll) walls of the epidermis. Plasmodesmata have been described not only in the anticlinal and the inner periclinal walls but also in the outer wall (Lambertz, 1954; Scott et al., 1958).

The most characteristic feature of the epidermal cells of the aerial parts of the plant is the presence of cutin, a fatty substance, as an incrustation of the outer walls and as a separate layer, the cuticle, on the outer surface of the cell (fig. 7.2, *E, F*). (The process of impregnation with cutin is called cutinization, and the formation of cuticle, cuticularization.) The structure of the cuticle is of particular interest to investigators concerned with problems of penetrability of the leaf surface by chemicals applied as mineral nutrients, fungicides, or herbicides. Although the cuticle appears to be highly impenetrable it may have areas of weakness during the expansion of the epidermal cells that would permit the entry of chemicals (Schiefferstein and Loomis, 1956). In many plants the cuticle is covered with a deposit of wax (bloom) in a variety of patterns (Mueller et al., 1954). Such deposits do not seem to play an important role in reducing the transpiration, but they reduce the wettability of the surface. The thickness of the cuticle is variable and is affected by environmental conditions. It not only occurs on the surface of the epidermal cells but often projects rib-like into the anticlinal walls (fig. 7.2, E).

The cutinized part of the epidermal wall beneath the cuticle has a complicated structure. It contains a cellulose framework and pectic compounds, cutin, waxes, and certain other compounds as incrusting substances. The source and the mode of migration of the cutin and waxes in the epidermal cells are still unsolved matters, although presence of plasmodesmata in the outer walls might be possibly significant in this regard. Some workers (Schiefferstein and Loomis, 1956) visualize an initial flooding of the outer surface of the epidermis with a substance resembling a drying and hardening oil. Under this primary cuticle further cuticular layers, containing a mixture of cutin, waxes, and other materials, may be deposited and produce thick, sometimes laminated cuticles. As the cutin and waxes migrate through the outer wall toward the surface they also impregnate this wall. The presence of a pectic layer beneath the cuticle has been demonstrated in some plants (Roelofsen, 1952; Scott et al., 1958) and probably explains why fungi often grow between the cuticle and the epidermal wall (Wood et al., 1952).

STOMATA

The stomata are openings (the stomatal pores, or apertures) in the epidermis bounded by two specialized epidermal cells, the guard cells, which by changes in shape bring about the opening and closure of the aperture (fig. 7.4). It is convenient to apply the term stoma to the entire unit, the pore and the two guard cells. The stoma may be surrounded by cells that do not differ from other ground cells of the epidermis. On the other hand, in many plants the stomata are flanked or surrounded by cells that differ in shape, and sometimes also in content, from the ordinary epidermal cells. These distinct cells are called subsidiary cells of the stoma (fig. 7.5, *B, C*). The subsidiary cells may or may not be closely related ontogenetically to the guard cells.

Stomata occur on all aerial parts of the plant, but they are most abundant on leaves. Roots usually lack stomata. Stomatal frequency varies greatly. It varies on different parts of the same leaf and on different leaves of the same plant and is influenced by environmental conditions. In leaves stomata may occur on both sides or mostly or only on one side, usually the lower. Stomata also vary in the level of their position in the epidermis. Some are even with the other epidermal cells; others are raised above or sunken below the surface.

The two features just discussed, the number of stomata per unit area and the positional level of the guard cells with respect to the other epidermal cells, are so variable that they are of little taxonomic value. The more frequently used taxonomic character is the appearance of the stomata as seen from the surface, especially with reference to the nature and orientation of the neighboring cells (Bondeson, 1952; Metcalfe and Chalk, 1950).

With regard to the neighboring cells, the dicotyledons show four principal stomatal types (as revised by Metcalfe and Chalk, 1950, p. XV): type A (fig. 7.5, *A*), no subsidiary cells are present, several ordinary epidermal cells irregularly surround the stoma (*anomocytic* or irregular-celled type); type B (fig. 7.5, *B*), three subsidiary cells, one distinctly smaller than the other two, surround the stoma (*anisocytic*

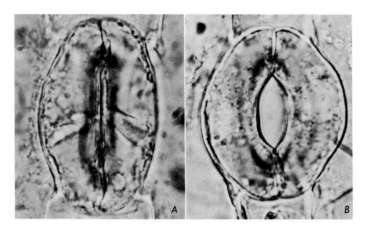

Fig. 7.4. Stomata of onion (*Allium Cepa*) in closed (*A*) and opened (*B*) states. (From Shaw, *New Phytol.,* 1954.)

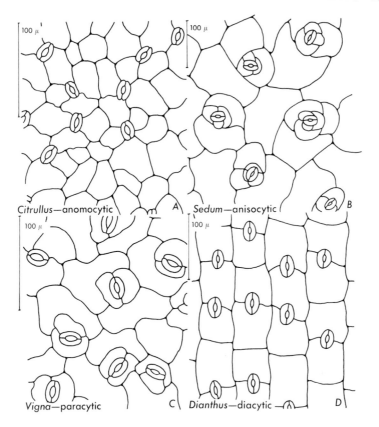

Fig. 7.5. Epidermis in surface view illustrating patterns formed by guard cells and surrounding cells.

or unequal-celled type); type C (fig. 7.5, *C*), one or more subsidiary cells flank the stoma parallel with the long axis of the guard cells (*paracytic* or parallel-celled type); type D (fig. 7.5, *D*), one pair of subsidiary cells, with their common walls at right angles to the long axis of the guard cells, surrounds the stoma (*diacytic* or cross-celled type).

The guard cells of the dicotyledons (fig. 7.6, *A–D*) are commonly crescent shaped, with rounded ends (kidney shaped), and have ledges of wall material on the upper or both the upper and the lower sides. They are covered with a cuticle which extends over the surfaces facing the stomatal pore and the substomatal chamber. The guard cells typically have unevenly thickened walls. This feature appears to play a role in the opening and closing of the stomatal pore (cf. Stålfelt, 1956). The thinner parts of the wall are more extensible and are affected more strongly by changes in the turgor of the cell than the thicker parts. As a result of this differential response the shape and volume of the guard cells—and concomitantly the size of the stomatal aperture—are changed when the turgor changes. The stomata are open at high turgor, closed at low turgor.

The changes in the turgor of the guard cells are commonly associated with the changes in the state of the carbohydrates in these cells (e.g., Yemm and Willis, 1954). Guard cells of many plants are known to have chloroplasts which show fluctuating amounts of starch. Some workers, however, question whether the changes in carbohydrates are the primary cause of guard-cell movement (e.g., Heath, 1952; Williams, 1954). At least these changes do not necessarily involve starch-sugar changes because starch may not occur in the guard cells. In the

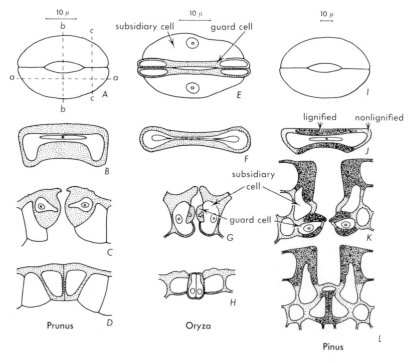

Fig. 7.6. Stomata in representatives of different groups of plants. Stomata in *A, E, I* are shown from the surface. Other drawings show sections of stomata made in the planes indicated in *A: B, F, J,* plane *aa; C, G, K,* plane *bb; D, H, L,* plane *cc.* *E,* guard cells are shown in a high focal plane so that the lumen is not visible in the narrow part of the cell. (*Prunus* and *Pinus,* after Esau, *Plant Anatomy,* John Wiley and Sons, 1953.)

onion, for example, these cells contain only small and pale chloroplasts and form no starch (Shaw, 1954).

Among the guard cells of the monocotyledons, those of the Gramineae and Cyperaceae constitute a rather special type (fig. 7.6, E–H). As seen from the surface, they are narrow in the middle and enlarged at both ends. The central narrow part has a very thick wall, the bulbous ends have thinner walls. The opening and closing of the stomata results from changes in the size of the bulbous ends. When they are swollen, the stoma is open. The guard cells of the Gramineae are associated with subsidiary cells, one on each long side of the stoma. These subsidiary cells are derived from epidermal cells flanking the guard cells although later they appear to be completely a part of the stoma.

The stomata of the gymnosperms (fig. 7.6, I–L) are commonly deeply sunken and sometimes appear as though suspended from subsidiary cells that overarch them. The characteristic feature of these stomata is that the walls of the guard cells and of the subsidiary cells are partly free of lignin. This combination of more and less rigid wall parts, the manner of connection between the guard cells and the subsidiary cells, and the presence of thin wall parts in the subsidiary cells all appear to be involved in the mechanism of opening and closing of the stomata. The subsidiary cells may or may not be related ontogenetically to the guard cells. In the *haplocheilic* type (e.g., cycads, conifers, *Ginkgo*) the subsidiary cells are not related to the guard cells; in the *syndetocheilic* type (e.g., *Gnetum, Welwitschia*) a protodermal cell divides into a guard-cell mother cell and two lateral cells, each of which either becomes a subsidiary cell or gives rise to subsidiary cells by division (Florin, 1951). In the establishment of typology of the gymnosperm stomata the origin and the arrangement of the subsidiary cells are taken into account.

In the development of stomata in angiosperms, the mother cell, or the precursor, of the guard cells (fig. 7.7, *A, D*) commonly originates by an unequal division of a protodermal cell and is the smaller of two cells resulting from such a division (Bondeson, 1952; Bünning and Biegert, 1953). It divides into the two guard cells (fig. 7.7, *A, B, E*) which, through differential expansion, acquire their characteristic shape. The intercellular substance between the guard cells swells (fig. 7.7, *B*), and the connection between the cells is weakened. They separate in their median parts, and the stomatal opening is thus formed (fig. 7.7, *C, F*). Various spatial readjustments occur between the guard cells and the adjacent subsidiary or other epidermal cells so that the guard cells may be elevated above or lowered below the sur-

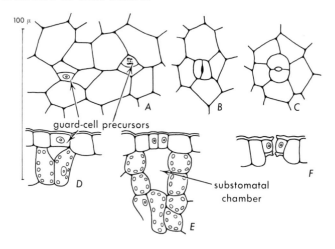

Fig. 7.7. Development of stomata. Guard-cell precursors (mother cells) have been formed by a division of a protodermal cell (*A, D*). The precursor has divided into two guard cells (*B, E*). Stomatal opening has been formed (*C, F*).

face of the epidermis. Finally, the adjacent cells may overarch the guard cells or grow under them into the substomatal chamber.

Stomata begin to develop in a leaf shortly before the main period of meristematic activity in the epidermis is completed and continue to develop through a considerable part of the later extension of the leaf by cell enlargement. In leaves with parallel venation, as in most monocotyledons, and with the stomata arranged in longitudinal rows, the formation of stomata begins at the apices of the leaves and progresses in the downward direction. In the netted-veined leaves, as in most dicotyledons, different developmental stages are mixed in a mosaic fashion.

TRICHOMES

Trichomes (fig. 7.8 and chapter 13) are highly variable appendages of the epidermis, including glandular (or secretory) and nonglandular hairs, scales, papillae, and the absorbing hairs of roots. They occur on all parts of the plant and may persist through the life of a plant part or may fall off early. Some of the persisting hairs remain alive; others die and become dry. Although trichomes vary widely in structure within larger and smaller groups of plants, they are sometimes remarkably uniform and may be used for taxonomic purposes (Cowan, 1950; Metcalfe and Chalk, 1950, pp. 1326–1329).

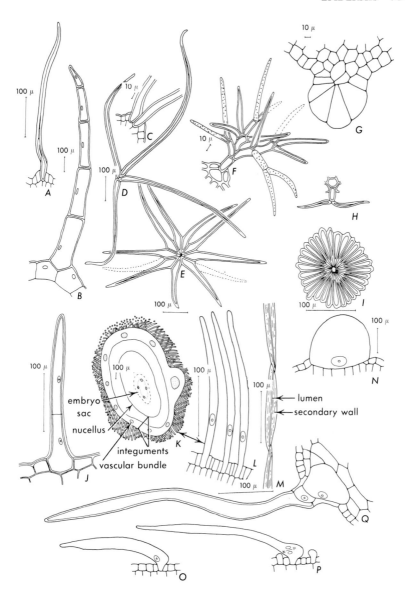

Fig. 7.8. Trichomes. *A,* simple hair from *Cistus* leaf. Structure resembling a short hair is included at the base. *B,* Uniseriate hair from *Saintpaulia* leaf. *C, D,* tufted hair from leaf of cotton (*Gossypium*). *E,* stellate hair from leaf of alkali mallow (*Sida*). *F,* dendroid hair from lavender leaf (*Lavandula*). *G,* short multicellular hair from leaf of potato (*Solanum*). *H, I,* peltate scale from leaf of olive (*Olea*). *J,* bicellular hair from stem of *Pelargonium*. *K–M,* cotton (*Gossypium*). Epidermal hairs from seed (*K*) in young stage (*L*) and mature, with secondary walls (*M*). *N,* water vesicle of *Mesembryanthemum.* *O–Q,* hairs in three stages of development from leaf of soybean (*Glycine*).

Trichomes are classified into morphological categories. Some of these categories are: (1) hairs, which may be unicellular or multicellular, glandular (chapter 13) or nonglandular (fig. 7.8, *A, B, J, Q*); (2) scales, or peltate hairs (Fig. 7.8, *H, I*); (3) water vesicles, which are enlarged epidermal cells (fig. 7.8, *N*); (4) root hairs (chapter 14). The hairs may be tufted (fig. 7.8, *C, D*), star shaped (stellate, fig. 7.8, *E*), or branched (dendroid, fig. 7.8, *F*). The hairs of cotton seeds are unicellular and develop secondary walls at maturity (fig. 7.8, *K–M*).

REFERENCES

Bondeson, W. Entwicklungsgeschichte und Bau der Spaltöffnungen bei den Gattungen *Trochodendron* Sieb. et Zucc., *Tetracentron* Oliv. und *Drimys* J. R. et G. Forst. *Acta Horti Bergiani* 16:169–218. 1952.

Bünning, E., and F. Biegert. Die Bildung der Spaltöffnungsinitialen bei *Allium Cepa. Ztschr. für Bot.* 41:17–39. 1953.

Cowan, J. M. *The Rhododendron leaf; a study of the epidermal appendages.* Edinburgh, Oliver and Boyd. 1950.

Florin, R. Evolution in cordaites and conifers. *Acta Horti Bergiani* 15:285–388. 1951.

Heath, O. V. S. Studies in stomatal behaviour. II: The role of starch in the light response of stomata. Part 2: The light response of stomata of *Allium cepa* L., together with some preliminary observations on the temperature response. *New Phytol.* 51:30–47. 1952.

Kaja, H. Untersuchungen über die Plastiden in den Epidermiszellen von *Spinacia oleracea* L. *Planta* 44:503–508. 1954.

Lambertz, P. Untersuchungen über das Vorkommen von Plasmodesmen in den Epidermisaussenwänden. *Planta* 44:147–190. 1954.

Metcalfe, C. R., and L. Chalk. *Anatomy of the dicotyledons.* 2 Vols. Oxford, Clarendon Press. 1950.

Mueller, L. E., P. H. Carr, and W. E. Loomis. The submicroscopic structure of plant surfaces. *Amer. Jour. Bot.* 41:593–600. 1954.

Roelofsen, P. A. On the submicroscopic structure of cuticular cell walls. *Acta Bot. Neerland.* 1:99–114. 1952.

Schiefferstein, R. H., and W. E. Loomis. Wax deposits on leaf surfaces. *Plant Physiol.* 31:240–247. 1956.

Scott, F. M., K. C. Hamner, E. Baker, and E. Bowler. Electron microscope studies of the epidermis of *Allium cepa. Amer. Jour. Bot.* 45:449–461. 1958.

Shaw, M. Chloroplasts in the stomata of *Allium cepa* L. *New Phytol.* 53:344–348. 1954.

Stålfelt, M. G. Die stomatäre Transpiration und die Physiologie der Spaltöffnungen. In: *Handbuch der Pflanzenphysiologie.* 3:350–426. 1956.

Weber, F., I. Thaler, and G. Kenda. Die Plastiden der *Cleome*-Epidermis. *Österr. Bot. Ztschr.* 102:84–88. 1955.

Williams, W. T. A new theory of the mechanism of stomatal movement. *Jour. Exp. Bot.* 5:343–352. 1954.

Wood, R. K. S., A. H. Gold, and T. E. Rawlins. Electron microscopy of primary cell walls treated with pectic enzymes. *Amer. Jour. Bot.* 39:132–133. 1952.

Yemm, E. W., and A. J. Willis. Stomatal movements and changes of carbohydrate in leaves of *Chrysanthemum maximum. New Phytol.* 53:373–396. 1954.

8. XYLEM:
GENERAL STRUCTURE AND CELL TYPES

THE XYLEM IS THE PRINCIPAL WATER-CONDUCTING TISSUE OF A vascular plant. It is usually spatially associated with the phloem (fig. 8.1), the principal food-conducting tissue. The two tissues together are called the vascular tissue or tissues. The combination of xylem and phloem forms a continuous vascular system throughout all parts of the plant, including all branches of stem and root.

Developmentally, it is convenient to distinguish between primary and secondary vascular tissues. The primary tissues differentiate during the formation of the primary plant body, that is, the body that originates in the embryo and is further elaborated through the activity of the apical meristems and their derivative meristematic tissues (chapter 3). The meristem directly concerned with the formation of the primary vascular tissues is the procambium, also called provascular tissue.

The secondary vascular tissues are produced during the second major stage of plant development, in which an increase in thickness results from lateral additions of new tissues to the primary body. This thickening is most pronounced in the axial parts of the plant (that is, stem and root) and their larger branches. It results from the activity of a special meristem, the vascular cambium (fig. 8.1). As was mentioned in chapter 3, secondary growth is absent in small annuals of the dicotyledons and in most monocotyledons.

The primary and the secondary xylem have histologic differences, but both are complex tissues containing at least water-conducting elements and parenchyma cells and usually also other types of cells,

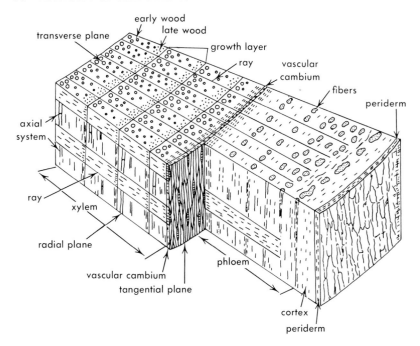

Fig. 8.1. Block diagram illustrating basic features of secondary vascular tissues and their spatial relation to one another and to vascular cambium and periderm.

especially supporting cells. The characteristics of these various types of cells and their interrelations in the tissue may be best introduced by a consideration of secondary xylem, or wood.

GROSS STRUCTURE OF THE SECONDARY XYLEM

Axial and radial systems

With the aid of low magnification, a study of a block of wood reveals the presence of two distinct systems of cells (fig. 8.1): the *axial* (longitudinal or vertical) and the *radial* (transverse or horizontal) or *ray* system. The axial system contains cells or files of cells with their long axes oriented vertically in the stem or root, that is, parallel to the axis of these organs (or their branches); and the *radial* is composed of files of cells oriented horizontally with regard to the axis of stem or root.

The two systems have their characteristic appearances in the three kinds of sections employed in the study of wood (figs. 9.1 and 9.4).

In the transverse section, that is, the section cut at right angles to the axis of the stem or root, the cells of the axial system are cut transversely and reveal their smallest dimensions. The rays, on the other hand, are exposed in their longitudinal extent in a cross section. When stems or roots are cut lengthwise, two kinds of longitudinal sections are obtained: the radial (parallel to a radius) and the tangential (perpendicular to a radius). Both show the longitudinal extent of the cells of the axial system, but they give strikingly different views of the rays. Radial sections expose the rays as horizontal bands lying across the axial system. When a radial section cuts a ray through its median plane it reveals the height of the ray. Tangential sections cut a ray approximately perpendicularly to its horizontal extent and reveal its height and width. In tangential sections, therefore, it is easy to measure the height of the ray—this is usually done in terms of number of cells—and to determine whether the ray is uniseriate (one cell wide, fig. 9.1) or multiseriate (two to many cells wide, fig. 9.4).

Growth layers

With little or no magnification the wood discloses the layering resulting from the presence of more or less sharp boundaries between successive growth layers—growth rings in transections (figs. 8.1 and 9.8, *B*). Each growth layer may be a product of one seasonal growth period, but various environmental conditions may induce the formation of more than one growth layer during one season. When conspicuous layering is present—commonly in woods of the temperate zone—each growth layer is divisible into early and late wood. The early wood is less dense than the late wood because wider cells with thinner walls predominate in the early wood, narrower cells with thicker walls in the late. The late wood forms a distinct boundary of a growth ring because of its sharp contrast to the early wood of the following season, but the change from early wood to the late wood of the same growth layer is more or less gradual. The relative amounts of early and late wood are affected by environmental conditions and specific differences. Adverse growth conditions, for example, increase the relative amount of late wood in a pine but decrease the relative amount of the same wood in an oak.

Sapwood and heartwood

The earlier increments of secondary xylem gradually become nonfunctional in conduction and storage. The relative amounts of the nonfunctioning wood, the heartwood, vary in different species and are

also affected by environmental conditions (Harris, 1954; Trendelenburg, 1955). The heartwood is generally darker in color than the
active wood, or sapwood. Formation of the heartwood involves removal of reserve materials or their conversion into heartwood substances and eventual death of protoplasts of the parenchymatous
members of the wood.

CELL TYPES IN THE SECONDARY XYLEM

The fine structure of the xylem is studied in preparations of the three
kinds of sections mentioned earlier and in macerated wood, that is,
wood dissociated into groups of cells or individual cells by treatments
that dissolve the middle lamella.

The principal cell types of the secondary xylem (figs. 8.2 and 8.3)
may be thus tabulated.

Cell Types	Principal Function
Axial system	
Tracheary elements	
Tracheids	Conduction of water
Vessel members	
Fibers	
Fiber-tracheids	Support; sometimes storage
Libriform fibers	
Parenchyma cells	
Ray system	Storage and translocation of ergastic substances
Parenchyma cells	
(Tracheids in some conifers)	

Tracheary elements

The tracheary elements are the most highly specialized cells of the
xylem and are concerned with the conduction of water and substances
dissolved in water. They are more or less elongated cells, nonliving
at maturity. They have lignified walls with secondary thickenings and
a variety of pits.

The two kinds of tracheary cells, the tracheids and vessel members,
differ from each other in that the tracheid is an imperforate cell
whereas the vessel member has perforations, one or more at each end
(fig. 8.2), sometimes also on a side wall (fig. 8.2, *F*). In tracheids the
passage of water from cell to cell is facilitated by the presence of pit-

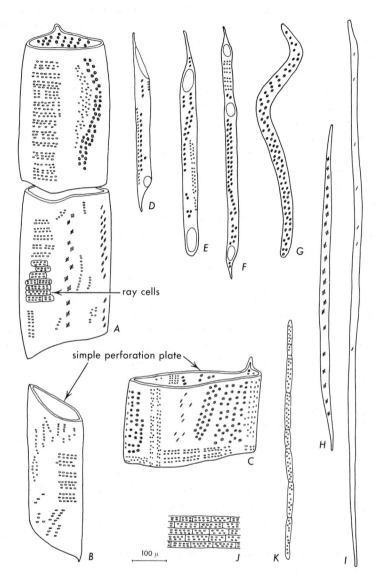

Fig. 8.2. Cell types in secondary xylem as illustrated by dissociated wood elements of *Quercus,* oak. Various pits appear on cell walls. *A–C,* wide vessel members. *D–F,* narrow vessel members. *G,* tracheid. *H,* fiber-tracheid. *I,* libriform fiber. *J,* ray parenchyma cells. *K,* axial parenchyma strand. (*A–I,* drawn from photographs in C. H. Carpenter and L. Leney, *91 Papermaking Fibers,* Tech. Publ. 74, College of Forestry at Syracuse, 1952.)

pairs, with thin primary pit membranes, in walls between superimposed or adjacent tracheids. In vessel members, on the other hand, the water moves freely from cell to cell through perforations (openings) in the wall.

Vessel members occur in longitudinal series which constitute the vessels. Vessels are not of indefinite lengths, although in some species with especially wide vessels in the early wood (ring-porous) they have been reported to extend through almost the entire height of a tree (Greenidge, 1952). Where a vessel ends, it is presumably connected with another through bordered pits. The question of length of vessels needs further study.

The perforated part of a wall of a vessel member is called the perforation plate. A plate may be simple, with only one perforation, or multiperforate, with more than one perforation (fig. 8.3). The multiperforate plates are scalariform if the perforations are elongated and arranged parallel to each other (fig. 8.3, *C*) and reticulate if the perforations form a net-like pattern.

The perforations in vessel-member walls develop during ontogeny. A young vessel member has a continuous primary wall (fig. 8.4, *A*). The middle lamella swells in the areas of future perforations (fig. 8.4, *B*). Subsequently the secondary wall develops, except over the pit membranes and the areas of future perforations (fig. 8.4, *C*). The primary walls and middle lamella of two contiguous vessel members are dissolved in the perforation areas (fig. 8.4, *D*), presumably by action of the still living protoplasts. The secondary wall may form a rim around the perforation.

Simple and bordered pits are encountered in the secondary walls of tracheids and vessel members (figs. 8.2 and 8.3). The number and arrangement of these pits are highly variable, even on different wall facets of the same cell, because they depend on the type of cell contiguous with the particular wall facet. Usually numerous bordered pit-pairs occur between contiguous tracheary elements (fig. 8.5, *A–G, P, Q*); few or no pit-pairs occur between tracheary elements and fibers; half-bordered or simple pit-pairs occur beetween tracheary elements and parenchyma cells. In half-bordered pit-pairs the border is on the side of the tracheary cell (fig. 8.5, *K*).

The bordered pit-pairs of conifer tracheids are large, particularly in early wood. As seen in front view, one of the common forms of a conifer pit-pair has a circular outline of the *border* and a circular *aperture* in the border (figs. 8.3, *A, B,* and 8.6, *B*). The border, which is formed by the secondary wall, overarches the *pit cavity* (fig. 8.6, *A*). At the bottom of the cavity is the *pit membrane* composed of the middle lamella and two layers of primary wall. The central thickened

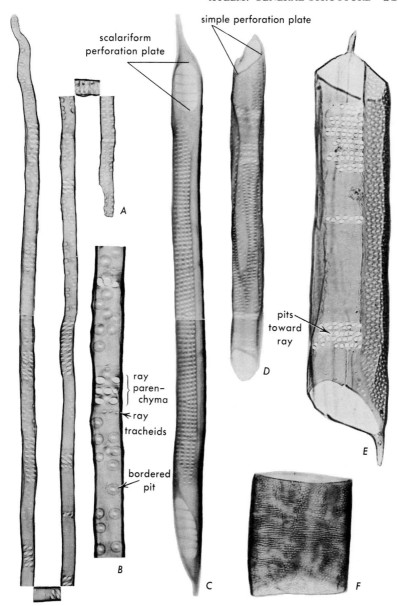

Fig. 8.3. Tracheary elements. *A,* early wood tracheid of sugar pine (*Pinus lambertiana*). *B,* enlarged part of *A. C–F,* vessel members of tulip tree, *Liriodendron tulipifera* (*C*), beech, *Fagus grandifolia* (*D*), black cottonwood, *Populus trichocarpa* (*E*), tree-of-heaven, *Ailanthus altissima* (*F*). (*A,* ×60; *B,* ×125; *C,* ×111; *D,* ×120; *E,* ×130; *F,* ×144. From C. H. Carpenter and L. Leney, *91 Papermaking Fibers,* Tech. Publ. 74, College of Forestry at Syracuse, 1952.)

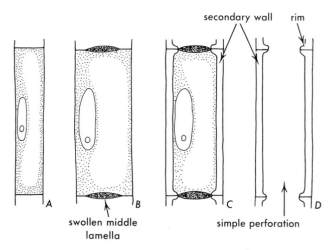

Fig. 8.4. Diagrams illustrating development of a vessel member. *A*, meristematic cell. *B*, future perforation area is thickened because of swelling of middle lamella. *C*, secondary thickening has been formed except over future perforation area. *D*, perforation has occurred and protoplast has disappeared.

part of the pit membrane is the *torus* (fig. 8.6, *A*). The margin of the torus appears in front view of a pit as a circle somewhat larger than that of the aperture (fig. 8.6, *B*). Above and below a pit thickenings of the middle lamella and primary wall may form *crassulae* (chapter 9).

The pit membrane has openings that are larger than plasmodesmatal pores (e.g., Eicke, 1954; Liese and Johann, 1954). According to electron-microscope studies on *Pinus sylvestris* (Frey-Wyssling et al., 1956), originally the membrane has a dispersed arrangement of microfibrils, but later the microfibrils are gathered into rather thick radial strands. Some tangentially oriented microfibrils link the radial strands. The resulting net has openings of about 0.3 micron in diameter, a sufficiently large opening to permit the passage of gold and carbon particles, which are 0.15 micron in diameter. The perforations in vessels are, of course, more penetrable.

In the active conifer wood the pit membrane with its torus occupies the median position in the pit-pair (fig. 8.6, *A*). In the heartwood, however, the pit membranes of most bordered pit-pairs are laterally displaced, and the tori block the apertures (fig. 8.6, *C*); in this condition the pits are said to be *aspirated* (Committee on Nomenclature, 1957). The aspiration of pits begins gradually in the sapwood and is thought to be related to the drying out of the central core of the wood; it seems that the displacement of pit membranes occurs where a tracheid wall lies between a tracheid containing gases and one containing water (Harris, 1954).

In walls between conifer tracheids and parenchyma cells the pit-pairs are *half-bordered*, with the border present on the side of the tracheid (fig. 8.6, *D, E*). No torus is present in such pit-pairs.

Fibers

The fibers are long cells with secondary, commonly lignified, walls. The walls vary in thickness but are usually thicker than the walls of tracheids in the same wood. Two principal types of xylem fibers are recognized, the fiber-tracheids (fig. 8.2, *H*) and the libriform fibers

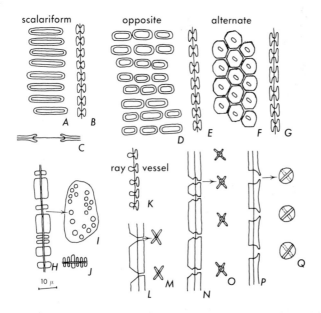

Fig. 8.5. Pits and patterns of pitting. *A–C*, scalariform pitting in surface (*A*) and side (*B, C*) views (*Magnolia*). *D–E*, opposite pitting in surface (*D*) and side (*E*) views (*Liriodendron*). *F–G*, alternate pitting in surface (*F*) and side (*G*) views (*Acer*). *A–G*, all bordered pit-pairs in vessel members. *H–J*, simple pit-pairs in parenchyma cells in surface (*I*) and side (*H, J*) views; *H*, in side wall; *J*, in end wall (*Fraxinus*). *K*, half-bordered pit-pairs between a vessel and a ray cell in side view (*Liriodendron*). *L, M*, simple pit-pairs with slit-like apertures in side (*L*) and surface (*M*) views (libriform fiber). *N, O*, bordered pit-pairs with slit-like inner apertures extended beyond the outline of the pit border; *N*, side view, *O*, surface view (fiber-tracheid). *P, Q*, bordered pit-pairs with slit-like inner apertures included within the outline of the pit border; *P*, side view, *Q*, surface view (tracheid).

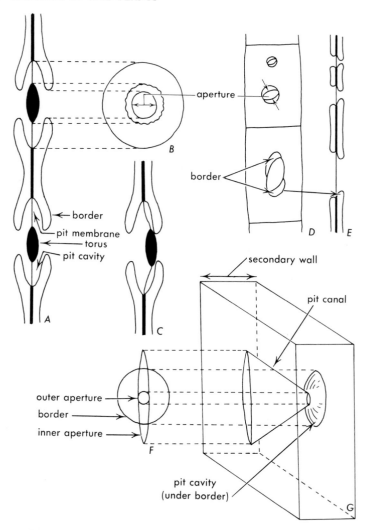

Fig. 8.6. Diagrams of bordered and half-bordered pit-pairs. *A,* two bordered pit-pairs, each with a torus, in side view. *B,* bordered pit in surface view. *C,* aspirated bordered pit-pair. *D, E,* half-bordered pit-pairs in surface (*D*) and side (*E*) views. *F, G,* bordered pit with extended inner aperture and reduced border. (*F, G,* after Record, *Timbers of North America,* John Wiley and Sons, 1934.)

(fig. 8.2, *I*). If both occur in the same wood, the libriform fiber is longer and has thicker walls than the fiber-tracheid. The fiber-tracheids have bordered pits with cavities smaller than the pit cavities of tracheids or vessels in the same wood (figs. 8.5, *N, O,* and 8.6, *F,*

G). These pits also have a distinct *pit canal* leading from the pit cavity into the cell lumen through the thick wall. The aperture from the cell lumen into the pit canal (*inner aperture*, fig. 8.6, *F*) is elongated and may be slit-like. It varies in length and commonly extends beyond the limit of the pit cavity (figs. 8.5, *N, O,* and 8.6, *F, G*). The aperture from the pit canal into the pit cavity (*outer aperture*) is circular (figs. 8.5, *N,* and 8.6, *F*). In three-dimensional aspect the canal appears like a flattened funnel (fig. 8.6, *G*).

The pit in a libriform fiber has a slit-like aperture toward the cell lumen, a canal resembling a much flattened funnel, but no pit cavity (fig. 8.5, *L, M*). In other words, the pit has no border; it is simple. The reference to the pits of libriform fibers as simple implies a sharper distinction between fibers and fiber-tracheids than actually exists. The fibrous xylem cells show a graduated series of pits between those with pronounced borders and those with vestigial borders or no borders. The intergrading forms with recognizable pit borders are placed, for convenience, in the fiber-tracheid category (Committee on Nomenclature, 1957).

Fibers of both categories may develop thin transverse walls across the cell lumen after the secondary walls are formed. The transverse walls of this kind are referred to as septa and the fibers as *septate fibers*. Septate fibers are widely distributed in the dicotyledons and usually retain their protoplasts in mature sapwood. Septate fibers, as well as some nonseptate fibers, are concerned with storage of reserve materials. Thus, the living fibers approach xylem-parenchyma cells in structure and function. The distinction between the two is particularly tenuous when the parenchyma cells develop secondary walls and septa. The retention of protoplasts by fibers is an indication of evolutionary advance (Bailey, 1953), and where living fibers are present the axial parenchyma is small in amount or absent (Money et al., 1950).

Another modification of fiber-tracheids and libriform fibers are the so-called gelatinous fibers. These fibers have a more or less unlignified inner wall with a gelatinous appearance. They are common components of the reaction wood (chapter 9) in the dicotyledons.

Phylogenetic specialization of tracheary and fiber cells

The lines of specialization of cells and tissues are better understood for the xylem than for any other tissue of the vascular plant. Among the individual lines, those pertaining to the evolution of the tracheary elements have been studied with particular thoroughness.

The specialization of tracheary elements was a concomitant of the separation of functions of conduction and strengthening in the vascular plant that occurred during the evolution of land plants (Bailey, 1953). In the less specialized state, conduction and support are combined in tracheids. With increasing specialization woods evolved with conducting elements—the vessel members—more efficient in conduction than in support. On the other hand, fibers evolved as principally strengthening elements. Thus, from primitive tracheids two lines of specialization diverged, one toward the vessels, the other toward the fibers.

Vessels evolved separately in several groups among the vascular plants. The widely accepted concept is that vessels evolved independently in lycopods, ferns, gymnosperms, monocotyledons, and dicotyledons and that they may have arisen several times independently in some of these groups, for example, in the dicotyledons (Cheadle, 1953). A large body of evidence suggests that in the dicotyledons vessels originated and underwent specialization first in the secondary xylem, then in the late primary xylem (metaxylem), and last in the early primary xylem (protoxylem). In the primary xylem of the monocotyledons, origin and specialization of vessels also occurred first in the metaxylem, then in the protoxylem; furthermore, vessels appeared first in the roots, then at progressively higher levels in the shoot (Cheadle, 1953; Fahn, 1954).

Vessels may undergo an evolutionary loss (Bailey, 1953; Cheadle, 1953). Absence of vessels in some aquatic plants, saprophytes, parasites, and succulents, for example, is interpreted as a result of a reduction in the xylem tissue. Reduction in this sense implies failure of potential xylem elements, including vessel members, to undergo typical ontogenetic differentiation and maturation. These vesselless plants are highly specialized as contrasted with the some ten known genera of primitively vesselless dicotyledons (*Trochodendron, Tetracentron, Drimys, Pseudowintera,* and others) belonging to the lowest taxonomic groups of dicotyledons (Bailey, 1953; Cheadle, 1953; Lemesle, 1956).

The evolutionary sequence of vessel members of the secondary xylem of dicotyledons began with long scalariformly pitted tracheids similar to those still found in some lower dicotyledons. These tracheids were succeeded by vessel members of long narrow shape with tapering ends (fig. 8.3, *C*). The cells shortened progressively, became wider, and their end walls became less inclined and finally transverse (fig. 8.3, *D–F*). In the more primitive state the perforation plate was scalariform, with numerous bars, resembling a wall with scalariformly arranged pits devoid of pit membranes. Increase in specialization resulted in a decrease in the number of bars (fig. 8.3, *C*) and finally their total elimination and the appearance of a simple perforation (fig. 8.3, *D–F*).

The pitting of vessel walls also changed during the evolution. In intervessel pitting, bordered pit-pairs in scalariform arrangement (fig. 8.5, *A*) were replaced by smaller bordered pit-pairs, first in opposite (fig. 8.5, *D*), later in alternate arrangement (fig. 8.5, *F*). The pit-pairs between vessels and parenchyma cells changed from bordered, through half-bordered, to simple.

The tracheids were not eliminated when vessels evolved, and they also underwent phylogenetic changes. They became shorter—not as short as the vessel members, however—and the pitting of their walls evolved parallel to that of the associated vessel members. They generally did not increase in width.

In the specialization of the xylem fibers the emphasis on mechanical function became apparent in the increase in wall thickness and decrease in cell width. Concomitantly the pits changed from elongated to circular, the borders became reduced (fig. 8.5, *N, O*) and eventually disappeared (fig. 8.5, *L, M*). The inner apertures of the pit became elongated and then slit-like. Thus, the evolutionary sequence was from tracheids, through fiber-tracheids, to libriform fibers.

The matter of evolutionary change in length of fibers is rather complex. The shortening of vessel members is correlated with a shortening of the fusiform cambial initials (chapter 10) from which the axial cells of the xylem are derived. Thus, in woods with shorter vessel members, the fibers are derived ontogenetically from shorter initials than in more primitive woods with longer vessel members. In other words, with increase in xylem specialization the fibers become shorter. Because, however, during ontogeny fibers undergo intrusive growth whereas vessel members do so only slightly or not at all, the fibers are longer than the vessel members in the mature wood, and, of the two categories of fibers, the libriform fibers become the longest. Nevertheless, the fibers of specialized woods are shorter than their ultimate precursors, the primitive tracheids.

The evolutionary lines in the xylem have been reconstructed from comparative studies of existing plants. The vascular plants now living show a wide range in the degree of specialization of their cells, tissues, and organs. These variations, incidentally, are useful in the identification of woods (chapter 9).

Parenchyma cells

The parenchyma of the secondary xylem is represented by the axial parenchyma and the ray parenchyma. Both tissues are fundamentally alike regarding wall structure and contents, and in both the cells may vary considerably in structure and contents (Chattaway, 1951; Wardrop and Dadswell, 1952). The parenchyma cells store starch, oils, and many other ergastic substances of unknown function. Tanniniferous compounds and crystals are common inclusions. The types of crystals and their arrangements may be sufficiently characteristic to serve in identification of woods (Chattaway, 1955).

The walls of radial and axial parenchyma cells may have secondary thickenings and be lignified (Wardrop and Dadswell, 1952). If secondary walls are present, the pit-pairs between the parenchyma cells may be bordered, half-bordered, or simple. Some parenchyma cells become sclerified by deposition of thick walls. These are sclerotic cells, or sclereids. Crystalliferous parenchyma cells frequently have lignified walls with secondary thickenings and may be chambered by septa, each chamber containing one crystal.

The axial parenchyma cells are derived from elongated fusiform cambial cells. If the derivative of such a cambial cell differentiates into a parenchyma cell without transverse (or oblique) divisions, a fusiform *parenchyma cell* results. If such divisions occur, a *parenchyma strand* is formed (fig. 8.2, *K*). Neither type undergoes intrusive growth.

The radial parenchyma cells are divided into categories according to their form. The two most common types are the *procumbent* and the *upright* ray cells (fig. 8.7). A ray cell whose longest diameter is oriented radially is procumbent; the cell elongated axially is upright. The two kinds of ray cells are often combined in the same ray, the upright cells typically appearing at the upper and lower margins of

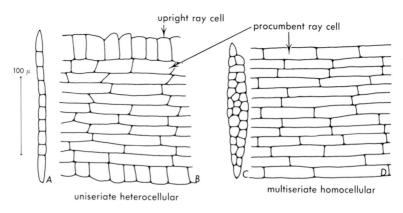

Fig. 8.7. Two types of rays as seen in tangential (*A, C*) and radial (*B, D*) sections. *A, B, Fagus grandifolia. C, D, Acer saccharum.*

the ray (fig. 8.7, *A*). Rays composed of one kind of cell are called *homocellular* (fig. 8.7, *C, D*), those containing procumbent and upright cells, *heterocellular* (fig. 8.7, *A, B*).

PRIMARY XYLEM

The primary xylem contains the same basic cell types as the secondary xylem: tracheary elements—both tracheids and vessel members—fibers, and parenchyma cells. It is, however, not organized into the combination of axial and radial systems, for it contains no rays. In stems and leaves, and floral parts, the primary xylem and the associated primary phloem commonly occur in strands, the vascular bundles (fig. 8.8). Panels of parenchyma, the interfascicular regions, occur between the vascular bundles in stems (chapter 16). These panels are often called medullary rays and are considered to be part of the ground tissue. In the root the primary xylem forms a core with or without parenchyma in the center (chapter 14).

Protoxylem and metaxylem

Developmentally, the primary xylem usually consists of an earlier part, the *protoxylem,* and a later part, the *metaxylem* (figs. 8.8 and 8.9, *B*). Although the two parts have some distinguishing characteristics, they intergrade so that the delimitation of the two can be made only approximately.

The protoxylem differentiates in the parts of the primary plant body that have not completed their growth and differentiation. In fact, in the shoot, the protoxylem matures among actively elongating tissues and is, therefore, subjected to stresses. Its mature nonliving tracheary elements are stretched and eventually destroyed. In the root they persist longer because here they mature beyond the region of maximum growth.

The metaxylem is commonly initiated in the still growing primary plant body, but it matures largely after the elongation is completed. It is, therefore, less affected by the primary extension of the surrounding tissues than the protoxylem.

The protoxylem usually contains only tracheary elements imbedded in parenchyma that is considered to be part of the protoxylem. When the tracheary elements are destroyed they may become completely obliterated by the surrounding parenchyma cells. In the shoot xylem of many monocotyledons the stretched nonfunctioning elements are

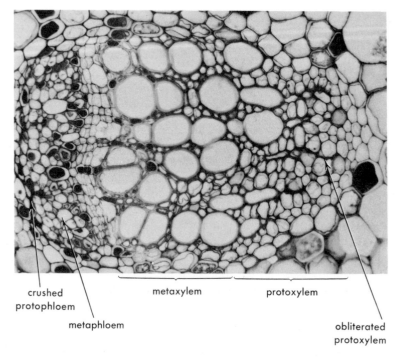

crushed   metaxylem   protoxylem
protophloem

metaphloem   obliterated
protoxylem

Fig. 8.8. Vascular bundle from petiole of grapevine (*Vitis*) leaf in cross section. Illustrates parts of primary xylem and primary phloem. (×290. From Esau, *Hilgardia*, 1948.)

partly collapsed but not obliterated; instead, open canals, the so-called protoxylem lacunae, surrounded by parenchyma cells appear in their place (fig. 11.7). If preserved in sectioning, the secondary walls of the nonfunctioning tracheary cells may be seen along the margin of the lacuna.

The metaxylem is somewhat more complex than the protoxylem and may contain fibers in addition to the tracheary elements and parenchyma cells. The parenchyma cells may be dispersed among the tracheary elements or may occur in radial rows simulating rays. Longitudinal sections reveal them as axial parenchyma cells. The radial seriation often encountered in the metaxylem, and also in the protoxylem, has given rise to a tendency in the literature to interpret the primary xylem of many plants as secondary because radial seriation is so characteristic of the secondary vascular tissues.

The tracheary elements of the metaxylem are retained after the primary growth is completed but become nonfunctioning after some secondary xylem is produced. In plants lacking secondary growth the metaxylem remains functional in mature plant organs.

Secondary wall in primary tracheary elements

The secondary wall thickenings of primary tracheary cells are highly characteristic; moreover, they appear in an orderly ontogenetic series of forms that clearly indicates a progressive increase in the extent of the primary wall area covered by secondary wall material (fig. 8.9). In the earliest tracheary elements the secondary walls may occur as rings (*annular* thickenings) not connected with one another. The elements differentiating next have *helical* (*spiral*) thickenings. Then follow cells with thickenings that may be characterized as helices with coils interconnected (*scalariform* thickenings). These are succeeded by cells with net-like, or *reticulate,* thickenings, and finally by *pitted* elements.

Not all types of secondary thickenings are necessarily represented in the primary xylem of a given plant or plant part, and the different types of wall structure intergrade. The annular thickenings may be interconnected here and there; annular and helical or helical and scalariform thickenings may be combined in the same cell; and the difference between scalariform and reticulate is sometimes so tenuous that the thickening may best be called scalariform-reticulate. The pitted elements also intergrade with the earlier ontogenetic type. The openings in a scalariform reticulum of the secondary wall may be comparable to pits, especially if a slight border is present. A border-like overarching of the secondary wall is common in the various types of secondary walls in the primary xylem. Rings, helices, and the bands of the scalariform-reticulate thickenings may be connected to the primary wall by narrow bases, so that the secondary wall layers widen out toward the lumen of the cell and overarch the exposed primary wall parts.

The intergrading nature of the secondary wall thickenings in the primary xylem makes it impossible to assign distinct types of wall thickenings to the protoxylem and the metaxylem with any degree of consistency. Most commonly the first tracheary elements to mature, that is, protoxylem elements, have the minimal amounts of secondary wall material. Annular and helical thickenings predominate. These types of thickenings do not hinder materially the stretching of the mature protoxylem elements during the extension growth of the primary plant body. The evidence that such stretching occurs is easily perceived in the increase in distance between rings in older xylem ele-

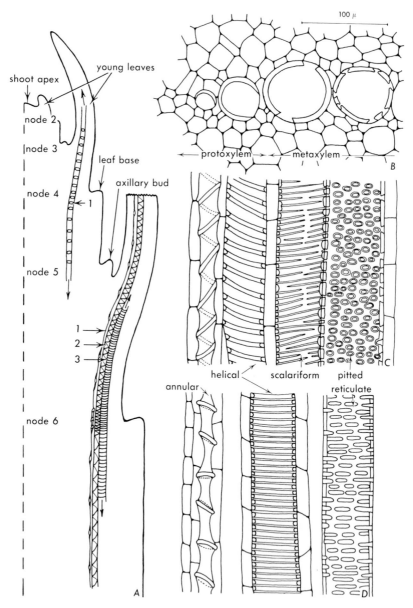

Fig. 8.9. Details of structure and development of primary xylem. *A*, diagram of a shoot tip showing stages in xylem development at the different levels. *B–D*, primary xylem of castor bean, *Ricinus,* in cross (*B*) and longitudinal (*C, D*) sections.

ments, the tilting of the rings, and in the uncoiling of the helices (fig. 8.9, *A*).

The metaxylem, in the sense of the xylem tissue maturing after the extension growth, may have helical, scalariform, reticulate, and pitted elements; one or more types of thickening may be omitted. If many elements with helical thickenings are present, the helices of the succeeding elements are less and less steep, a condition suggesting that some stretching occurs during the development of the earlier metaxylem elements.

Convincing evidence exists that the type of wall thickening in primary xylem is strongly influenced by the internal environment in which these cells differentiate. Annular thickenings develop when the xylem begins to mature before the maximum extension of the plant part occurs, as, for example, in the shoots of normally elongating plants (fig. 8.9, *A*, nodes 3–5); they may be omitted if the first elements mature after this growth is largely completed, as is common in the roots. If the elongation of a plant part is suppressed before the first xylem elements mature, one or more of the early ontogenetic types of thickenings are omitted. On the contrary, if elongation is stimulated, as for example by etiolation, more than the usual number of elements with annular and helical thickening will be present.

The intergrading of the different types of thickenings of tracheary elements is not limited to the primary xylem. The delimitation between the primary and the secondary xylem may also be vague. To recognize the limits of the two tissues it is necessary to consider many features, among these the length of tracheary cells—the last primary elements are typically longer than the first secondary—and the organization of the tissue, particularly the appearance of the combination of ray and axial systems characteristic of the secondary xylem.

In the primary xylem the protoxylem elements may be the narrowest, but not necessarily so. The metaxylem elements sometimes increase in width in the successively developed tissue layers. The secondary xylem may also have rather narrow cells in the beginning and thus be distinct from the wide-celled metaxylem. On the whole, however, no simple method is available to distinguish between developmental categories of tissues and of their meristems.

REFERENCES

Bailey, I. W. Evolution of the tracheary tissue of land plants. *Amer. Jour. Bot.* 40:4–8. 1953.

Chattaway, M. M. Morphological and functional variations in the rays of pored timbers. *Austral. Jour. Sci. Res. Ser. B, Biol. Sci.* 4:12–29. 1951.

Chattaway, M. M. Crystals in woody tissues. Part I. *Trop. Woods* 1955 (102):55–74. 1955.

Cheadle, V. I. Independent origin of vessels in the monocotyledons and dicotyledons. *Phytomorphology* 3:23–44. 1953.

Committee on Nomenclature, International Association of Wood Anatomists. *International glossary of terms used in wood anatomy*. *Trop. Woods* 1957 (107):1–36. 1957.

Eicke, R. Beitrag zur Frage des Hoftüpfelbaues der Koniferen. *Deut. Bot. Gesell. Ber.* 67:213–217. 1954.

Fahn, A. Metaxylem elements in some families of the Monocotyledoneae. *New Phytol.* 53:530–540. 1954.

Frey-Wyssling, A., H. H. Bosshard, and K. Mühlethaler. Die submikroskopische Entwicklung der Hoftüpfel. *Planta* 47:115–126. 1956.

Greenidge, K. N. H. An approach to the study of vessel length in hardwood species. *Amer. Jour. Bot.* 39:570–574. 1952.

Harris, J. M. Heartwood formation in *Pinus radiata* (D. Don). *New Phytol.* 53:517–524. 1954.

Lemesle, R. Les éléments du xyléme dans les Angiospermes à caractères primitifs. *Soc. Bot. de France Bul.* 103:629–677. 1956.

Liese, W., and I. Johann. Experimentelle Untersuchugen über die Feinstruktur der Hoftüpfel bei den Koniferen. *Naturwiss.* 41:579. 1954.

Money, L. L., I. W. Bailey, and B. G. L. Swamy. The morphology and relationships of the Monimiaceae. *Arnold Arboretum Jour.* 31:372–404. 1950.

Trendelenburg, R. *Das Holz als Rohstoff.* 2nd ed. Revised by H. Mayer-Wegelin. München, Carl Hauser. 1955.

Wardrop, A. B., and H. E. Dadswell. The cell wall structure of xylem parenchyma. *Austral. Jour. Sci. Res. Ser. B, Biol. Sci.* 5:223–236. 1952.

9. XYLEM:
VARIATIONS IN WOOD STRUCTURE

WOODS ARE USUALLY CLASSIFIED IN TWO MAIN GROUPS, THE SOFT-woods and the hardwoods. The term softwood is applied to gymnosperm wood, that of hardwood to the dicotyledon wood. The two kinds of woods show basic structural differences, but they are not necessarily distinct in degree of density and hardness. The gymnosperm wood is homogeneous in structure—with long straight elements predominating—and, therefore, easily workable. It is highly suitable for papermaking. Many commercially important dicotyledon woods are especially strong, dense, and heavy because of high proportion of fiber tracheids and libriform fibers (e.g., *Quercus, Carya, Eucalyptus, Acacia*), but some are light and soft (the lightest and softest is balsa, *Ochroma*). Among the gymnosperms only the conifers are an important source of commercial timber, and among the angiosperms only the dicotyledons. The monocotyledons that have secondary growth do not produce a commercially important homogeneous body of secondary xylem.

CONIFER WOOD

The secondary xylem of the conifers is relatively simple in structure (figs. 9.1 and 9.2; Greguss, 1955), simpler than that of most of the dicotyledons. One of its outstanding features is the lack of vessels. The tracheary elements are imperforate and are mainly tracheids.

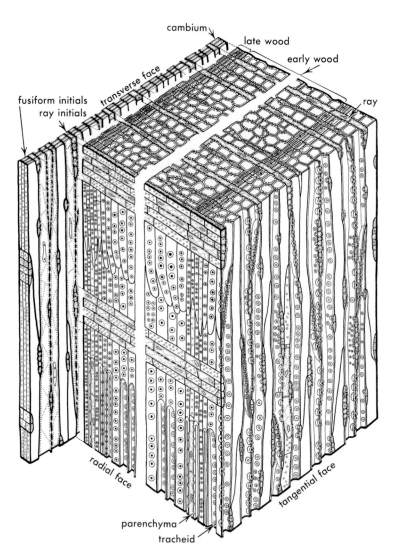

Fig. 9.1. Block diagram of vascular cambium and wood of *Thuja occidentalis* (white cedar), a conifer. The axial system is composed of tracheids and some parenchyma cells. The rays contain only parenchyma cells. (From Esau, *Plant Anatomy,* John Wiley and Sons, 1953. Courtesy of I. W. Bailey. Drawn by Mrs. J. P. Rogerson under the supervision of L. G. Livingston. Redrawn.)

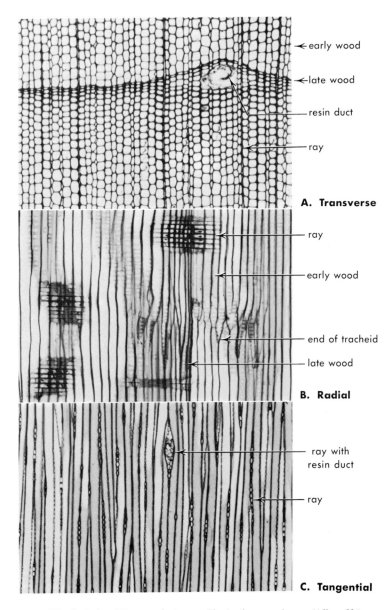

← early wood

← late wood

— resin duct

— ray

A. Transverse

— ray

— early wood

— end of tracheid

— late wood

B. Radial

— ray with resin duct

— ray

C. Tangential

Fig. 9.2. Wood of pine (*Pinus strobus*), a conifer in three sections. (All, ×53.)

Fiber-tracheids may occur in the late wood, but libriform fibers are absent. The tracheids are narrow elongated cells averaging 2 to 5 millimeters in length (fig. 8.3, *A;* Trendelenburg, 1955). Their overlapping ends may be curved and branched because of intrusive growth. Basically the ends are wedge shaped, with the truncated end of the wedge exposed in the radial section (fig. 9.1).

The early-wood tracheids have circular bordered pits with circular inner apertures (fig. 9.3, *D*). The late-wood tracheids (or fiber-tracheids) have somewhat reduced borders with oval inner apertures. This difference in pit structure is a concomitant of the increase in wall thickness in the late-wood cells. The pit-pairs between tracheids are usually with tori (fig. 9.3, *D, F*). Throughout most of a growth layer the pits are restricted to the radial walls (fig. 9.1); only in the late wood may the tangential walls bear pits. The pit-pairs are abundant on the overlapping ends including the parts that were added by intrusive growth. The pits are typically in one row. In the Taxodiaceae and Pinaceae some wide early-wood tracheids may have two or more rows of pits in opposite arrangement, and in the Araucariaceae pits occur in alternate arrangement (Phillips, 1948). In addition to the pitted secondary wall layers, conifer tracheids may have helical thickenings (fig. 9.3, *B*).

Axial parenchyma may or may not be present in conifer wood (Īatsenko-Khmelevskiĭ, 1954; Phillips, 1948). In Podocarpaceae, Taxodiaceae, and Cupressaceae parenchyma is prominent in the wood (figs. 9.1 and 9.3, *C*). It is scantily developed or absent in Araucariaceae, Pinaceae, and Taxaceae. In some genera axial parenchyma is restricted to that associated with resin ducts (*Pinus, Picea, Larix, Pseudotsuga*). Resin ducts (figs. 9.2 and 9.3, *A*) appear as a constant feature of some woods (Pinaceae), but they also develop as a result of injury (traumatic resin ducts). They occur in the axial and in the radial systems.

The rays of conifers are mostly one cell wide (figs. 9.1 and 9.2), occasionally biseriate (fig. 9.3, *C*), and from one to twenty or even to fifty cells high. Presence of resin ducts makes the normally uniseriate rays appear multiseriate (fig. 9.2, *C*). The rays consist of parenchyma cells or may also contain ray tracheids. These tracheids resemble parenchyma cells in shape but are devoid of protoplasts at maturity and have secondary walls with bordered pits (fig. 9.3, *E*). Ray tracheids are normally present in most Pinaceae, occasionally in *Sequoia* and the Cupressaceae (Phillips, 1948). The ray tracheids commonly occur along the margins of rays, one or more cells in depth.

Each axial tracheid is in contact with one or more rays (fig. 9.1). The pit-pairs between the axial tracheids and ray parenchyma cells are

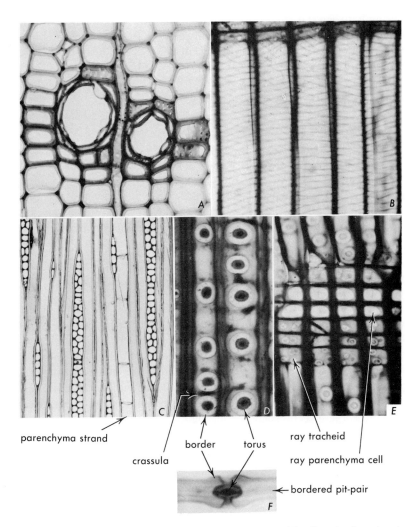

parenchyma strand

border   torus   ray tracheid

crassula   ray parenchyma cell

bordered pit-pair

Fig. 9.3. Details of conifer wood. *A, B, Pseudotsuga taxifolia,* Douglas fir. *A,* resin ducts with thick-walled epithelial cells, in cross section. *B,* tracheids with spiral inner-most layer of secondary wall in radial section. *C, Sequoia sempervirens* wood in tangential section: tracheids, axial parenchyma, uniseriate and biseriate rays. *D, E, F,* pine wood in radial (*D, E*) and tangential (*F*) sections. Only axial tracheids in *D, F;* ray with ray tracheids and parenchyma cells in *E.* (*A, B, E,* ×200; *C,* ×70; *D,* ×280; *F,* ×850.)

half-bordered, with the border on the side of the tracheid (chapter 8); those between the axial and the ray tracheids are fully bordered. The pitting between the ray parenchyma cells and the axial tracheids form such characteristic patterns in radial sections that the cross-field, or the rectangle formed by the radial wall of a ray cell against an axial tracheid (fig. 9.2, *B*), is utilized in classification and in phylogenetic studies of conifer woods.

DICOTYLEDON WOOD

The wood of the dicotyledons is more varied than that of the gymnosperms. The wood of the primitively vesselless dicotyledons is relatively simple, but that of the vessel-containing species is usually complex. Wood of the latter species may have both vessels and tracheids, one or more categories of fibers (chapter 8), axial parenchyma, and rays of one or more kinds (figs. 9.4, 9.5, and 9.6).

Storied and nonstoried wood

In transverse sections the secondary xylem shows more or less orderly radial seriation of cells—a result of the origin of cells from tangentially dividing cambial cells. In the homogeneous conifer wood this seriation is pronounced (fig. 9.2); in vessel-containing dicotyledons it may be somewhat obscured by the ontogenetic enlargement of the vessel members and the consequent displacement of adjacent cells (figs. 9.5 and 9.6). Radial sections also reveal the radial seriation; moreover, these sections show that the radial series of the axial system are superimposed one upon the other in horizontal strata (figs. 9.4 and 9.6). The tangential sections, however, are more varied in their appearance in different woods. In some, the cells of one stratum unevenly overlap those of another; in others the horizontal strata are as regular in tangential sections as they are in the radial sections. Thus, some woods are nonstratified, or nonstoried, in tangential sections (e.g., fig. 9.7, *A*; *Castanea, Fraxinus, Juglans, Quercus*), others stratified, or storied (e.g., fig. 9.7, *B; Aesculus, Cryptocarya, Ficus, Tilia*, and numerous Leguminosae). The storied condition is especially pronounced when the height of the ray matches that of a horizontal stratum of the axial system. From the evolutionary aspect the storied woods are more highly specialized than the nonstoried. They are derived from vascular cambia with short fusiform initials. Many intermediate patterns are found between the strictly stratified woods and the

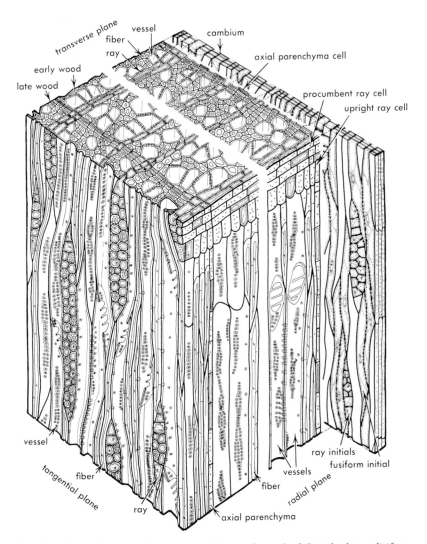

Fig. 9.4. Block diagram of vascular cambium and wood of *Liriodendron tulipifera* (tulip tree), a dicotyledon. The axial system consists of vessel members with scalariform perforation plates, fiber-tracheids, and axial xylem parenchyma strands in terminal arrangement. (Courtesy of I. W. Bailey. Drawn by Mrs. J. P. Rogerson under the supervision of L. G. Livingston. Redrawn.)

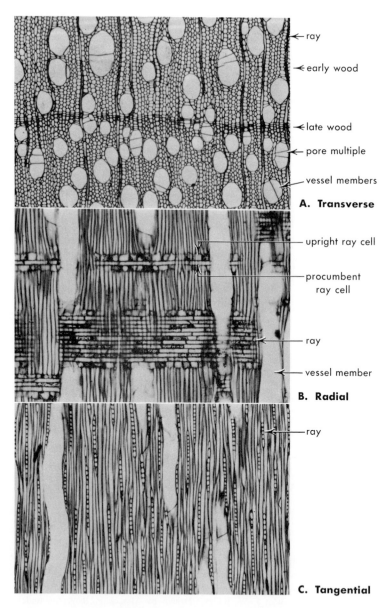

← ray

← early wood

← late wood

← pore multiple

vessel members

A. Transverse

— upright ray cell

— procumbent ray cell

— ray

— vessel member

B. Radial

— ray

C. Tangential

Fig. 9.5. Wood of willow (*Salix nigra*), a dicotyledon, in three sections. Diffuse-porous nonstoried wood with uniseriate heterocellular rays. (All, ×53.)

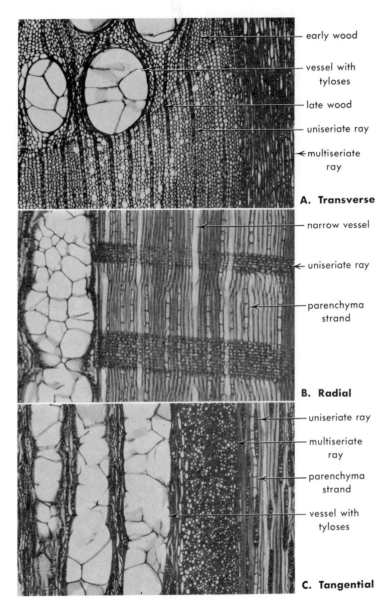

- early wood
- vessel with tyloses
- late wood
- uniseriate ray
- multiseriate ray

A. Transverse

- narrow vessel
- uniseriate ray
- parenchyma strand

B. Radial

- uniseriate ray
- multiseriate ray
- parenchyma strand
- vessel with tyloses

C. Tangential

Fig. 9.6. Wood of oak (*Quercus alba*), a dicotyledon, in three sections. Ring-porous nonstoried wood with high multiseriate and low uniseriate rays. The large vessels are occluded by tyloses. (All, ×53.)

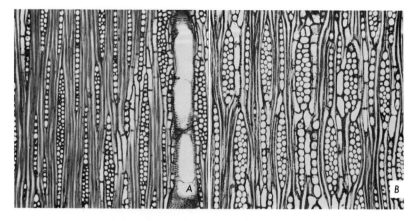

Fig. 9.7. *A*, nonstoried wood of pecan (*Hicoria pecan*). *B*, storied wood of persimmon (*Diospyros virginiana*). (Both tangential sections, ×70.)

strictly nonstratified woods derived from cambia with long fusiform initials.

Distribution of vessels

The wood anatomist refers to a vessel in cross section as a pore. Two principal types of woods are recognized on the basis of the distribution of pores in a growth layer: diffuse-porous wood with pores rather uniform in size and distribution throughout a growth ring (e.g., fig. 9.5 and species in the following genera: *Acer, Betula, Carpinus, Fagus, Juglans, Liriodendron, Platanus, Populus, Pyrus*); ring-porous wood with pores distinctly larger in the early wood than in the late wood (e.g., figs. 8.1 and 9.6 and species in the following genera: *Castanea, Catalpa, Celtis, Fraxinus, Gleditsia, Morus, Quercus, Robinia, Ulmus*). Intergrading patterns occur between the two types of patterns. The ring-porous condition appears to be an indication of evolutionary specialization and occurs in comparatively few species, nearly all characteristic of the north temperate zone.

Within the main distributional patterns of vessels, minor variations occur in the spatial relation of the pores to each other. A pore is called solitary when the vessel is completely surrounded by other types of cells (fig. 9.6). A group of two or more pores appearing together form a pore multiple (fig. 9.5, *A*). This may be a radial pore multiple, with pores in a radial file, or a pore cluster, with an irregular grouping of pores.

Distribution of axial parenchyma

The distribution of the axial xylem parenchyma shows many inter-grading patterns. An agreement on their best classification has not been reached. The spatial relation to vessels, as seen in transections, serves for the division in two main patterns: *apotracheal*, parenchyma not definitely associated with the vessels; *paratracheal*, parenchyma consistently associated with the vessels (fig. 9.8, *D*). The apotracheal parenchyma is further subdivided into: *diffuse*, single parenchyma cells or parenchyma strands scattered among fibers (fig. 9.6); apotracheal *banded* (figs. 9.8, *C*, and 9.9, *B*); *boundary* parenchyma (Jane, 1956),

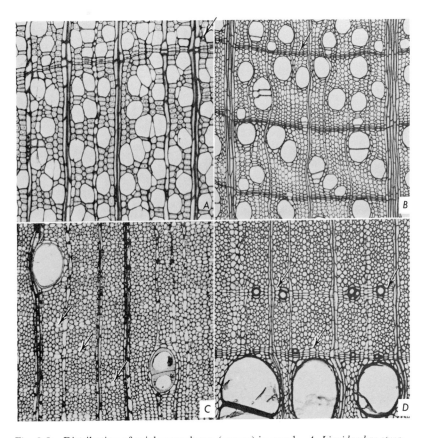

Fig. 9.8. Distribution of axial parenchyma (arrows) in wood. *A, Liquidambar styraciflua,* parenchyma very sparce. *B, Acer saccharum,* boundary parenchyma. *C, Hicoria pecan,* apotracheal banded parenchyma. *D, Fraxinus* sp., paratracheal and boundary parenchyma. (All cross sections, ×70.)

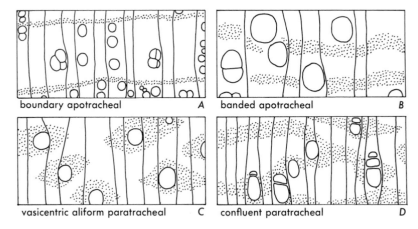

Fig. 9.9. Distribution of axial parenchyma (stipples) in wood of *A, Michelia; B, Saccopetalum; C,* a leguminous species; *D, Terminalia.* (Drawn from photographs in Record, *Timbers of North America,* John Wiley and Sons, 1934.)

singly or in a band at the end (terminal) or at the beginning (initial) of a growth layer (figs. 9.8, *B, D,* and 9.9, *A*). Diffuse apotracheal parenchyma may be scanty (fig. 9.8, *A*). The paratracheal parenchyma appears in the following forms: *vasicentric,* forming a complete sheath around vessels; *aliform,* vasicentric with wing-like tangential extensions (fig. 9.9, *C*); and *confluent,* coalesced aliform forming irregular tangential or diagonal bands (fig. 9.9, *D*). Paratracheal parenchyma may be scanty. If septate fibers instead of axial parenchyma occur in the xylem, they show distributional patterns similar to those assumed by the axial xylem parenchyma. From the evolutionary aspect the apotracheal and diffuse patterns are primitive (Money et al., 1950).

Structure of rays

In contrast to the predominately uniseriate rays of conifers, those of the dicotyledons may be one to many cells wide, that is, they may be uniseriate (fig. 9.5) or multiseriate (figs. 9.4 and 9.7), and range in height from one to many cells (from a fraction to 1 inch or more). The multiseriate rays frequently have uniseriate margins (fig. 9.7, *A*). Small rays may be grouped so as to appear to be one large ray. Such groups are called aggregate rays (*Carpinus*).

The appearance of rays in radial and tangential sections can be used as a basis for their classification. Individual rays may be homo-cellular, that is, composed of cells of one form only (figs. 8.7, *C, D,* and 9.6), either procumbent or upright, or heterocellular, that is, com-posed of two morphological types, procumbent and upright (figs. 8.7, *A, B,* 9.4, and 9.5). The entire ray system of a wood may consist of either homocellular or heterocellular rays or of combinations of the two types of rays. On this basis the ray tissue system is classified into homogeneous, rays all homocellular (procumbent cells only), or heterogeneous, rays all heterocellular or combinations of homocellular and heterocellular (Jane, 1956, p. 121). Further variations between homogeneous and heterogeneous ray tissues result from combinations of uniseriate and multiseriate rays or absence of multiseriate rays.

The different ray combinations have a phylogenetic significance. The primitive ray tissue may be exemplified by that of the Winteraceae (e.g., *Drimys*). The rays are of two kinds: one homocellular, uniseriate composed of upright cells; the other heterocellular, multiseriate com-posed of radially elongated or nearly isodiametric cells in the mul-tiseriate part and upright cells in the uniseriate marginal parts Both kinds of rays are many cells in height. From such primitive ray structure other ray systems, more specialized, have been derived. For example, multiseriate rays may be eliminated (*Aesculus hippocastanum*) or increased in size (*Quercus*), or both multiseriate and uniseriate rays may be decreased in size (*Fraxinus*).

The evolution of rays strikingly illustrates the maxim that phylo-genetic changes depend on successively modified ontogenies. Even in a given wood the specialized ray structure may appear gradually. The earlier growth layers may have a more primitive ray structure than the later because the vascular cambium commonly undergoes successive changes before it begins to produce a ray pattern of a more specialized type.

Tyloses

In many species, axial and ray parenchyma cells located next to the vessels form outgrowths through the pit cavities into the lumina of the vessels when the latter become inactive (fig. 9.6). These out-growths are called tyloses. The growth of a tylosis appears to involve surface growth of the pit membrane of a pit-pair between a parenchyma cell and a vessel. The nucleus and part of the cytoplasm of the paren-chyma cell commonly migrate into the tylosis. Tyloses store ergastic substances and may develop secondary walls, or even differentiate into sclereids. It seems that tylosis development is possible only if the pit aperture on the vessel side is no less than 10 microns wide (Chattaway, 1949). Examples of woods with abundant development of tyloses are those of *Quercus* (White Oak species), *Robinia, Vitis, Morus, Catalpa, Juglans nigra, Maclura.*

Tyloses block the lumina of vessels and reduce the permeability of the wood. Technically this phenomenon is important in the treatment of wood with preservatives and in its selection for tight coopering. With regard to conduction in the xylem, the significance of tyloses is not clearly understood. They are known to block vessels in the heartwood, and in the sapwood beneath wounds and in connection with some diseases.

Intercellular canals and cavities

Intercellular canals similar to the resin ducts of gymnosperms occur in the dicotyledon woods. They are often called gum ducts although they may contain resins. They occur in both the axial and the radial systems and may be normal or traumatic. Intercellular canals vary in extent, and some may be more appropriately called intercellular cavities. Intercellular canals and cavities may be schizogenous, but gummosis of surrounding cells may also occur. Canals associated with gummosis are well known in such genera as *Amygdalus* and *Prunus.*

REACTION WOOD

A more or less distinctive wood is formed on the lower sides of branches and leaning or crooked stems of conifer trees, and on the upper sides of similar structures in dicotyledon trees. This wood is called reaction wood (compression wood in conifers and tension wood in the dicotyledons) because its development is assumed to result from the tendency of the branch or stem to counteract the force inducing the inclined position (Sinnott, 1952). The extent to which gravity is involved in the formation of the reaction wood is controversial (Scott and Preston, 1955).

The reaction wood differs from the normal in both anatomy and chemistry (Dadswell et al., 1958). The compression wood of conifers is typically denser and darker than the surrounding tissue, its tracheids are shorter than those in normal wood, and the cell walls are heavily lignified. The inner layer of the usually three-layered secondary wall is missing. In the tension wood of dicotyledons the vessels are reduced in width and number and the fibers have a thick highly refractive inner layer—the so-called gelatinous layer—consisting largely of cellulose. The walls of these fibers may be two to four layered; the gelatinous layer is always the innermost.

KEY FOR IDENTIFICATION OF WOODS

An exercise in identification of woods serves two useful purposes: first, practice in the use of a key gives a motivation for seeing a variety of woods and for reviewing structural features and terminology; second, it can illustrate an important feature in the evolutionary specialization of woods: whereas generally the various structures in woods tend to specialize together, the rates of specialization of the structures in a given species differ, and this difference makes identification possible.

To establish presence or absence of characters mentioned in a key, it is often necessary to look at all three kinds of sections of wood. When one is in doubt about making a choice, one should follow out both alternatives before making the final decision on the identity of a specimen.

This key was constructed by the use of conifer (softwoods) and dicotyledon (hardwoods) species listed below. Keys in Jane (1956), Stover (1951), and especially Record (1934) were consulted.

Conifer woods: *Abies balsamea* (Balsam Fir), *Larix occidentalis* (Western Larch), *Pinus strobus* (White Pine), *Pseudotsuga taxifolia* (Douglas Fir), *Sequoia sempervirens* (Redwood), *Thuja occidentalis* (American Arbor-vitae).

Dicotyledon woods: *Acer saccharum* (Sugar Maple), *Alnus rubra* (Red Alder), *Betula papyrifera* (Paper Birch), *Carpinus caroliniana* (Blue Beech), *Castanea dentata* (Chestnut), *Celtis occidentalis* (Hackberry), *Diospyros virginiana* (Persimmon), *Fagus grandifolia* (American Beech), *Fraxinus americana* (White Ash), *Hicoria pecan* (Pecan), *Juglans nigra* (Black Walnut), *Liquidambar styraciflua* (Sweet Gum), *Liriodendron tulipifera* (Tulip Tree, Yellow Poplar), *Platanus occidentalis* (Sycamore), *Populus deltoides* (Cottonwood), *Prunus serotina* (Wild Black Cherry), *Quercus alba* (White Oak), *Quercus borealis* (Red Oak), *Robinia pseudoacacia* (Black Locust), *Salix nigra* (Black Willow), *Tilia americana* (American Linden, Basswood), *Ulmus americana* (American Elm).

Vesselless dicotyledon wood: *Drimys winteri*.

I WOOD NONPOROUS (Vessels absent)

A1. Rays mostly uniseriate, occasionally biseriate; wider when resin ducts are present in them. (Conifers.)

B1. Axial resin ducts normally present

C1. Epithelial cells of resin ducts mostly with thin walls. Ray tracheids present.

PINUS
(fig. 9.2)

C2. Epithelial cells of resin ducts mostly with thick walls.

D1. Tracheids with helical thickenings, at least in early wood. No ray tracheids present.

PSEUDOTSUGA
(fig. 9.3, *A, B*)

D2. Tracheids rarely with helical thickenings. Ray tracheids present.

LARIX

B2. Axial resin ducts normally absent; may be traumatic, then in tangential rows or groups.

C1. Axial wood parenchyma present and conspicuous. End walls of ray cells not conspicuously pitted. (Wood without aromatic scent.)

SEQUOIA
(fig. 9.3, *C*)

C2. Axial wood parenchyma absent or sparce.

D1. End walls of ray cells conspicuously pitted (nodular). (Wood with aromatic scent.)

THUJA

D2. End walls of ray cells not conspicuously pitted. (Wood without aromatic scent.)

ABIES

A2. Rays multiseriate and high-celled uniseriate in the same wood. (Dicotyledons.)

DRIMYS

II WOOD POROUS (Vessels present)
Dicotyledons.

1. WOOD RING-POROUS. Pores of the early wood distinctly larger than those of the late wood.

A1. Late wood with radial lines or patches of small pores, tracheids, and parenchyma cells, usually light in color. Fine bands of metatracheal parenchyma.

B1. Rays all uniseriate.

CASTANEA

B2. Rays of two distinct sizes, the large ones multiseriate and very high, the small ones uniseriate and low.

    C1. Pores individually distinct in late wood, usually crowded in early wood. Transition between early and late wood relatively gradual. Vessels usually without tyloses.

        QUERCUS (Red Oak)

    C2. Pores rarely individually distinct in late wood, usually not crowded in early wood. Transition between early and late wood abrupt. Vessels usually with tyloses in heartwood.

        QUERCUS (White Oak)

        (fig. 9.6)

A2. Late wood without distinct radial lines or patches of pores and other cells.

    B1. Pores in late wood numerous. Small pores with rather thin walls. Ground mass of libriform fibers.

        C1. Pores in late wood small and numerous, in tangential, more or less curved bands. Paratracheal and boundary axial parenchyma.

            D1. Rays 1–6 seriate. Ray tissue homogeneous.

                ULMUS

            D2. Rays 1–13 seriate. Ray tissue heterogeneous.

                CELTIS

        C2. Pores in late wood variable in size and occur in clusters. Paratracheal, often confluent, axial parenchyma.

            ROBINIA

    B2. Pores in late wood few, solitary or in small multiples. Small pores with thick walls. Groundmass usually of fiber-tracheids, but libriform fibers may be present.

        C1. Wood parenchyma in late wood paratracheal, often aliform and becoming confluent. Pores in late wood all much smaller than those in early wood.

            FRAXINUS

            (fig. 9.8, *D*)

        C2. Wood parenchyma in late wood in numerous fine metatracheal bands; also boundary parenchyma. Pores in late wood sometimes as wide as those in early wood. Comparatively few, irregularly arranged pores in early wood.

            D1. Storied wood. Bands of axial parenchyma much or somewhat finer than the rays. Tyloses absent.

                DIOSPYROS

                (fig. 9.7, *B*)

            D2. Nonstoried wood. Bands of axial parenchyma as distinct as the rays. Tyloses often present.

                HICORIA

                (figs. 9.7, *A*, and 9.8, *C*)

2. WOOD DIFFUSE POROUS. Pores throughout growth ring fairly uniform in size or only gradually changing in size and distribution from early to late wood.

A1. Pores variable in size, the large ones readily visible without magnification, not crowded.

JUGLANS

A2. Pores small to minute, often not distinct without lens; sometimes few and scattered but mostly crowded, although well distributed throughout growth ring.

B1. Wide aggregate rays present.

C1. Perforation plates in vessels mostly simple.

CARPINUS

C2. Perforation plates exclusively scalariform.

ALNUS

B2. No wide aggregate rays.

C1. Pores solitary or in small multiples, not crowded.

D1. Rays narrower than the pores, inconspicuous. Scalariform perforations in vessels.

BETULA

D2. Large rays as wide or wider than the pores, conspicuous. Simple perforations in vessels.

ACER

(fig. 9.8, *B*)

C2. Pores very numerous, usually crowded.

D1. Largest rays as wide as or wider than the pores. (See also tangential section.)

E1. Largest pores and rays differ little in width. Vessels with gum plugs but without tyloses. Gum ducts (gummosis type) frequently present.

PRUNUS

E2. Largest rays much wider than the pores. Vessels with tyloses but without gum plugs. Gum ducts absent.

F1. Rays nearly all wide, numerous, uniformly spaced.

PLATANUS

F2. Rays of several widths, the wider not numerous and irregularly spaced.

FAGUS

D2. Rays narrower than the pores.

E1. Scalariform perforations in vessels.

F1. Axial parenchyma at boundary.

LIRIODENDRON

F2. Axial parenchyma paratracheal or apotracheal diffuse or both, the cells scattered. Frequently very scanty.

LIQUIDAMBAR
(fig. 9.8, *A*)

E2. Simple perforations in vessels.

F1. Rays of two sizes, uniseriate and conspicuous multiseriate.

TILIA

F2. Rays all uniseriate.

G1. Rays homocellular.

POPULUS

G2. Rays heterocellular.

SALIX
(fig. 9.5)

REFERENCES

Chattaway, M. M. The development of tyloses and secretion of gum in heartwood formation. *Austral. Jour. Sci. Res. B, Biol. Sci.* 2:227–240. 1949.

Dadswell, H. E., A. B. Wardrop, and A. J. Watson. The morphology, chemistry and pulp characteristics of reaction wood. In: *Fundamentals of Papermaking Fibers,* pp. 187–219. British Paper and Board Makers' Association. 1958.

Greguss, P. *Identification of living gymnosperms on the basis of xylotomy.* Budapest, Akademiai Kiado. 1955.

Îatsenko-Khmelevskiĭ, A. A. *Drevesiny Kavkaza.* [Woods of Caucasus.] Erevan, Akad. Nauk Armîan. SSSR. 1954.

Jane, F. W. *The structure of wood.* New York, The Macmillan Company. 1956.

Money, L. L., I. W. Bailey, and B. G. L. Swamy. The morphology and relationships of the Monimiaceae. *Arnold Arboretum Jour.* 31:372–404. 1950.

Phillips, E. W. J. The identification of softwoods by their microscopic structure. *Dept. Sci. and Indust. Res. Forest Products Res. Bul.* 22. London. 1948.

Record, S. J. *Identification of the timbers of temperate North America.* New York, John Wiley and Sons. 1934.

Scott, D. R., and S. B. Preston. Development of compression wood in eastern white pine through the use of centrifugal force. *Forest Sci.* 1:178–182. 1955.

Sinnott, E. W. Reaction wood and the regulation of tree form. *Amer. Jour. Bot.* 39:69–78. 1952.

Stover, E. L. *An introduction to the anatomy of seed plants.* Boston, D. C. Heath and Company. 1951.

Trendelenburg, R. *Das Holz als Rohstoff.* 2nd ed. Revised by H. Mayer-Wegelin. München, Carl Hauser. 1955.

10. VASCULAR CAMBIUM

THE VASCULAR CAMBIUM—THE MERISTEM THAT PRODUCES THE SEC-
ondary xylem and phloem—is commonly called a lateral meristem to
distinguish it from the apical meristem, because it occupies a lateral
position in stem and root. In the three-dimensional aspect, the cambium is a continuous sheath about the xylem of stem and root and
their branches and extends in the form of strips into leaves if the
leaves have secondary growth.

ORGANIZATION OF CAMBIUM

The cells of the vascular cambium do not fit the usual concept of
meristematic cells, that is, cells with dense cytoplasm, large nuclei,
and of approximately isodiametric shape. The cambial cells are highly
vacuolated and occur in two forms. One type of cell, the *fusiform
initial* (fig. 10.1, *A*), is several to many times longer than wide; the
other, the *ray initial* (fig. 10.1, *B*), is slightly elongated to nearly isodiametric. The term fusiform implies that the cell is shaped like a
spindle. A fusiform initial, however, is an approximately prismatic
cell in its middle part and wedge shaped at the ends. The pointed end
of the wedge is seen in tangential sections, the truncated in radial. The
tangential sides of the cell are wider than the radial.

In their relative arrangement the two kinds of initials duplicate the
cell arrangement in the secondary xylem. The fusiform initials consti-

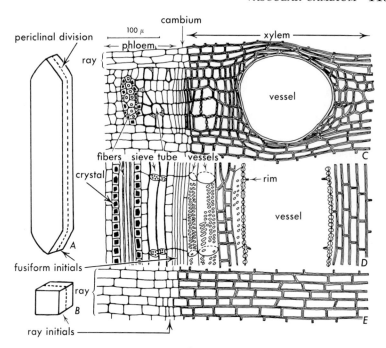

Fig. 10.1. Vascular cambium in relation to derivative tissues. *A,* diagram of fusiform initial; *B,* of ray initial. In both, orientation of division concerned with formation of phloem and xylem cells (periclinal division) is indicated by broken lines. *C, D, E, Robinia pseudoacacia;* sections of stem include phloem, cambium, and xylem. *C,* transverse; *D,* radial (axial system only); *E,* radial (ray only).

tute the axial system of the zone of cambial initials; the ray initials, the radial system (figs. 10.1, *C–E,* and 10.2). The axial system of the secondary xylem is derived from the fusiform initials; the ray system, from the ray initials. Like the secondary xylem, the vascular cambium may be stratified or nonstratified depending on whether or not, as seen in tangential sections, the cells are arranged in horizontal strata (fig. 10.3). In a stratified cambium the fusiform initials are shorter and less extensively overlapping than in a nonstratified cambium.

When the cambial initials produce secondary xylem and phloem cells they divide periclinally (fig. 10.1, *A, B*). At one time a derivative cell is produced toward the xylem, at another time toward the phloem, although not necessarily in alternation. Thus, each cambial initial produces radial files of cells, one toward the outside, the other toward the inside, and the two files meet at the cambial initial (figs. 10.1, *C,* and 10.2, *A*).

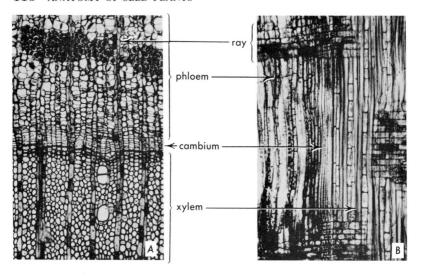

Fig. 10.2. Vascular tissues and cambium of walnut (*Juglans hindsii*) in cross (*A*) and radial (*B*) sections. (Both, ×90.)

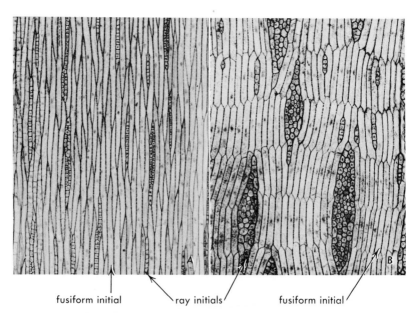

Fig. 10.3. Arrangement of cells in vascular cambium as seen in tangential sections. *A*, nonstoried cambium of *Rhus typhina*. *B*, storied cambium of *Wisteria* sp. (Both, ×90.)

During the height of cambial activity, cell addition occurs so rapidly that older cells are still meristematic when new cells are added to them. Thus, a wide zone of more or less undifferentiated cells accumulates. Within this zone only one cell in a given radial file is an initial in the sense that it continues to divide periclinally and to produce derivatives without itself becoming differentiated into a phloem or a xylem cell. The initials are difficult to distinguish from their recent derivatives, especially because these derivatives divide periclinally one or more times before they begin to differentiate into xylem and phloem cells. It is convenient, therefore, to use the word cambium loosely to include the cambial initials as well as the undifferentiated periclinally dividing derivatives (Bannan, 1955, 1957) and to refer to these derivatives as phloem initials (or phloem mother cells) and xylem initials (or xylem mother cells).

The cambium thus constitutes a more or less wide stratum of periclinally dividing cells organized into axial and ray systems. In the approximately median plane (it is more often an off median) of this stratum is a single layer of cambial initials flanked on both sides by initials of the two vascular tissues.

DEVELOPMENTAL CHANGES IN THE INITIAL LAYER

As the core of the secondary xylem increases in thickness, the cambium is displaced outwardly, and its circumference increases. This increase is accomplished by division of cells, but it also involves complex phenomena of intrusive growth, elimination of initials, and formation of ray initials from fusiform initials. In cambia with short fusiform initials, the divisions are radial anticlinal (fig. 10.4, *A*). Thus, two cells appear side by side where one was present formerly, and each enlarges tangentially. Long fusiform initials divide by anticlinal walls of various degrees of inclination (figs. 10.4, *B–D*, and 10.5, *A*), and each new cell elongates by apical intrusive growth (fig. 10.4, *E, F*). As a result of this growth the new sister cells come to lie side by side in the tangential plane (fig. 10.4, *F*), and they thus increase the circumference of the cambium. During the intrusive growth the ends of the cells may fork (fig. 10.4, *G, H*). The ray initials also divide radially anticlinally if the plant has biseriate or multiseriate rays.

The formation of ray initials from fusiform initials, or their segments, is a common phenomenon. If one compares growth layers in the xylem near the pith with those farther outward, a relative constancy in the ratio between the rays and the axial components may be observed (Braun, 1955). This constancy results from the addition of new rays

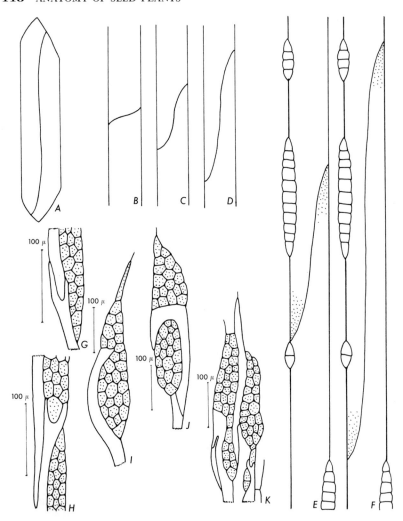

Fig. 10.4. Division and growth of cambial initials. *A*, fusiform initial divided by a radial anticlinal wall; *B–D*, by various oblique anticlinal walls. *E, F*, oblique anticlinal division is followed by apical intrusive growth (growing apices are stippled). *G–H*, forking of fusiform initials during intrusive growth (*Juglans*). *I–K*, intrusion of fusiform initials into rays (*Liriodendron*). (All tangential views.)

as the column of xylem increases its girth; that is, new ray initials appear in the cambium. These new ray initials are derived from fusiform initials.

According to some workers (Braun, 1955), the initials of new uni-seriate rays of the conifers arise as unicellular segments cut out from fusiform initials at their apices or in the middle parts. Studies on *Thuja* (Bannan, 1953) have shown, however, that the origin of rays may be a highly complicated process involving a transverse subdivision of fusiform initials into several cells, loss of some of the products of these divisions, and the transformation of others into ray initials. The loss, or elimination, of initials is a displacement of these cells toward the xylem or phloem and eventual maturation into a xylem or a phloem cell.

It has been observed in conifers and dicotyledons that new uniseriate rays begin as rays one or two cells high and only gradually attain the height typical of the species (Braun, 1955). The increase in height occurs through transverse divisions of the established ray initials and through fusion of rays located one above the other. In the formation of multiseriate rays radial anticlinal divisions and fusions of laterally approximated rays are involved. Indications are that in the process of fusion some fusiform initials intervening between rays are converted into ray initials by transverse divisions; others are displaced toward the xylem or the phloem and are thus lost from the initial zone.

The reverse process, a splitting of wide rays, also occurs. Two phenomena account for such splitting. Some ray initials within a multiseriate ray grow out into fusiform initials by intrusive growth and thus break the continuity of the panel of ray initials. The other method is a separation of such a panel by a fusiform initial intruding among the ray initials (figs. 10.4, *H–K,* and 10.5, *C*).

The divisions that increase the number of fusiform cambial initials and those that are involved in the production of new ray initials occur toward the end of the maximal growth concerned with the seasonal production of xylem and phloem (Bannan, 1956; Braun, 1955). In plants with nonstratified cambia this timing in divisions means that the cambium contains, on the average, shorter fusiform initials at the end of the season than earlier. Subsequently the new cells elongate—provided that they are not eliminated from the cambium—so that the average length of the initials increases until a new period of divisions ensues at the end of the growth season.

The periodic changes in the length of the fusiform initials are reflected in the variation in the length of the resulting xylem cells. In both gymnosperms and angiosperms the length of the elongated types of cells (e.g., tracheids, fibers) rises from the first-formed early wood to the last-formed late wood (Bisset and Dadswell, 1950; Bosshard, 1951). There is also an over-all increase in the length of the fusiform initials from the beginning of secondary growth and through the successive years until the length is more or less stabilized or perhaps reduced (Bosshard, 1951).

The foregoing discussion of developmental transformations in the initial region of the cambium clearly indicates that this meristem is in continuous state of change. The concept of cambial initials must take into account this lack of stability. The elimination of initials is a particularly significant feature in this regard. The initials have no continuing individuality, but the function of initiation of new cells is continuing and is "inherited" by one cell after another (Newman, 1956).

The cytokinesis, or formation of new cells, in the cambium is of especial interest when the cells divide longitudinally and the new wall is formed along the long diameter of the cell. In such a division, the diameter of the initial phragmoplast originating during telophase (fig. 10.6, *A*) is very much shorter than the long diameter of the cell. The phragmoplast and the cell plate reach the longitudinal walls of the cambial cell soon after nuclear division (fig. 10.6, *E*), but the progress of the cell plate toward the ends of the cell is an extended process (fig. 10.6, *A–C*). Before the side walls are reached the phragmoplast appears as a circular halo in front view (fig. 10.6, *D*). After these walls are intersected by the cell plate—but before the ends are reached—the phragmoplast in the same view forms two bars intersecting the side walls (figs. 10.5, *B*, and 10.6, *E*).

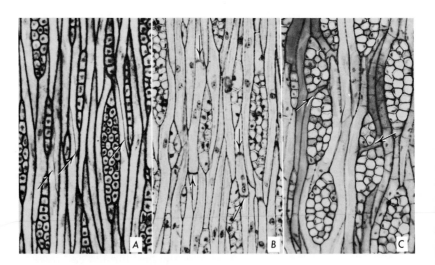

Fig. 10.5. Division and growth of fusiform initials. *A*, cambium of *Juglans* with three fusiform initials recently divided by oblique anticlinal walls (arrows). *B*, cambium of *Cryptocarya;* two periclinally dividing fusiform initials with phragmoplasts (arrows), which indicate extent of cell plates. *C*, phloem of *Liriodendron* with two rays penetrated by axial cells, as a result of intrusive growth, while the tissue was still in cambial state. (All, tangential sections; *A*, *B*, ×140; *C*, ×100.)

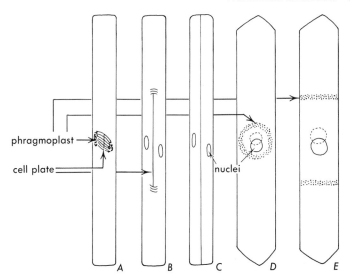

Fig. 10.6. Cell division in fusiform initials. *A–C*, three stages in formation of cell plate as seen in radial sections. *D, E,* two stages of cell-plate formation as seen in tangential sections. The cell plate has extended through about one-third of the cell in *B* and *E*. All views illustrate tangential divisions.

REFERENCES

Bannan, M. W. Further observations on the reduction of fusiform cambial cells in *Thuja occidentalis* L. *Canad. Jour. Bot.* 31:63–74. 1953.

Bannan, M. W. The vascular cambium and radial growth in *Thuja occidentalis* L. *Canad. Jour. Bot.* 33:113–138. 1955.

Bannan, M. W. Some aspects of the elongation of fusiform cambial cells in *Thuja occidentalis* L. *Canad. Jour. Bot.* 34:175–196. 1956.

Bannan, M. W. The relative frequency of the different types of anticlinal divisions in conifer cambium. *Canad. Jour. Bot.* 35:875–884. 1957.

Bisset, I. J. W., and H. E. Dadswell. The variation in cell length within one growth ring of certain angiosperms and gymnosperms. *Austral. Forestry* 14:17–29. 1950.

Bosshard, H. H. Variabilität der Elemente des Eschenholzes in Funktion der Kambiumtätigkeit. *Schweiz. Ztschr. für Forstwesen.* 12:648–665. 1951.

Braun, H. J. Beiträge zur Entwicklungsgeschichte der Markstrahlen. *Botan. Studien* No. 4:73–131. 1955.

Newman, I. V. Pattern in meristems of vascular plants. I. Cell partition in living apices and in the cambial zone in relation to the concept of initial cells and apical cells. *Phytomorphology* 6:1–19. 1956.

11. PHLOEM

THE PHLOEM, OR FOOD-CONDUCTING TISSUE OF A VASCULAR PLANT IS associated with the xylem in the vascular system. Like the xylem, the phloem consists of several types of cells and may be classified, developmentally, into a primary and a secondary tissue. The primary phloem is derived from procambium. The secondary phloem originates in the vascular cambium and reflects the organization of this meristem in that it has an axial and a ray system. The rays are continuous through the cambium with those of the xylem.

Thus, the over-all development and structure of the phloem tissue parallel those of the xylem, but the distinct function of the phloem is associated with characteristics peculiar to this tissue. It is a less sclerified and less persisting tissue than the xylem. Because of its usual position near the periphery of stem and root the phloem becomes much modified in relation to the increase in circumference of the axis and is eventually cut off by the periderm. In contrast, the old xylem remains relatively unchanged in its basic structure.

CELL TYPES

The primary and secondary phloem contain the same categories of cells, but unlike the secondary phloem the primary is not organized into the two systems, the axial and the radial; it has no rays. The summary illustration (fig. 11.1) and the list of phloem cells below are based on the secondary phloem.

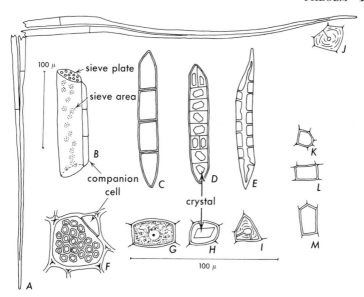

Fig. 11.1. Cell types in secondary phloem of a dicotyledon, *Robinia pseudoacacia*. *A–E*, longitudinal views; *F–J*, cross sections. *A, J*, fiber. *B*, sieve-tube member and, *F*, sieve plate. *C, G*, phloem parenchyma cells (parenchyma strand in *C*). *D, H*, crystal-containing parenchyma cells. *E, I*, sclereid. *K, L, M*, ray cells in tangential (*K*), radial (*L*), and cross (*M*) sections of phloem.

Cell Types	Principal Function
Axial system	
Sieve elements	Conduction, especially longitudinal,
Sieve cells	of food materials
Sieve-tube members	
(with companion cells)	
Sclerenchyma cells	Support; sometimes storage
Fibers	
Sclereids	
Parenchyma cells	Storage and translocation of food
Radial system	substances
Parenchyma cells	

Sieve Elements

The sieve elements are the most highly specialized cells in the phloem. Their principal morphologic characteristics are the sieve areas (modified pits) in their walls and the absence of nuclei in mature protoplasts.

Wall and Sieve Areas. The sieve elements are commonly described as having primary walls except in some conifers. The walls, however, are highly variable in thickness, and in many families the wall thickening is remarkably prominent (fig. 11.14, *D*), sometimes to the extent of almost occluding the lumen of the cell (Esau and Cheadle, 1958). The significance of the extremely thick wall with relation to the function of the cell is still obscure.

The sieve areas are wall areas with pores penetrated by strands that connect the protoplasts of adjoining cells (figs. 11.2, *A, B, D, E,* and 11.3, *B, C*). Throughout the vascular plants, the connecting strands vary in thickness between those that are comparable to plasmodesmata and those several microns in diameter. In the lower vascular plants and up to and including the gymnosperms the strands are commonly thin and rather uniform in size in the sieve areas on different walls of the same cell. In the angiosperms the size of the pores and the thickness of the strands vary considerably, even on different walls of the same cell (fig. 11.4, *A, B;* Esau and Cheadle, 1959). Sieve areas with the larger pores and strands usually occur on the end walls (fig. 11.2, *A, B*), occasionally on the side walls. The wall parts bearing the more highly differentiated sieve areas, that is, areas with compara-

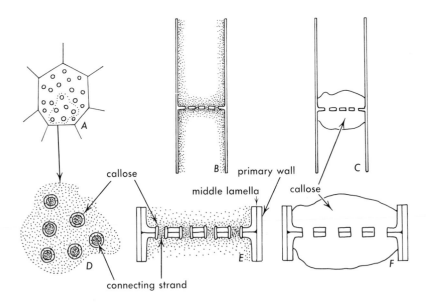

Fig. 11.2. Details of a sieve plate. *A, D,* surface views. *B, C, E, F,* side views. *A, B, D, E,* sieve plate at functioning stage; *C, F,* at cessation of function or during dormancy.

tively large pores and prominent connecting strands, are called sieve plates (figs. 11.2, 11.3, *B, C,* and 11.4, *A*). This term parallels the designation perforation plate used in describing walls of vessel members having perforations.

The question has been raised in the literature whether the connecting strands in the sieve areas, especially the thinner kind, are continuous from cell to cell. The larger connecting strands are undoubtedly continuous, and available electron micrographs show open pores in sieve plates (e.g., Frey-Wyssling and Müller, 1957; Preston, 1958).

Each connecting strand is surrounded by callose (fig. 11.2, *D, E*), a carbohydrate that stains blue with aniline blue and resorcin blue and gives glucose upon hydrolysis (Eschrich, 1956; Kessler, 1958). Callose forms first a thin layer around a strand (fig. 11.2, *E*), but, as the sieve element ages, more callose accumulates. The layer around the strand thickens, and some callose appears also on the surface of the sieve area. The connecting strand is gradually constricted and then completely obliterated when the sieve element becomes dormant or dies. At this stage the callose forms a pad on the sieve area (fig. 11.2, *C, F,* and 11.3, *D*). In old completely inactive sieve elements callose is absent, and open perforations are exposed in the sieve areas. If the phloem is only dormant, connecting strands again appear in the callose during the reactivation of the tissue in the spring (fig. 11.3, *E*), and the callose decreases in amount.

Sieve areas occur between laterally (fig. 11.4, *B, C*) and vertically contiguous sieve elements (fig. 11.3, *A*). If a sieve element adjoins a parenchyma cell, sieve areas on the sieve element side are complemented by ordinary pits on the parenchyma side. Similar structures occur between sieve elements and companion cells.

Sieve cells and sieve-tube members. The degree of specialization of the sieve areas and differences in their distribution on the walls of a given cell serve for the classification of sieve elements into sieve cells and sieve-tube members. In sieve cells the sieve areas are not highly specialized and are not markedly aggregated on restricted wall parts into sieve plates (fig. 11.5, *A*). In sieve-tube members the more highly differentiated sieve areas occur on restricted wall parts—the sieve plates—usually at ends of the cells (fig. 11.5, *B–H*). Moreover, the sieve-tube members form conspicuous vertical series—the sieve tubes (fig. 11.3, *A*)—interconnected through the sieve plates. Comparative studies carried out thus far suggest that gymnosperms and lower vascular plants have sieve cells; angiosperms, both dicotyledons and monocotyledons, have sieve-tube members and sieve tubes.

The sieve areas and sieve plates of angiosperms vary considerably

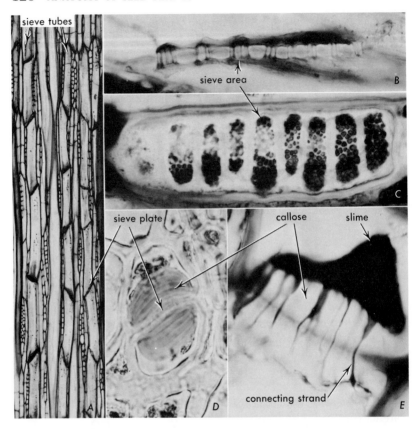

Fig. 11.3. *A,* tangential section of phloem of *Campsis radicans* with sieve tubes. *B–E,* sieve plates of grapevine, *Vitis vinifera. B,* side view (tangential section) and, *C,* surface view (radial section) of compound sieve plates of functioning sieve elements. Sieve areas show connecting strands. *D,* sieve plate with massive dormancy callose. *E,* sieve plate during reactivation: connecting strands are being reformed within dormancy callose. *D, E,* cross sections of phloem. (*A,* ×90; *B, C,* ×750; *D,* ×760; *E,* ×1200; *D, E,* from Esau, *Hilgardia,* 1948.)

in differentiation and arrangement. To a degree these differences are related to the length and form of cells (Esau and Cheadle, 1959). Long sieve-tube members with much inclined end walls commonly have compound sieve plates (fig. 11.5, *B–D*), that is, sieve plates composed of several sieve areas in scalariform or reticulate arrangement. The ends of such elements are wedge shaped, and the sieve areas are borne on the oblique face of the wedge which is part of the radial side of the cell (fig. 11.3, *C*).

The highly compound sieve plates—a concomitant of much inclined end walls—often have relatively thin connecting strands (fig. 11.4, *D*) and are not strikingly different from sieve areas on the lateral walls of the same cells. Such sieve-tube members are interpreted as being rather primitive for angiosperms. Increasing specialization is characterized by a decrease in the inclination of the end walls, reduction in the number of sieve areas in the sieve plates, increase in the thickness of strands in the sieve plates, and the concomitant increase in the difference in specialization between the lateral sieve areas and those of the sieve plates (fig. 11.5, *B–H*). Considerable evidence suggests that the most highly specialized sieve element has simple sieve plates with large pores on transverse end walls and lateral sieve areas of low degree of specialization (fig. 11.5, *H*).

The foregoing discussion suggests an evolutionary sequence for

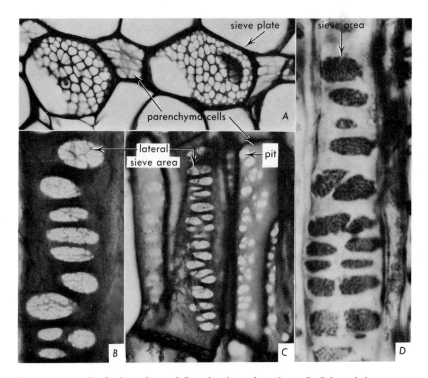

Fig. 11.4. *A*, simple sieve plates of *Cucurbita* in surface view. *B*, *C*, lateral sieve areas in sieve elements and pits in parenchyma cells of *Cucurbita* in surface view. *D*, compound sieve plate of *Pyrus*, with numerous sieve areas. (*A*, *C*, ×320; *B*, ×820; *D*, ×840; from Esau, Cheadle, and Gifford, *Amer. Jour. Bot.*, 1953.)

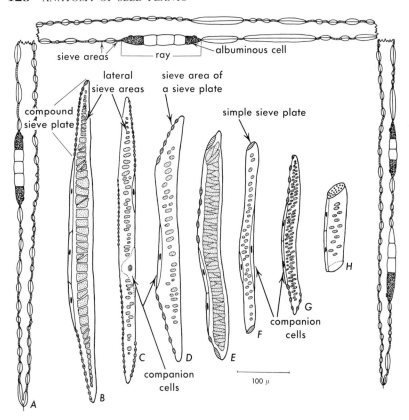

Fig. 11.5. Variations in structure of sieve elements. *A,* sieve cell of *Pinus pinea,* with associated rays, as seen in tangential section. Others are sieve-tube members with companion cells from tangential sections of phloem of the following species: *B, Juglans hindsii; C, Pyrus malus; D, Liriodendron tulipifera; E, Acer pseudoplatanus; F, Cryptocarya rubra; G, Fraxinus americana; H, Wisteria* sp. In *B–G,* the sieve plates appear in side views and their sieve areas are thicker than the intervening wall regions because of deposition of callose.

sieve elements similar in many respects to that of the vessel members in the xylem. The length of cells, which is such a useful criterion in the evaluation of the evolutionary level of vessel members, cannot be used, however, with the same ease in comparable studies of sieve-tube members. Although short fusiform cambial cells give rise to short sieve-tube members, the derivatives of long fusiform initials on the phloem side may become divided so that the potential length of sieve-tube members is reduced (Esau and Cheadle, 1955).

Protoplast. The sieve-element protoplast undergoes a profound change during ontogeny (fig. 11.6). The nucleus disintegrates, the delimitation between the vacuole and the parietal cytoplasm seems to vanish, and features appear that indicate a lowering of the metabolic rate (Esau et al., 1957). Nevertheless, the enucleate protoplast continues to be plasmolyzable (Currier et al., 1955) and capable of depositing callose and again removing it under certain conditions. Plastids elaborating a carbohydrate related to starch may be present. The sieve elements of dicotyledons and conifers contain proteinaceous material called slime which originates in the form of discrete bodies in the cytoplasm of young cells (fig. 11.6, C) but disperses throughout the protoplast in older cells, probably in connection with the breakdown of the tonoplast. In cut sections the slime usually appears near the sieve areas, especially those of the sieve plates, as the so-called slime plug (fig. 11.3, A). No such accumulations appear to be present in undisturbed cells.

The nature of the sieve-element protoplast is undoubtedly reflected in the structure of the connecting strands between the sieve elements. If the protoplast is denatured to the extent that a delimitation between the cytoplasm and vacuole is lost, the connecting strands may be not purely cytoplasmic in composition. At least the slime, which is part of the vacuolar material of mature sieve elements, has been recognized in connecting strands. It is also possible that the connecting strands are highly permeable. The behavior of callose supports this idea. When inactivation of a sieve element approaches or when the cell is

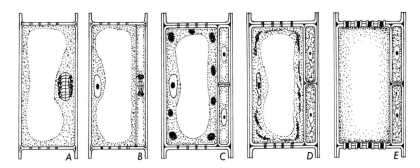

Fig. 11.6. Diagrams illustrating differentiation of a sieve-tube member. *A,* mother cell of sieve-tube member in division. *B,* after division: sieve-tube member and dividing mother cell of companion cells. *C,* slime bodies in sieve-tube member; two companion cells. *D,* slime bodies, in sieve-tube member, are dispersing and nucleus is degenerating. *E,* mature enucleate sieve-tube member; callose incloses connecting strands. The diagrams suggest that each connecting strand is derived from a group of plasmodesmata.

injured the connecting strands are constricted and even obliterated by callose. This reaction suggests a protective sealing off of highly permeable areas (Currier, 1957).

An understanding of the nature of sieve-element protoplast and of the connecting strands in the sieve areas is prerequisite to a satisfactory explanation of the mechanism of translocation of food materials in the phloem. The known characteristics of the protoplast do not wholly support any of the hypotheses advanced so far with regard to that mechanism, and it is likely, moreover, that the formulation of a generally acceptable theory of translocation will not be forthcoming until the relation between the sieve elements and the associated nucleate cells is understood.

Companion cells

The relation between the sieve elements and the companion cells suggests an interdependence between nucleate and enucleate cells of the phloem. The sieve elements and the companion cells are ontogenetically related: one or more companion cells are cut off from the cell that afterwards differentiates into a sieve-tube member (fig. 11.6). The two kinds of cells also appear to have a close physiological relation because cessation of function in a sieve element is accompanied by death of the associated companion cells, and the walls between the two kinds of cells are either very thin or densely pitted.

The presence of companion cells is usually cited together with the presence of sieve plates as a distinguishing characteristic of sieve-tube members. In the gymnosperms and lower vascular plants, cells strictly comparable to the companion cells have not been found. In conifers, however, certain cells in the rays, and sometimes also in the axial system, are connected with the sieve cells by sieve area–pit combinations and die when the sieve cells cease to function. These cells are called *albuminous cells* (fig. 11.5, *A*) and are regarded as being similar to the companion cells in their physiologic relation to the sieve elements (Esau et al., 1953).

Sclerenchyma cells

Fibers (fig. 11.1, *A*) are common components of both primary and secondary phloem. In the primary phloem fibers occur in the outermost part of the tissue; in the secondary, in various distributional patterns among the other phloem cells of the axial system. The fibers may be nonseptate or septate and may be living or nonliving at maturity. In many species the primary and secondary fibers are long cells with thick walls and are used as a commercial source of fiber (e.g., *Linum, Cannabis, Hibiscus*).

Sclereids (fig. 11.1, *E*) are also frequently found in the phloem. They may occur in combination with the fibers or alone, and they may be present in both the axial and the radial systems of the secondary phloem. Sclereids typically differentiate in older parts of the phloem as a result of sclerification of parenchyma cells. This sclerification may or may not be preceded by intrusive growth of the cells. During this growth the sclereids may become branched or considerably elongated. The distinction between fibers and sclereids is not always sharp, especially if the sclereids are long and slender. Sometimes it is appropriate to classify the intermediate cell types as fiber-sclereids.

Parenchyma cells

Parenchyma cells containing various ergastic substances, such as starch, tannins, and crystals, are regular components of the phloem. In the secondary phloem they are classified into axial parenchyma cells (fig. 11.1, *C, D*) and ray parenchyma cells (fig. 11.1, *K–M*). The axial cells may occur in parenchyma strands or as single fusiform parenchyma cells. The strand results from division of a precursor cell in two or more cells.

Crystal-forming parenchyma cells may become subdivided into small cells, each containing a single crystal (fig. 11.1, *D*). Such cells are commonly associated with fibers or sclereids and have lignified walls with secondary thickenings.

The companion cells may be regarded as specialized parenchyma cells. The distinction between the companion cells and the other phloem parenchyma cells is not necessarily sharp, however. In many families of dicotyledons sieve-tube members and some parenchyma cells may be derived from the same phloem initial, and these parenchyma cells may die at the same time as the associated sieve elements (Esau and Cheadle, 1955; Cheadle and Esau, 1958). Thus, it appears that parenchyma cells in the phloem of dicotyledons may range in their degree of ontogenetic and physiologic relation to the sieve elements; the companion cells are the most highly evolved in this respect.

PRIMARY PHLOEM

The primary phloem is classified into *protophloem* and *metaphloem* on the same basis that the primary xylem is classified into protoxylem and metaxylem. The protophloem matures in plant parts that are still undergoing extension growth, and its sieve elements are stretched and become nonfunctional. Eventually they are completely obliterated (fig. 11.7). The metaphloem differentiates later and, in plants without secondary growth, constitutes the only conducting phloem in adult plant parts.

The protophloem sieve elements of angiosperms are usually narrow and inconspicuous, but they are enucleate and have sieve areas with callose. They may or may not have companion cells. They may appear in groups or singly among parenchyma cells or, in many dicotyledons, among conspicuously elongated living cells. In many species these elongated cells are fiber primordia. They undergo considerable elongation and, after the sieve elements are obliterated, mature as fibers. Such fibers on the periphery of the phloem of numerous

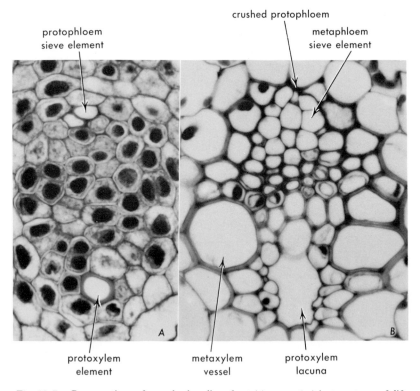

Fig. 11.7. Cross sections of vascular bundles of oat (*Avena sativa*) in two stages of differentiation. *A,* first elements of protophloem and protoxylem have matured. *B,* metaphloem and metaxylem are mature; protophloem is crushed; protoxylem is replaced by a lacuna. (Both, ×850. From Esau, *A, Amer. Jour. Bot.,* 1957; *B, Hilgardia,* 1958.)

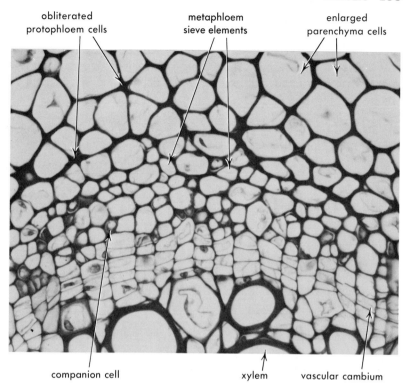

Fig. 11.8. Primary phloem from leaf of sugar beet (*Beta vulgaris*). Metaphloem is mature, sieve elements and companion cells of protophloem have been crushed. A few secondary phloem and xylem elements have been formed. (×470. From Esau, *Jour. Agr. Res.*, 1944.)

dicotyledon stems are often called pericyclic fibers (chapters 6 and 16). Protophloem fibers occur in roots also.

The metaphloem has more and commonly wider sieve elements than the protophloem. Companion cells are regularly present in the metaphloem of angiosperms (fig. 11.8), but fibers are usually absent. Parenchyma cells may become sclerified after the phloem ceases to conduct.

SECONDARY PHLOEM

The secondary phloem constitutes a much less prominent part of a branch, a tree trunk, or a root than the secondary xylem. The amount of phloem produced by the vascular cambium is usually smaller than that of xylem, the old phloem becomes more or less conspicuously crushed, and eventually the nonfunctional phloem is separated from the axis by the periderm. Thus, whereas the successive increments of the xylem accumulate in the branch, trunk, or root, the phloem remains restricted in amount.

Under the term bark the phloem is often included with the tissues located outside the vascular cambium (chapter 12). The functional phloem constitutes the innermost part of the bark of woody stems and roots.

Conifer phloem

Phloem in the conifer is generally simpler and less variable among species than it is in the dicotyledons. The axial system contains sieve cells, parenchyma cells, and frequently fibers (figs. 11.9 and 11.10). Sclereids also may occur. The rays are uniseriate and contain parenchyma cells and albuminous cells, if these cells are present in the given species. The albuminous cells are located at the margins of the rays as a rule (fig. 11.5, *A*). Resin ducts may be present in both systems.

The sieve elements are long cells that have numerous sieve areas, commonly restricted to the radial faces (fig. 11.5, *A*). The parenchyma cells form strands (fig. 11.10, *B, C*) or are single cells. Fibers are regularly absent in *Pinus* (fig. 11.11) and regularly present in Taxaceae, Taxodiaceae, and Cupressaceae. When they are present the fibers occur, as a rule, in uniseriate tangential bands (fig. 11.10, *A*) alternating with similar bands of parenchyma cells and sieve cells (Chang, 1954).

In a given section of conifer phloem only a narrow band, approximately one growth layer, may be in active state; the rest is no longer conducting. If fibers are absent, collapse of the sieve cells gives the tissue a distorted appearance, especially because the rays assume a wavy course (fig. 11.11). The parenchyma cells are enlarged in the nonfunctional phloem and remain alive until they are cut off by the periderm (figs. 11.10 and 11.11). Ray parenchyma cells also remain active but not the albuminous cells. These collapse in the nonfunctional phloem (fig. 11.11, *A*).

Dicotyledon phloem

The secondary phloem of dicotyledons varies with regard to composition, arrangement, and size of cells and the characteristics of the nonfunctional phloem in both the axial and the ray systems (fig. 11.12 and 11.13). Sieve tubes, companion cells, and parenchyma cells are constant elements of the axial system, but fibers may be absent (*Aristolochia*). When they are present, the fibers may be scattered (*Campsis,* fig. 11.14, *C, Cephalanthus, Laurus*) or may appear in tangential bands in parallel arrangement (*Fraxinus,* fig. 11.13, *A, Liriodendron, Magnolia, Robinia, Tilia,* fig. 11.14, *A*) or somewhat scattered (*Ostrya,* fig. 11.14, *B*). Fibers may be so abundant that the sieve tubes and parenchyma cells appear as small groups scattered

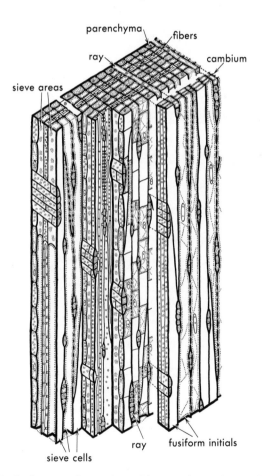

Fig. 11.9. Block diagram of secondary phloem and vascular cambium of *Thuja occidentalis* (white cedar), a conifer. (From Esau, *Plant Anatomy,* John Wiley and Sons, 1953. Courtesy of I. W. Bailey. Drawn by Mrs. J. P. Rogerson under the supervision of L. G. Livingston. Redrawn.)

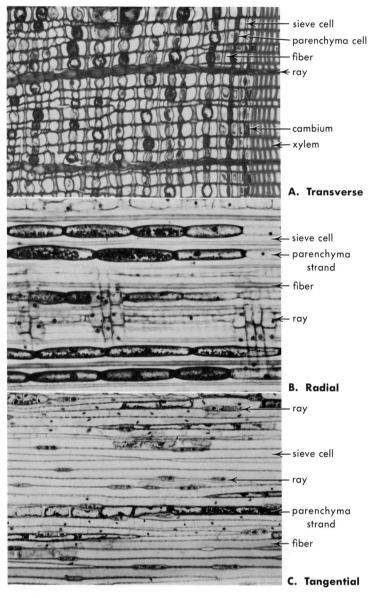

Fig. 11.10. Secondary phloem of *Thuja occidentalis* in three kinds of sections. A conifer phloem with fibers. (All, ×150.)

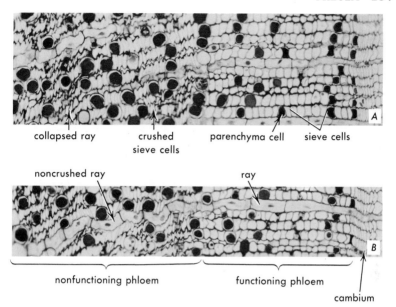

| collapsed ray | crushed sieve cells | parenchyma cell | sieve cells | A |

noncrushed ray .... ray

nonfunctioning phloem .... functioning phloem

B

cambium

Fig. 11.11. Phloem of *Pinus pinea* in cross sections. A conifer phloem without fibers. (Both, ×130.)

among the fibers (*Carya*). In some species sclerenchyma cells, usually sclereids or fiber-sclereids, differentiate only in the nonfunctional part of the phloem (*Prunus*). The septate fibers of *Vitis* are living cells concerned with storage of starch.

Depending on the characteristics of the cambium, the secondary phloem may be stratified (*Robinia*) or nonstratified (*Betula, Quercus, Populus, Tilia, Liriodendron, Juglans*). The long sieve-tube members in nonstratified phloem typically have inclined end walls with compound sieve plates. In more or less definitely storied phloem the sieve elements have slightly inclined or almost transverse end walls, and their sieve plates have few sieve areas or only one.

The rays resemble the xylem rays of the same plant and may be uniseriate or multiseriate, high or low; and different kinds of rays may be represented in the same tissue. The rays are composed of parenchyma cells (fig. 11.13) but may contain sclereids or sclerified parenchyma cells with crystals. In older parts of the phloem the rays may become dilated in response to the increase in circumference of the stem or the root (fig. 11.14, *A*). Radial anticlinal cell division and tangential cell enlargement bring about the dilatation. Sometimes the divisions are restricted to the median position in a ray, and the tissue appears like a meristem (Holdheide, 1951; Schneider, 1955). Usually only some rays become dilated; others remain as wide as they were at the time of origin in the cambium (fig. 11.14, *A*). In some species of *Eucalyptus* wide wedges of tissue develop in the dilatating phloem by division of phloem parenchyma cells (Chattaway, 1955).

The nonfunctional phloem assumes different aspects depending on the types of cells present and on their behavior. The dilatation of rays is one of the characteristics of the old phloem. Sieve tubes may be completely crushed, or they may remain open and become filled

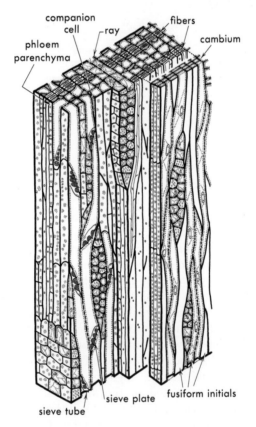

Fig. 11.12. Block diagram of secondary phloem and vascular cambium of *Liriodendron tulipifera* (tulip tree), a dicotyledon. (From Esau, *Plant Anatomy*, John Wiley and Sons, 1953. Courtesy of I. W. Bailey. Drawn by Mrs. J. P. Rogerson under the supervision of L. G. Livingston. Redrawn.)

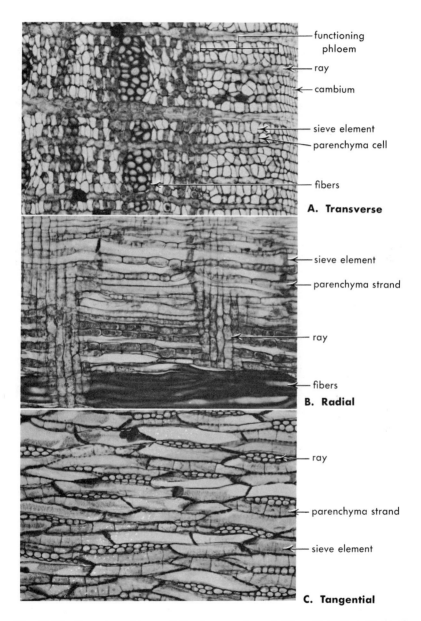

A. **Transverse** — functioning phloem, ray, cambium, sieve element, parenchyma cell, fibers

B. **Radial** — sieve element, parenchyma strand, ray, fibers

C. **Tangential** — ray, parenchyma strand, sieve element

Fig. 11.13. Secondary phloem of *Fraxinus americana* (white ash) in three kinds of sections. (All, ×130.)

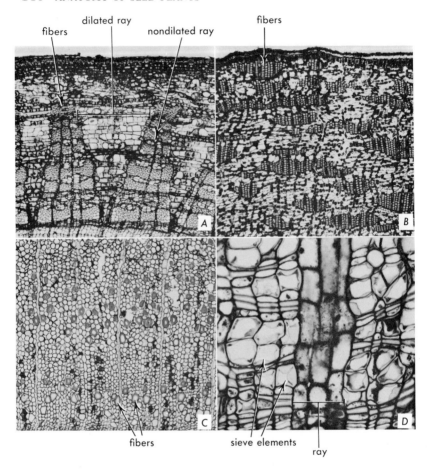

Fig. 11.14. Secondary phloem of dicotyledons in cross sections. *A, Tilia,* fibers in parallel tangential bands. *B, Ostrya,* fibers in groups and bands. *C, Campsis,* fibers scattered singly. *D, Liriodendron,* sieve elements with nacreous walls. (*A, C,* ×54; *D,* ×45; *E,* ×200.)

with gases. Parenchyma cells may enlarge and thus crush the sieve tubes. If the tissue shrinks because of collapse of cells, the rays may be bent. The parenchyma cells in the nonfunctional phloem continue to store starch until they are cut off by the periderm.

The amount of the functional phloem is commonly limited to one growth increment because the sieve elements originating from the cambium in the spring usually cease to conduct and die in the fall. There are exceptions to this sequence, however. As was mentioned earlier, in *Vitis* the sieve elements of one year become dormant in the winter but resume activity in the following spring. How many other species might show such reactivation remains to be investigated. The amount of the nonfunctional phloem is highly variable. If periderm is formed repeatedly at short intervals, the old phloem does not accumulate. In some species growth layers may be detected in the phloem—the early sieve elements may be wider than those formed in the late phloem, or there may be a band of sclerenchyma formed in the late phloem—but generally the delimitation of the layers is obscured by the changes affecting the nonfunctioning phloem.

REFERENCES

Chang, Y. P. Bark structure of North American conifers. *U. S. Dept. Agric. Tech. Bul.* 1095. 1954.

Chattaway, M. M. The anatomy of bark. VI. Peppermints, boxes, ironbarks and other eucalypts with cracked and furrowed barks. *Austral. Jour. Bot.* 3:170–176. 1955.

Cheadle, V. I., and K. Esau. Secondary phloem of Calycanthaceae. *Calif. Univ., Pubs., Bot.* 24:397–510. 1958.

Currier, H. B. Callose substance in plant cells. *Amer. Jour. Bot.* 44:478–488. 1957.

Currier, H. B., K. Esau, and V. I. Cheadle. Plasmolytic studies of phloem. *Amer. Jour. Bot.* 42:68–81. 1955.

Esau, K., and V. I. Cheadle. Significance of cell divisions in differentiating secondary phloem. *Acta Bot. Neerl.* 4:348–357. 1955.

Esau, K., and V. I. Cheadle. Wall thickening in sieve elements. *Natl. Acad. Sci. Proc.* 44:546–553. 1958.

Esau, K., and V. I. Cheadle. Size of pores and their contents in sieve elements of dicotyledons. *Natl. Acad. Sci. Proc.* 45:156–162. 1959.

Esau, K., V. I. Cheadle, and E. M. Gifford, Jr. Comparative structure and possible trends of specialization of the phloem. *Amer. Jour. Bot.* 40:9–19. 1953.

Esau, K., H. B. Currier, and V. I. Cheadle. Physiology of phloem. *Annu. Rev. Plant Physiol.* 8:349–374. 1957.

Eschrich, W. Kallose. *Protoplasma* 47:487–530. 1956.

Frey-Wyssling, A., and H. R. Müller. Submicroscopic differentiation of plasmodesmata and sieve plates in *Cucurbita*. *Jour. Ultrastr. Res.* 1:38–48. 1957.

Holdheide, W. Anatomie mitteleuropäischer Gehölzrinden. In: H. Freund. *Handbuch der Mikroskopie in der Technik.* Vol. 5, Part 1:195–367. 1951.

Kessler, G. Zur Charakterisierung der Siebröhrenkallose. *Schweiz. Bot. Gesell. Ber.* 68:5–43. 1958.

Preston, R. D. The physiological significance of electron microscopic investigations of plant cell walls. *Science Progress* 46:593–605. 1958.

Schneider, H. Ontogeny of lemon tree bark. *Amer. Jour. Bot.* 42: 893–905. 1955.

12. PERIDERM

THE PERIDERM IS A PROTECTIVE TISSUE OF SECONDARY ORIGIN replacing the epidermis in stems and roots that have continual secondary growth. Woody dicotyledons and the gymnosperms furnish the best examples of periderm development. Leaves usually produce no periderm although scales of winter buds may do so. Periderm occurs in herbaceous dicotyledons, especially in the oldest parts of stem and root. Some monocotyledons have periderm; others a different kind of secondary protective tissue.

The periderm develops along surfaces that are exposed after abscission of plant parts, such as leaves and branches. Periderm formation is also an important phase in the development of protective layers near injured or dead (necrosed) tissues (wound periderm or wound cork; fig. 12.1), whether resulting from mechanical wounding (Morris and Mann, 1955) or invasion of parasites (Struckmeyer and Riker, 1951). In several families of dicotyledons, periderm is formed in the xylem—interxylary cork—sometimes in relation to a normal dying back of annual shoots or a splitting of perennating roots and stems into strands (Moss and Gorham, 1953).

The nontechnical term bark must be distinguished from the term periderm. Although the word bark is employed loosely, and often inconsistently, it is a useful term if properly defined. Bark may be used most appropriately to designate all tissues outside the vascular cambium. In the secondary state, bark includes the secondary phloem, the primary tissues that may still be present outside the secondary

142

wound periderm   dead surface cells

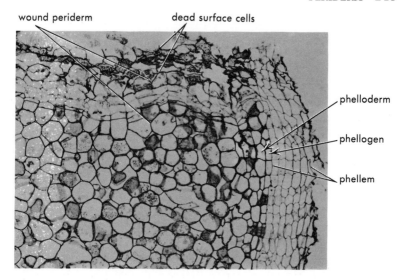

phelloderm

phellogen

phellem

Fig. 12.1. Periderm of root of sweet potato (*Ipomoea batatas*). Natural periderm to the right, wound periderm above. The development of wound periderm is part of the process of wound healing on broken ends of sweet potato at end of curing period. (×63. From Morris and Mann, *Hilgardia*, 1955.)

phloem, the periderm, and the dead tissues outside the periderm. The death of cells isolated outside the periderm brings about a distinction between the outer nonliving and the inner living bark. The functional phloem is the innermost part of the living bark. The term bark is sometimes used for stems in primary state of growth. It then includes the primary phloem, the cortex, and the epidermis. Because of the radially alternate arrangement of xylem and phloem in roots in primary state, in such roots the phloem could not be conveniently included with the cortex under the term bark.

STRUCTURE OF PERIDERM AND RELATED TISSUES

The periderm consists of the *phellogen* (*cork cambium*), the meristem that produces the periderm; the *phellem* (commonly called *cork*), the protective tissue formed outwardly by the phellogen; and the *phelloderm*, a living parenchyma tissue formed inwardly by the meristem (fig. 12.1). The death of the tissues lying outside the periderm results from the insertion of the nonliving cork between these tissues and the living inner tissues of the plant.

The phellogen is relatively simple in structure. In contrast to the vascular cambium it has only one type of cell. In cross section the phellogen appears as a continuous tangential layer (lateral meristem) of rectangular, radially flattened cells, each with the derivatives in a radial file that extends outwardly through the cork cells and inwardly through the phelloderm cells (fig. 12.1). In longitudinal sections the phellogen cells are rectangular or polygonal in outline, sometimes somewhat irregular.

The cork cells are often approximately prismatic in shape (fig. 12.2, A, B) although they may be rather irregular in the tangential plane (fig. 12.2, F). They may be elongated vertically (fig. 12.2, E, F), radially (fig. 12.2, B–E), or tangentially (fig. 12.2, A, narrower cells). They are usually arranged compactly, that is, the tissue lacks intercellular spaces. The cells are nonliving at maturity.

The cork cells are characterized by suberization of their walls. The suberin, a fatty substance, usually occurs as a distinct lamella that covers the original primary cellulose wall, which may be lignified. The walls of cork cells vary in thickness. In thick-walled cells a lignified cellulose layer occurs on the inside of the suberin lamella, which thus may be imbedded between two cellulose layers. The walls of cork cells may be colored brown or yellow, or the lumina may contain colored resinous or tanniniferous materials. Frequently, however, cork cells lack visible contents.

Cork used commercially as bottle cork has thin walls and lumina filled with air. It is highly impervious to water (effect of suberin) and resistant to oil. It is light in weight and has thermal insulating qualities. Mature cork of this type is also a compressible, resilient tissue. The commercially valuable properties, imperviousness to water and insulating qualities, make the cork also effective as a protective layer on the plant surface. The dead tissue that becomes isolated by periderm adds to the insulating effect of the cork.

In many species the phellem consists of cork cells and of nonsuberized cells called *phelloid cells.* Like the cork cells, the nonsuberized cells may be thin walled or thick walled, the latter often differentiated as sclereids (fig. 12.2, D).

The phelloderm cells resemble cortical parenchyma cells and may be distinguished from the latter by their position in the same radial files as the phellem cells (fig. 12.1).

Polyderm. A special type of protective tissue called polyderm occurs in roots and underground stems of Hypericaceae, Myrtaceae, Onagraceae, and Rosaceae (Luhan, 1955; Nelson and Wilhelm, 1957). It consists of alternating tissue layers: layers one cell deep of partly

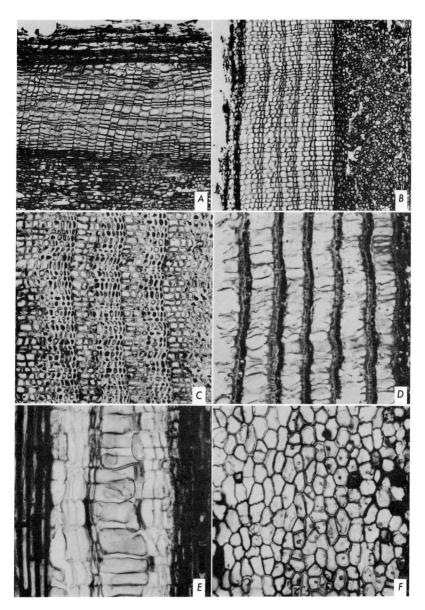

Fig. 12.2. Variation in structure of phellem in stems. *A, B, Rhus typhina.* Phellem in transverse (*A*) and radial (*B*) sections of stem shows growth layers because of alternation of narrower and wider cells. *C*, birch (*Betula populifolia*). Phellem with thick cell walls and conspicuous growth layers; in radial section. *D, Rhododendron maximum.* Heterogeneous phellem consisting of cells of different sizes; sclereids compose some of the layers of small cells; in radial section. *E, F, Vaccinium corymbosum.* Phellem in radial (*E*, light-colored cells in the middle) and tangential (*F*) sections. (*A, B,* ×54; *C, E, F,* ×200; *D,* ×150.)

suberized cells and layers several cells deep of nonsuberized cells (fig. 12.3). The polyderm may become twenty or more layers in total thickness, but only the outermost layers are dead. In the living part the nonsuberized cells function as storage cells.

Rhytidome. As a tree grows older, periderms often arise at successively greater depths and thus cause an accumulation of dead tissues on the surface of stem and root (fig. 12.4). This dead part of the bark, composed of layers of tissues isolated by the periderms and of layers of no longer growing periderms, is called rhytidome. Rhytidome thus

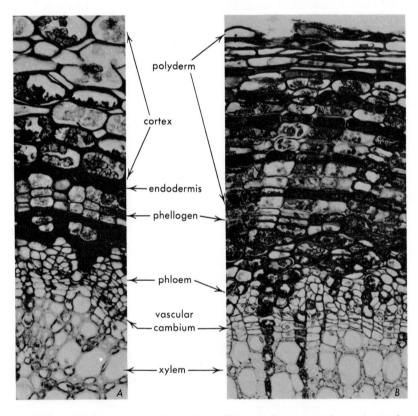

polyderm

cortex

endodermis

phellogen

phloem

vascular cambium

xylem

A   B

Fig. 12.3. Polyderm of root of strawberry (*Fragaria*) in cross sections. *A,* root in early stage of secondary growth. Phellogen has been initiated, but the cortex is still intact. *B,* older root. Wide layer of polyderm has been formed by the phellogen. The cells composing the darkly stained bands in the polyderm are suberized. These cells alternate with nonsuberized cells. Both kinds of cells are living. Nonliving suberized cells form the other covering. No cortex is present. (*A,* ×285; *B,* ×248. From Nelson and Wilhelm, *Hilgardia,* 1957.)

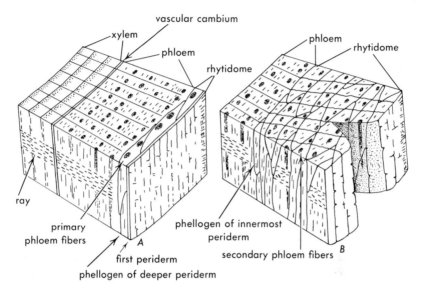

Fig. 12.4. Diagrams showing an earlier (*A*) and a later (*B*) stage in the development of rhytidome. In *A* cortex and primary phloem are included in the rhytidome; in *B*, many layers of secondary phloem. The earlier layers of rhytidome have been shed in *B*.

constitutes the outer bark and is especially well developed in older stems and roots of trees. In shrubs early exfoliation of older bark is common and precludes accumulation of thick rhytidome.

DEVELOPMENT OF PERIDERM

The first periderm commonly appears during the first year of growth of stem or root. The subsequent, deeper periderms may be initiated later the same year or may not appear for many years (species of *Abies, Carpinus, Fagus, Quercus*) or may never appear. In addition to specific differences, environmental conditions influence the appearance of the sequent periderms. Exposure to sunlight, for example, hastens the development of deep periderms (Zeeuw, 1941).

The first periderm of a stem originates most commonly in the sub-epidermal layer (fig. 12.5, *A*), occasionally in the epidermis. In some species, however, the first periderm appears rather deep in the stem (*Berberis, Ribes, Vitis;* fig. 12.6), usually in the primary phloem. In most roots the first periderm originates in the pericycle (chapter 14), but it may appear near the surface as, for example, in some trees and perennial herbaceous plants in which the root cortex serves for food storage. The sequent periderms appear in successively deeper layers beneath the first (fig. 12.4) and thus, eventually, originate from parenchyma of the phloem, including ray cells.

The first phellogen is initiated either uniformly around the circumference of the axis or in localized areas and becomes continuous by a lateral spread of meristematic activity. The sequent periderms appear most commonly as discontinuous but overlapping layers (figs. 12.4 and 12.7, *C, D*). These approximately shell-shaped layers originate beneath cracks of overlying periderms. The sequent periderms may also be continuous around the circumference or at least for considerable parts of the circumference (fig. 12.7, *B*).

The phellogen, the first or the sequent, is initiated by divisions of cells of various kinds. Depending on the position of the phellogen these may be cells of the epidermis, subepidermal parenchyma or collenchyma, or parenchyma of the pericycle or phloem, including that of the phloem rays. Usually these cells are indistinguishable from other cells of the same categories; all are living cells and, therefore, potentially meristematic. The initiating divisions may begin in the presence of chloroplasts and ergastic substances, such as starch and tannins, and while the cells still have thick primary walls, as in collenchyma. Eventually the chloroplasts, ergastic substances, and wall thickenings disappear. Sometimes the subepidermal cells, in which phellogen is to arise, have no collenchymatous thickenings and show an orderly and compact arrangement.

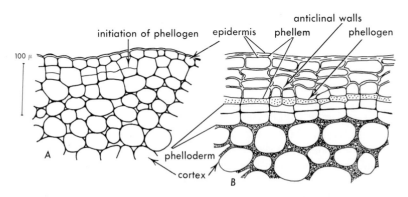

Fig. 12.5. Origin of periderm in *Pelargonium* stem as seen in cross sections. *A*, periclinal divisions in subepidermal layer have produced phellogen cells toward the outside and phelloderm cells toward the inside, one of each to a divided cell. *B*, periderm is established.

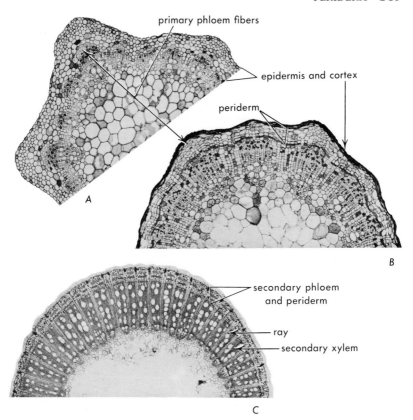

Fig. 12.6. Origin of periderm in grapevine (*Vitis vinifera*) as seen in cross sections. *A,* seedling stem without periderm. *B,* older seedling stem with periderm that originated in the primary phloem and caused the death and collapse of cortex. Primary phloem fibers also appear outside the periderm. *C,* one-year-old cane with periderm outside the secondary phloem. (*A,* ×50; *B,* ×49; *C,* ×10. From Esau, *Hilgardia,* 1948.)

The phellogen is initiated by periclinal divisions, and it produces the phellem and the phelloderm by means of periclinal divisions (fig. 12.5). The phellogen keeps pace with the increase in circumference of the axis by periodic division of its cells in the radial anticlinal plane (fig. 12.5, *B*).

The exact sequence of divisions initiating the periderm is somewhat variable, even in plants of the same species growing under different environmental conditions. There may be some preparatory divisions before the phellogen becomes defined. This sequence is especially

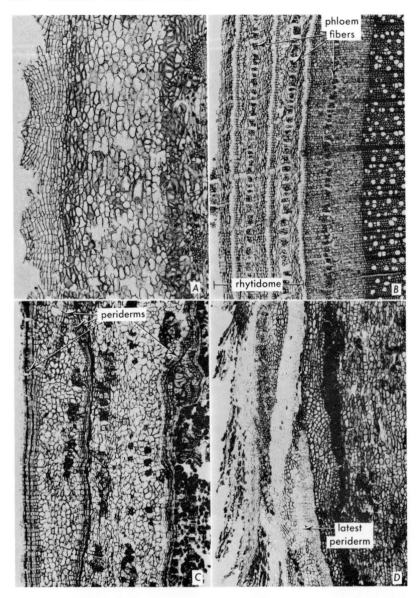

Fig. 12.7. Periderm and rhytidome in cross sections of stems. *A, Talauma.* Periderm with deep cracks. *B, Lonicera tartarica.* Rhytidome in which periderm layers alternate with those derived from secondary phloem and which contains phloem fibers. *C, Quercus alba* (oak). Rhytidome with narrow layers of periderm and wide layers of dead phloem tissue. *D, Gymnocladus dioica.* Rhytidome with wide layers of periderm (light stain). (*A, D,* ×45; *B, C,* ×43.)

common in roots. Phellogen, however, may become restricted to a single layer of cells, usually the outer of the two layers formed by the first periclinal division.

The phellogen of the first periderm produces most of the cells outwardly. The phelloderm is commonly small in amount, sometimes restricted to the single layer of cells left on the inner side of the phellogen after the first periclinal division. The deeper seated, sequent periderms also have phelloderm. In general a given phellogen cell produces only a few cork cells yearly. In many species the yearly increments are not discernible; in some, however, the early formed cork cells are wider and have thinner walls than those formed later (*Betula,* fig. 12.2, *C; Prunus, Robinia*); also the later cork may have dark contents (Wutz, 1955).

Wound periderm. Natural and wound periderms are basically alike in method of origin and growth and may have the same cellular composition (Morris and Mann, 1955). The difference between them is mainly in timing of origin and restriction of the wound periderm to the place of injury. Moreover, as a result of the injury, necrosed cells (scar tissue) occur exterior to the cork cells (fig. 12.1). Successful development of wound periderm is important in horticultural practice when plant parts used for propagation are apt to be injured by handling or must be cut (e.g., potato tubers, sweet potato roots). Environmental conditions markedly influence the proper development of wound periderm (Morris and Mann, 1955). The ability to develop wound periderm in response to invasion by parasites may distinguish resistant from susceptible plants (Struckmeyer and Riker, 1951).

A wound reaction occurs when the periderm is peeled off to the living cells underneath. The newly exposed cells die, and a new periderm arises below them. This reaction is utilized in the production of commercial cork from the cork oak. The first cork, which is of inferior quality, is removed to the phellogen and the new phellogen developing beneath the scar tissue produces massive cork of superior quality.

Protective tissue in monocotyledons. Some monocotyledons produce a periderm similar to that in the dicotyledons (species of *Aloe;* coconut and royal palms). The majority of monocotyledons that form a secondary protective tissue at all show, however, a special method of development of the protective tissue. Parenchyma cells in successively deeper positions divide several times periclinally, and the products of these divisions become suberized. Because of its storied appearance in transections, this tissue is called storied cork (fig. 12.8).

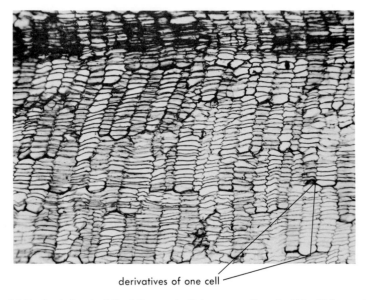

derivatives of one cell

Fig. 12.8. Storied cork of *Cordyline terminalis* in cross section. (×110. Slide, courtesy of V. I. Cheadle.)

OUTER ASPECT OF BARK IN RELATION TO STRUCTURE

The external features of the periderm and rhytidome vary in relation to the structure and development of periderm and the kinds of tissues isolated by the periderms. If there is only a superficial periderm with thin cork, the surface is smooth. Massive cork is usually cracked and fissured (fig. 12. 7, *A*). When the yearly production of cork occurs in isolated positions, the outer cork layers are sloughed off in these positions so that the surface resembles that produced by some scaly rhytidomes. The stems of some species produce a so-called winged cork, a form that results from a symmetrical longitudinal splitting of the cork in relation to the uneven expansion of different sectors of the stem (Smithson, 1954).

The rhytidome presents a variety of aspects. On the basis of manner of formation two forms are distinguished, scale bark and ring bark. Scale bark occurs when the sequent periderms develop in restricted overlapping strata, each cutting out a "scale" of tissue (figs. 12.4 and 12.7, *C, D; Pinus, Pyrus*). Ring bark is less common and results from the formation of successive periderms approximately concentrically around the axis (*Vitis, Clematis, Lonicera;* fig. 12.7, *B*).

With regard to the nonperidermal tissue enclosed in the rhytidome, the fibrous tissue lends a characteristic aspect to the bark (Holdheide, 1951). If fibers are absent, the bark breaks into individual scales or shells (*Pinus, Acer pseudoplatanus*). In fibrous bark a net-like pattern of splitting occurs (*Tilia, Fraxinus*). The scaling off of the bark may have different structural bases. If thin-walled cells of cork or of phelloids are present in the periderms of the rhytidome, the scales may "exfoliate" along these cells (fig. 12.2, *D*). Breaks in the rhytidome can occur also through cells of the nonperidermal tissues; in *Eucalyptus*, for example, through phloem parenchyma cells (Chattaway, 1953). Cork is frequently a strong tissue and renders the bark persistent, even if deep cracks develop (species of *Betula*, fig. 12.2, *C; Pinus, Quercus, Robinia, Salix, Sequoia*). Such barks wear off without forming scales.

LENTICELS

A lenticel may be defined as a limited part of the periderm in which the phellogen is more active than elsewhere and produces a tissue which, in contrast to the phellem, has numerous intercellular spaces (Wutz, 1955). The lenticel phellogen itself also has intercellular spaces. Because of this relatively open arrangement of cells, the lenticels are assumed to be structures permitting the entry of air through the periderm.

Lenticels are usual components of periderm of stems and roots. Outwardly, a lenticel often appears as a vertically or horizontally elongated mass of loose cells that protrudes above the surface through a fissure in the periderm (fig. 12.9, *B*). Lenticels vary in size from structures barely visible without magnification to those 1 centimeter and more in length. They may occur singly or in rows. Vertical rows of lenticels frequently occur opposite the wide vascular rays, but in general there is no constant positional relation between lenticels and rays (Wutz, 1955).

The phellogen of a lenticel is continuous with that of the corky periderm but usually bends inward so that it appears more deeply situated (fig. 12.9). The loose tissue formed by the lenticel phellogen toward the outside is the *complementary* or *filling tissue* (Wutz, 1955); the tissue formed toward the inside is the phelloderm.

The degree of difference between the filling tissue and the neighboring phellem varies in different species. In the gymnosperms this tissue is composed of the same types of cells as the phellem. The

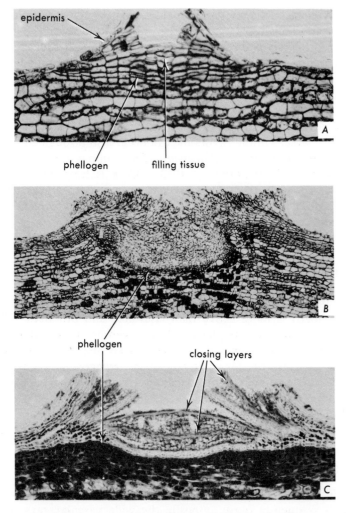

Fig. 12.9. Lenticels in cross sections of stems. *A, B,* avocado (*Persea americana*). Young lenticel in *A,* older in *B.* No closing layers are present. *C,* beech (*Fagus grandifolia*). Lenticel with closing layers. (*A, C,* ×110; *B,* ×43.)

main difference between the two is that the tissue of the lenticel has intercellular spaces. Lenticel cells may also have thinner walls and be radially elongated, instead of radially flattened like the phellem cells of so many species.

In the dicotyledons three structural types of lenticels are recognized (Wutz, 1955). The first and simplest, exemplified by species of

Liriodendron, Magnolia, Malus, Populus, Pyrus, and *Salix,* has a complementary tissue composed of suberized cells. The tissue, though having intercellular spaces, may be more or less compact and may show annual growth layers, with thinner-walled looser tissue appearing earlier, and thicker-walled more compact tissue later.

Lenticels of the second type, as found in species of *Fraxinus, Quercus, Sambucus,* and *Tilia,* consist mainly of a mass of more or less loosely arranged nonsuberized complementary tissue, succeeded at the end of the season by a more compactly arranged layer of suberized cells. The third type, illustrated by lenticels of species of *Betula, Fagus* (fig. 12.9, *C*), *Prunus,* and *Robinia,* shows the highest degree of specialization. The filling tissue is layered because loose nonsuberized tissue regularly alternates with compact suberized tissue. The compact tissue forms the closing layers, each one to several cells in depth, that hold together the loose tissue, usually in layers several cells deep. Several strata of each kind of tissue are produced yearly. The closing layers are successively broken by the new growth, but one closing layer on the outside is always intact.

The first lenticels frequently appear under stomata. The parenchyma cells beneath a stoma undergo some preparatory divisions; then a phellogen becomes established deeply in this new tissue. Growth from this phellogen pushes the overlying cells outward and ruptures the epidermis (fig. 12.9, *A*).

Lenticels are maintained in the periderm as long as the periderm continues to grow, and new ones arise from time to time by change in the activity of the phellogen from formation of phellem to that of lenticel tissue. The deeper periderms also have lenticels. They usually appear at bottom of cracks in the rhytidome. The lenticels of the rhytidome are basically similar to those of the initial periderm, but their phellogen is less active, and, therefore, they are not as well differentiated.

REFERENCES

Chattaway, M. M. The anatomy of bark. I. The genus *Eucalyptus. Austral. Jour. Bot.* 1:402–433. 1953.

Holdheide, W. Anatomie mitteleuropäischer Gehölzrinden. In: H. Freund. *Handbuch der Mikroskopie in der Technik.* Vol. 5, Part 1:195–367. 1951.

Luhan, M. Das Abschlussgewebe der Wurzeln unserer Alpenpflanzen. *Deut. Bot. Gesell. Ber.* 68:87–92. 1955.

Morris, L. L., and L. K. Mann. Wound healing, keeping quality, and compositional changes during curing and storage of sweet potatoes. *Hilgardia* 24:143–183. 1955.

Moss, E. H., and A. L. Gorham. Interxylary cork and fission of stems and roots. *Phytomorphology* 3:285–294. 1953.

Nelson, P. E., and S. Wilhelm. Some aspects of the strawberry root. *Hilgardia* 26:631–642. 1957.

Smithson, E. Development of winged cork in *Ulmus* x *hollandica* Mill. *Leeds Philos. and Lit. Soc., Sci. Sec., Proc.* 6:211–220. 1954.

Struckmeyer, B. E., and A. J. Riker. Wound periderm formation in white-pine trees resistant to blister rust. *Phytopathology* 41:276–281. 1951.

Wutz, A. Anatomische Untersuchungen über System und periodische Veränderungen der Lenticellen. *Bot. Studien* No. 4:43–72. 1955.

Zeeuw, C. De. Influence of exposure on the time of deep cork formation in three northeastern trees. New York State Col. Forestry, *Syracuse Univ. Bul.* 56. 1941.

13. SECRETORY STRUCTURES

MANY SUBSTANCES IN THE PLANT ARE SEPARATED FROM THE protoplast and deposited in nonliving cells or vacuoles of living cells, in cavities, or in canals. The phenomenon of separation of substances from the protoplast is commonly referred to as secretion. The secreted substances may be products of dissimilatory metabolism, that is, substances that are no longer utilized by the plant (e.g., terpenes, resins, tannins, various crystals), or they may be substances having a special physiologic function after they are secreted (e.g., enzymes, hormones). Strictly speaking, the secretion of substances eliminated from the metabolism is excretion. However, no sharp line is commonly drawn between secretion and excretion in the plant, and the same containers usually accumulate varieties of substances, some being waste products, others substances that are again utilized. Furthermore, the role of many of the secreted substances, perhaps of most, is not known. In this chapter, therefore, the term secretion covers both secretion in the strict sense and excretion.

EXTERNAL SECRETORY STRUCTURES

Trichomes and glands

The secreted substances may remain within cells, internal cavities, or canals; or they may emerge from superficial secretory cells to the surface of the plant. The external secretory structures assume many forms. Part of the epidermis itself is sometimes secretory, or there are secretory epidermal appendages of various degrees of complexity, or the secretory appendages are derived from both the epidermis and subepidermal layers. The relatively highly differentiated secretory structures involving many cells are referred to as glands (fig. 13.1, *G*). The adjective glandular is applied to the less complex secretory structures such as the glandular hairs (fig. 13.1, *D, E*) or the glandular epidermis. The simple secretory trichomes intergrade with the glands, and, therefore, the distinction between glandular trichomes and glands is not strict.

The common feature of the various external secretory structures is that they have cells capable of secreting substances to the surface. These substances are highly variable (cf. Stocking, 1956*b*). Some secretory structures produce oils and resins. Floral nectaries secrete

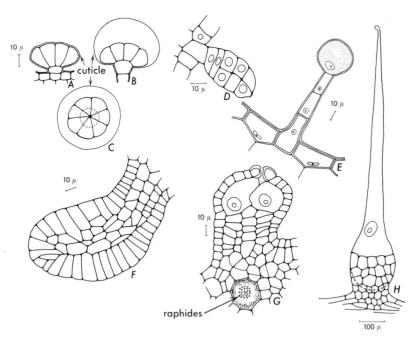

Fig. 13.1. Secretory trichomes. *A–C*, glandular hairs from leaf of lavender (*Lavandula vera*) with cuticle undistended (*A*) and distended (*B, C*) by glandular accumulation. *D*, glandular hair from leaf of cotton (*Gossypium*). *E*, glandular hair with unicellular head from stem of *Pelargonium*. *F*, colleter from young leaf of *Pyrus*. *G*, pearl gland from leaf of grapevine (*Vitis vinifera*). *H*, stinging hair of nettle (*Urtica urens*).

a sugary liquid. In many plants of saline habitats glands secrete salts (Arisz et al., 1955). Glands of certain insectivorous plants secrete nectar, mucilages, or digestive juices. Water glands—a type of hydathode—secrete water.

According to many observations, the secretion released from the glandular cells first accumulates between the cell wall and the cuticle (fig. 13.1, *A–C*). Later the cuticle is ruptured and the secretion released. Apparently this process may occur only once in some species (Stahl, 1953), or the cuticle may be regenerated and the subcuticular accumulation repeated (Trapp, 1949).

In certain glandular hairs and more elaborate glands, cells with cutinized walls resembling endodermal cells of roots (chapter 14) occur below the secretory cells (cf. Stocking, 1956*b*). These cells are assumed to play some role in the movement and accumulation of materials in the glandular structures.

The glandular hairs have a unicellular or multicellular head composed of cells producing the secretion and borne on a stalk of nonglandular cells (fig. 13.1, *E*). Among the glandular trichomes, the stinging hair of the nettle (*Urtica urens*) has a remarkable structure (fig. 13.1, *H*). The bladder-like end of the hair is embedded in epidermal cells, which are raised above the surface. The upper part of the hair resembles a fine capillary tube closed at the top by means of a spherical tip. When the hair comes in contact with the skin, the spherical tip breaks off along a predetermined line and leaves a sharp edge. This edge readily penetrates the skin, and the pressure upon the bulbous part forces the liquid into the wound.

In many woody plants (e.g., *Aesculus, Betula, Carya, Malus*) glandular hairs and more complex appendages called colleters (fig. 13.1, *F*) develop on young leaf primordia and produce a sticky secretion that both permeates and covers the entire bud. When the bud opens and the leaves expand, the glandular appendages commonly dry up and fall off. They seem to provide a protective coating for the dormant buds.

Nectaries

The nectaries are external structures that secrete a sugary liquid. They occur on flowers (floral nectaries) and on vegetative parts of the plant (extrafloral nectaries). The nectaries may have the form of glandular surfaces (fig. 13.2, *D, E*), or they may be differentiated into specialized structures (fig. 13.2, *A–C, F–H*). Both kinds are called nectaries or glands. The floral nectaries occupy various positions on the flower: on sepals, petals, stamens, ovaries, or the receptacle (Fahn, 1952, 1953). The extrafloral nectaries occur on stems, leaves, stipules, and pedicels of flowers.

The secretory tissue of a nectary may be restricted to the epidermis, or it may be several layers of cells deep (fig. 13.2). The secretory tissue is covered on the outside with a cuticle. Vascular tissue occurs more or less close to the secretory tissue. Sometimes this vascular tissue is merely a trace to some other part of the flower (fig. 13.2, B, D), but some nectaries have their own vascular bundles (fig. 13.2, G), often consisting of phloem only (Frei, 1955). A close relation exists

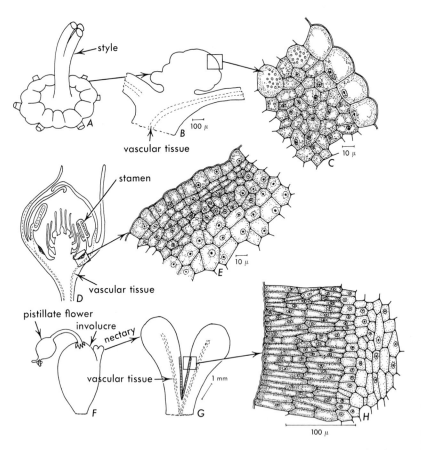

Fig. 13.2. Floral nectaries. *A–C, Ceanothus.* Nectary is a lobed disc inserted at base of gynoecium (*A*). *D, E*, strawberry (*Fragaria*). Nectary tissue lines floral tube beneath stamens (*D*). *F–H*, poinsettia (*Euphorbia pulcherrima*). Lobed nectary (*G*) is attached to involucre investing inflorescence.

between the relative amount of phloem in the vascular tissue supplying the nectary and the concentration of sugar in the nectar. If phloem predominates, the nectar may have up to 50 per cent sugar; at the opposite extreme—xylem predominating in the vascular supply—the content of sugar may fall to as low as 8 per cent (Frey-Wyssling, 1955). Nectaries with xylem predominating in their vascular supply intergrade physiologically with hydathodes.

Hydathodes

Hydathodes discharge water from the interior of the leaf to its surface. This process is called guttation. Some hydathodes are glands in the sense that they have a tissue actively secreting water. Others are merely parts of leaves with pathways along which water flowing from the endings of the xylem to the surface of the leaf meets with little resistance. The water is forced out by root pressure, flows through intercellular spaces of a modified mesophyll (epithem), and leaves the leaf through openings in the epidermis. These openings are frequently modified stomata (fig. 13.3) that are incapable of closing and opening movements (Reams, 1953; Stevens, 1956). It has been suggested that the term hydathode be restricted to the structures from which the water is forced by root pressure, that is, not actively secreting structures (Stocking, 1956a). If the term is so restricted, the actively secreting hydathodes should be called water glands.

INTERNAL SECRETORY STRUCTURES

Secretory cells

Internal secretory cells have a wide variety of contents. The secretory cells often appear as specialized cells dispersed among other, less specialized cells. They are then called idioblasts, more specifically excretory idioblasts (cf. Foster, 1956) if their contents seem to be waste products. The secretory cells may be much enlarged, especially in length, and are then called sacs or tubes. The secretory cells are usually classified on the basis of their contents, but many secretory cells contain mixtures of substances, and in many the contents have not been identified. Nevertheless, the secretory cells, as well as the secretory cavities and canals, are useful for diagnostic purposes in taxonomic work (Metcalfe and Chalk, 1950, pp. 1346–1349).

Some families, as, for example, Calycanthaceae, Lauraceae,

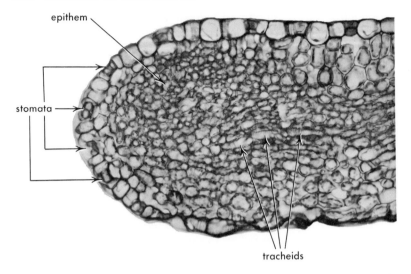

epithem

stomata

tracheids

Fig. 13.3. Hydathode from cabbage leaf. Section shows a strand of tracheids ending at a small-celled tissue, the epithem. Guttation water passes through intercellular spaces of epithem and is released through stomata. (×200. From Esau, *Plant Anatomy*, John Wiley and Sons, 1953.)

Magnoliaceae, Simarubaceae, and Winteraceae, have secretory cells with oily contents. These cells appear like enlarged parenchyma cells (fig. 13.4, *A*) and are known to occur in vascular and the ground tissues of stem and leaf. Cells similar in appearance to the oil cells but with unspecified contents occur in many other families and are often referred to as oil cells (e.g., Guttiferae, Hypericaceae, Rutaceae, Tetracentraceae, Trochodendraceae, and others). Some dicotyledon families contain resiniferous (e.g., Meliaceae), others mucilaginous cells (e.g., Cactaceae, Lauraceae, Magnoliaceae, Malvaceae, Tiliaceae). Mucilage cells often contain raphid crystals (fig. 13.4, *B*). Cells containing the enzyme myrosin have been identified in such families as Capparidaceae, Cruciferae, and Resedaceae. The myrosin cells may be elongated and even branched.

Some secretory cells have tannin as the most conspicuous inclusion. Tannin is a common ergastic substance in parenchyma cells (chapter 4), but some cells contain this material in great abundance, and, in addition, such cells may be conspicuously enlarged. The tannin cells often form connected systems and may be associated with vascular bundles. Tanniniferous idioblasts occur in many families (e.g., Ampelidaceae, Crassulaceae, Ericaceae, Leguminosae, Myrtaceae,

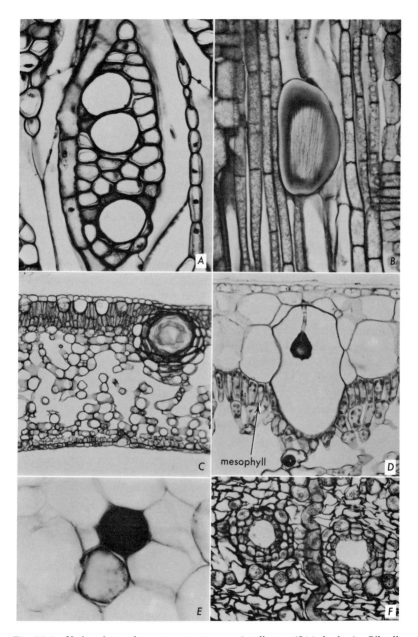

Fig. 13.4. Various internal secretory structures. *A*, tulip tree (*Liriodendron*). Oil cells in phloem ray cut tangentially. *B, Hydrangea paniculata.* Idioblast containing mucilage and raphides in a radial section of phloem. *C*, lemon (*Citrus*). Lysigenous oil cavity in upper part of leaf to the right. *D*, rubber plant (*Ficus elastica*). Enlarged epidermal cell contains a cystolith—calcium carbonate precipitation on a stalk of cellulose. The cell is part of multiple epidermis of leaf (all cells above mesophyll in *D*). Cystolith is cut off median and does not show base of stalk. *E*, elderberry (*Sambucus*). Tannin sacs in pith of stem in cross section. *F, Rhus typhina.* Schizogenous secretory canals in cross section of nonfunctioning phloem. (*A, B, D–F*, ×210; *C*, ×125.)

Rosaceae, and others). Easily procured examples are the tannin cells in the leaves of *Sempervivum tectorum* and species of *Echeveria* and tube-like tannin cells 1 and more centimeters long in the pith and phloem of stems of *Sambucus* (fig. 13.4, *E*). The tannin compounds in the tanniniferous cells are oxidized to brown and reddish-brown phlobaphenes which are easily perceived under the microscope. Cells in the ground tissue of the fruit of *Ceratonia siliqua* contain solid tannoids, inclusions of tannins combined with other substances.

Certain investigators include among the excretory idioblasts cells containing crystals (cf. Foster, 1956). Crystal-containing cells often do not differ from other parenchyma cells, but they also may be more or less specialized in form and contents. Striking examples of such specializations are the cystolith-containing cells in *Ficus elastica* leaves and the raphid cells. The cystoliths are structures combining wall material, including cellulose and callose (Eschrich, 1954), with calcium carbonate. In *Ficus elastica* the cystoliths occur singly in epidermal cells, and each is attached by means of a cellulose stalk to the outer epidermal wall (fig. 13.4, *D*). The raphides are often found in long sac-like cells filled with mucilage. In the Amaryllidaceae and Commellinaceae of the monocotyledons, raphides occur in long mucilage tubes, the origin of which still needs a critical study. In some plants raphid idioblasts are much elongated single cells (Kowalewicz, 1956). In the secondary vascular tissues a cell forming crystals often becomes subdivided into small cells, each depositing one crystal. In another modification the crystal is walled off by cellulose from the living part of the protoplast.

Secretory cavities and canals

The cavities and canals differ from secretory cells in that they are spaces resulting from either a dissolution of cells (lysigenous spaces) or a separation of cells (schizogenous spaces). The origin of the second type of secretory spaces resembles that of the ordinary air spaces characteristic of mature ground tissue. Lysigeny and schizogeny may be combined in the formation of secretory spaces. In the lysigenous spaces, partly disintegrated cells appear along the periphery of the space (fig. 13.4, *C*). The schizogenous spaces are usually lined with intact cells (fig. 13.4, *F*).

Lysigenous secretory cavities may be observed in *Citrus, Eucalyptus,* and *Gossypium.* The secretion is formed in cells that eventually break down and release the substances into the cavity resulting from the breakdown. The secretions in the genera mentioned above are commonly characterized as oily, although their composition is not known exactly (Metcalfe and Chalk, 1950, p. 1348). An example of lysigenous mucilaginous canals is found in the bud scales of *Tilia cordata*. Schizogenous canals with resiniferous contents occur in the Compositae and with unknown contents in the Umbelliferae. The copal-yielding secretory canals of certain tropical Leguminosae also arise as schizogenous spaces (Moens, 1955).

The best-known schizogenous canals are the gum ducts and the resin ducts in the conifers and the woody dicotyledons. Ducts in the conifers are called resin ducts, those in the dicotyledons gum ducts (chapter 9), but both kinds of ducts may occur in both plant groups. The resin ducts of the conifers occur in the vascular and the ground tissues of all plant organs and are actually long intercellular spaces lined with resin-producing epithelial cells. They occur in both the rays and the axial system in the secondary vascular tissues (chapter 9).

Secretory cavities and canals resulting from normal development may be difficult to distinguish from canals and cavities arising under the stimulus of injury. Resin and gum ducts and pockets are frequently traumatic formations in the wood of gymnosperms and dicotyledons, but their development and contents may parallel those constituting normal features of these woods.

Laticifers

The laticifers are cells or series of connected cells that contain latex, a fluid of complex composition. The laticifers are discussed in the same chapter with the various secretory structures because they are depositories of substances of which some may be classified as excretions (e.g., terpenes, resins) and others as secretions in the narrow sense (e.g., enzymes). Furthermore, in some plant groups, laticifers intergrade with structures interpreted as secretory or excretory idioblasts.

The laticifers may be simple or compound by origin. The simple laticifers are single cells; the compound laticifers are derived from series of cells. In a more highly specialized state the series of cells in a compound laticifer become united by dissolution of intervening walls. Because of this junction of cells the compound laticifers are commonly called *articulated* laticifers. In contrast, the simple laticifers are called *nonarticulated*. Both kinds of laticifers may be branched or unbranched.

Laticifers occur in various tissues of the plant and may permeate all tissues of a given plant. Such ubiquitous distribution results from

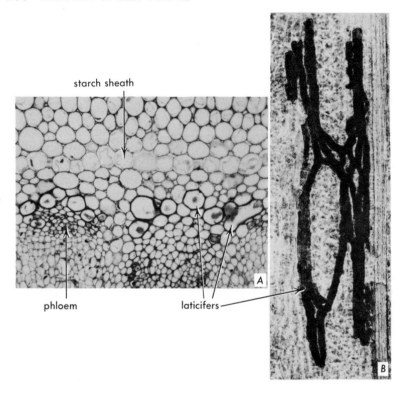

starch sheath

phloem

laticifers

A

B

Fig. 13.5. Articulated anastomosing laticifers of *Lactuca scariola. A,* cross section of stem. Laticifers appear outside the phloem. *B,* longitudinal view of laticifers in partly macerated tissue from stem. (*A,* ×180; *B,* ×70.)

the mode of development of the laticifers. The articulated laticifers (fig. 13.5) extend into new tissues by addition of cells from these tissues; that is, certain cells in newly formed tissues become laticiferous cells in juxtaposition with the older laticiferous cells. The nonarticulated laticifers originate as single cells in the embryo (fig. 13.6, *A*) and follow the growth of the plant by penetrating into tissues newly formed by the apical meristems. A combination of coordinated and apical intrusive growth is involved in the development of the single-cell laticifers. A similar individualistic behavior of cells occurs in a comparatively modest degree among sclereids (chapter 6); it is rather spectacular in the branched nonarticulated laticifers, which grow among other cells like hyphae of a successful intercellular fungal parasite (fig. 13.6, *B, C*). In plants with secondary growth laticifers develop in the secondary tissues also. Complicated developmental features are involved when a nonarticulated laticifer penetrates the

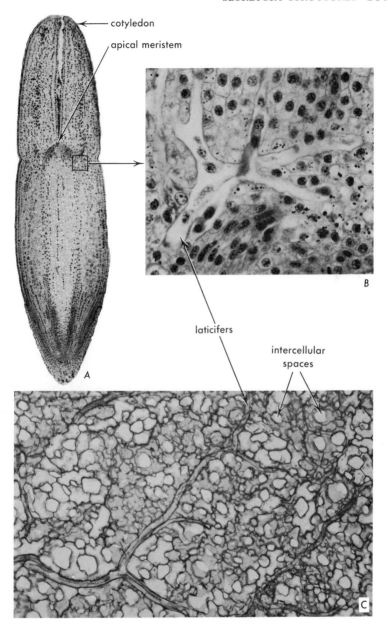

Fig. 13.6. Nonarticulated branching laticifers of *Euphorbia* sp. *A*, embryo. Rectangle indicates a locus of origin of laticifers. *B*, section through laticifers showing multinucleate condition. *C*, laticifers branching within spongy parenchyma as seen in a paradermal section of leaf. (*A*, ×41; *B*, ×320; *C*, ×150. Slide for *A* and *B*, courtesy of K. C. Baker.)

vascular cambium and maintains its continuity during the subsequent growth of tissues from the cambium (Vreede, 1949).

The laticifers have primary nonlignified walls variable in thickness. In mature laticifers the latex is said to be enclosed by the protoplast. The articulated laticifers become multinucleate when series of cells fuse by dissolution of walls. Nonarticulated laticifers also become multinucleate as they grow. In these laticifers the multinucleate condition is attained by nuclear division. The relation between the protoplast and the latex is just as difficult to assess as the relation between the cytoplasm and the vacuole in sieve elements. The latex is commonly regarded as vacuolar material, but there is no clear demarcation between the cytoplasm and the vacuole in mature laticifers.

The latex is highly variable in appearance and composition. It is frequently milky but may be colorless or brown or orange. Latex often contains rubber—a member of the organic substances called terpenes—and plants having large amounts of this substance in the laticifers (e.g., *Hevea brasiliensis, Ficus elastica*) are the source of natural commercial rubber. The rubber occurs as particles in colloidal suspension. The particles vary in size in different plants, have a complex structure, and appear to be coated with proteinaceous material (Schoon and Phoa, 1956). Many other substances occur in latices (plural of latex), e.g., alkaloids (opium poppy, *Papaver somniferum*), sugar (Compositae), waxes, proteins, enzymes (the proteolytic enzyme papain, for example, in *Carica papaya*), crystals, tannins, and starch, which is often in the form of large grains of unusual shape (chapter 4).

Laticifers occur in several families and in certain genera of some families of dicotyledons and monocotyledons. The familiar plants with latex are representatives of the spurge family, Euphorbiaceae, which includes plants of commercial source of rubber (e.g., *Hevea* and *Manihot*); several genera of the composite tribe Cichorieae (e.g., dandelion, *Taraxacum;* sow thistle, *Sonchus;* lettuce, *Lactuca*); and the Moraceae, which include the Indian rubber plant, *Ficus elastica.*

REFERENCES

Arisz, W. H., I. J. Camphuis, H. Heikens, and A. J. Van Tooren. The secretion of the salt glands of *Limonium latifolium* Ktze. *Acta Bot. Neerl.* 4:322–338. 1955.

Eschrich, W. Ein Beitrag zur Kenntnis der Kallose. *Planta* 44:532–542. 1954.

Fahn, A. On the structure of floral nectaries. *Bot. Gaz.* 113:464–470. 1952.

Fahn, A. The topography of the nectary in the flower and its phylogenetic trend. *Phytomorphology* 3:424–426. 1953.

Foster, A. S. Plant idioblasts: remarkable examples of cell specilization. *Protoplasma* 46:184–193. 1956.

Frei, E. Die Innervierung der floralen Nektarien dikotyler Pflanzenfamilien. *Schweiz. Bot. Gesell. Ber.* 65:60–114. 1955.

Frey-Wyssling, A. The phloem supply to the nectaries. *Acta Bot. Neerl.* 4:358–369. 1955.

Kowalewicz, R. Zur Kenntnis von *Epilobium* und *Oenothera.* I. Über die Raphidenschläuche. 2. Über intergenerische Transplantation. *Planta* 47:501–509. 1956.

Metcalfe, C. R., and L. Chalk. *Anatomy of the dicotyledons.* 2 Vols. Oxford, Clarendon Press. 1950.

Moens, P. Les formations sécrétrices des copaliers congolais. Étude anatomique, histologique et histogénétique. *Cellule* 57:33–64. 1955.

Reams, W. M., Jr. The occurrence and ontogeny of hydathodes in *Hygrophila polysperma* T. Anders. *New Phytol.* 52:8–13. 1953.

Schoon, T. G. F., and K. L. Phoa. Morphology of the rubber particles in natural latices. *Arch. Rubber Cult.* 33:195–215. 1956.

Stahl, E. Untersuchungen an den Drüsenhaaren der Schafgarbe (*Achillea millefolium* L.) *Ztschr. für Bot.* 41:123–146. 1953.

Stevens, A. B. P. The structure and development of hydathodes of *Caltha palustris* L. *New Phytol.* 55: 339–345. 1956.

Stocking, C. R. Guttation and bleeding. In: *Handbuch der Pflanzenphysiologie.* 3:489–502. 1956a.

Stocking, C. R. Excretion by glandular organs. In: *Handbuch der Pflanzenphysiologie.* 3:503–510. 1956b.

Trapp, I. Neuere Untersuchungen über den Bau und Tätigkeit der pflanzlichen Drüsenhaare. *Oberhess. Gesell. für Nat. und Heilk. Giessen Ber., Naturw. Abt.* 24:182–205. 1949.

Vreede, M. C. Topography of the laticiferous system in the genus *Ficus. Ann. Jard. Bot. Buitenzorg.* 51:125–149. 1949.

14. THE ROOT:
PRIMARY STATE OF GROWTH

TYPES OF ROOTS

The first root of a seed plant develops from the root promeristem (apical meristem) of the embryo. This is the taproot, usually called primary root. In the gymnosperms and dicotyledons the taproot and its variously branched lateral roots constitute the root system. In the monocotyledons the first root commonly lives only a short time, and the root system of the plant is formed by adventitious roots that arise on the shoot, often in connection with axillary buds. The formation in grasses of numerous axillary shoots and associated roots is known as tillering. The adventitious roots also become branched but form a relatively homogeneous system referred to as fibrous root system. A taproot system generally penetrates the soil more deeply than a fibrous root system, but the latter binds the superficial layers of the soil more firmly. The main components of a taproot system undergo secondary growth. The small absorbing roots on such a system, however, remain in the primary state and are often ephemeral. The components of an adventitious root system may or may not have secondary growth.

The two types of root systems just described are the most common in seed plants, and both are concerned with anchorage, absorption, storage, and conduction. Some roots or root parts, however, are highly specialized with reference to one particular function. The fleshy parts of the roots of the carrot (*Daucus*), radish (*Raphanus*), beet (*Beta*), sweet potato (*Ipomoea*), and others are specialized as storage organs. Some fleshy storage roots show peculiar forms of secondary growth in their development (chapter 15). The prop roots of the mangrove plants are mainly supporting structures. In a minor form prop roots (brace roots) are encountered in some Gramineae as in maize; large ones occur in *Pandanus*. Some vines and epiphytes develop aerial roots capable of attaching themselves to the surface on which the shoot of the plant may be growing.

The anchorage of plants may assume a specialized aspect in the form of contractile roots. Root contraction is widely distributed among monocotyledons and herbaceous perennial dicotyledons. The contraction occurs in taproots, in their lateral roots, and in adventitious roots. In all instances the contraction pulls the shoot promeristem closer to the ground or, in bulbous plants, deeper into the ground. In some plants, and especially in bulbous ones, certain roots are specialized as contractile roots; others have the usual form (e.g., Chan, 1952).

PRIMARY STRUCTURE

The internal organization of the root is variable but relatively simple compared with that of the stem; phylogenetically it is considered to be more primitive than that of the stem. The root is a simple axial structure with no leaf-like organs and with no division into nodes and internodes. Correspondingly, in the root the arrangement of tissues shows relatively little difference from level to level, whereas in the stem the connection of the axis with the leaves results in differences in structure between nodes and internodes and even between different levels of a given internode.

A transverse section through a root in the primary state of growth shows a clear separation between the usual three tissue systems, the epidermis (dermal tissue system), the cortex (ground tissue system), and the vascular tissue system (fig. 14.1). The vascular tissues form a solid cylinder (fig. 14.5, *A–C*) or, if pith is present, a hollow cylinder (fig. 14.3, *A*). Each system has some structural features that are characteristic of roots (see below). The rootcap, which covers the promeristem of the root, may be considered also a part of the primary body of the root.

Epidermis

In young roots the epidermis is specialized as an absorbing tissue and usually bears root hairs. The root hairs are tubular extensions of

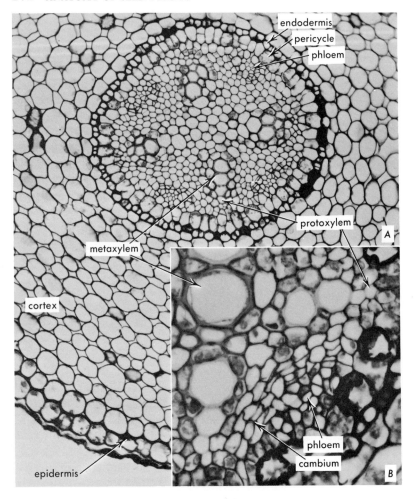

Fig. 14.1. Cross sections of root of strawberry (*Fragaria*) in primary state of growth (*A*) and at initiation of vascular cambium (*B*). (*A*, ×225; *B*, ×605. From Nelson and Wilhelm, *Hilgardia*, 1957.)

the epidermal cells. Some or most of the epidermal cells develop root hairs. In some plants the protodermal cells that give rise to root hairs (trichoblasts) are smaller than those that do not; in others no such differentiation occurs. The occurrence and nonoccurrence of this differentiation seems to be of taxonomic importance (Row and Reeder, 1957). The root hairs reach their full development in the region where the xylem is at least partly mature. The absorbing function, however, is not restricted to the root hairs. Epidermal cells lacking hairs also absorb. Root hairs markedly extend the absorbing surface of the root. Certain calculations based on the root system of a rye plant (Rosene, 1955) suggest that comparatively few of the total number of root hairs can supply all the water necessary for transpiration and growth of the plant. Such efficiency would be particularly important when the available moisture is irregularly distributed in the soil.

Root hairs arise as small papillae in the root zone located behind the zone of active cell division. The cytoplasm concentrates in the growing papilla, and the nucleus migrates toward its tip (Bouet, 1954). The wall of the growing tip is softer than the proximal part of the wall. It appears to contain pectic and callose substances, and also cellulose (cf. Cormack, 1949; Currier, 1957; Ekdahl, 1953). The hardening of the tip of the root hair at the end of growth is variously ascribed to calcification of pectic substances or to changes in the cellulosic component of the wall. The root hairs usually die off in older parts of the root but may be rather persistent.

A thin cuticle has been identified on the epidermis, including the root hairs, in the young part of the root (chapter 7). If the epidermis is persistent it may show eventually a conspicuous cutinization. In fact, in some herbaceous perennials the epidermis remains for a long time or permanently as a protective tissue (Luhan, 1955). Its walls increase in thickness, and the lumina sometimes become filled with deeply colored substances.

In air roots of the tropical Orchidaceae and the epiphytic Araceae the epidermis develops into a multiseriate tissue (multiple epidermis) called velamen. The velamen consists of nonliving cells, compactly arranged and often bearing secondary wall thickenings. The velamen is commonly interpreted as absorptive tissue, but some physiological studies on orchid velamen indicate that the principal roles of this tissue are mechanical protection and reduction in loss of water from the cortex (Dycus and Knudson, 1957).

Cortex

The root cortex is often composed of parenchyma cells only, but if it is persistent it may develop sclerenchyma, or it may become collenchymatous. The presence of conspicuous intercellular spaces is characteristic of the root cortex (fig. 14.1, A). The spaces may be large lacunae which involve a breakdown of cells in their formation (aerenchyma). Such lacunae are encountered in plants growing in moist habitats (e.g., rice) and also in relatively arid regions (cf. Beckel, 1956). Chloroplasts are usually absent in root cortex, but starch is often present. The innermost layer is differentiated as an endodermis, and one or more layers at the periphery may develop into an exodermis.

Endodermis. The endodermis receives much attention in discussions of movement of materials absorbed by the root. In the young absorbing region the endodermal wall contains suberin in a band-like structure extending completely around the cell within the radial and transverse walls (fig. 14.2). This band, called *Casparian strip* or band, is not merely a wall thickening but part of the primary wall. Moreover, the suberin deposition is continuous across the middle lamella. The protoplast is relatively firmly attached to the Casparian strip. The significance of this relation seems to be that materials moving radially through the endodermis cannot pass through the endodermal wall because of the suberized condition of the strip; neither can they move between the wall and the cytoplasm because of the connection between the protoplast and the strip. Thus, the only available path is through cytoplasm, which has a controlling influence upon the movement. This possibility is often considered in the formulation of hypotheses pertaining to the complex problem of movement of materials in roots and their release by living cells into the nonliving conducting cells of the xylem (e.g., Arnold, 1952).

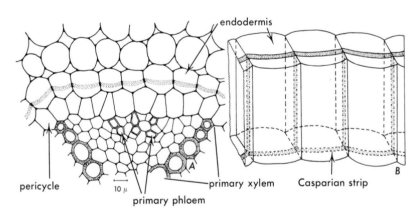

Fig. 14.2. Structure of endodermis. *A,* cross section of part of root of morning-glory (*Convolvulus arvensis*) showing position of endodermis with regard to xylem and phloem. The endodermis is shown with transverse walls bearing Casparian strips in focus. *B,* diagram of three connected endodermal cells oriented as they are in *A;* Casparian strip occurs in transverse and radial walls (that is, in all anticlinal walls) but is absent in tangential walls.

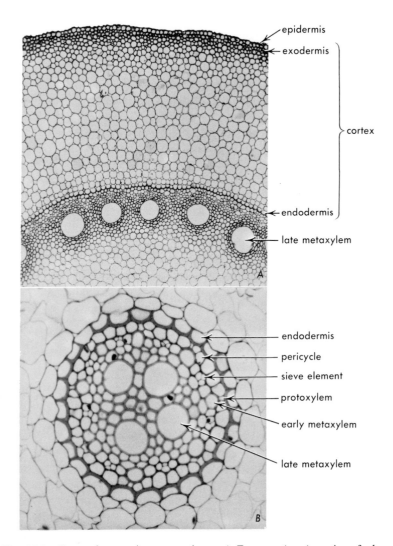

Fig. 14.3. Roots of grasses in cross sections. *A, Zea mays* (corn), section of a large root with pith. *B,* vascular cylinder of *Bromus* root without pith. In both, endodermis with secondary cell wall thickenings. Sieve element in *B* is associated with two parenchyma cells. (*A,* ×60; *B,* ×320. *A,* from J. E. Sass, *Botanical Microtechnique,* 3rd ed., The Iowa State College Press, 1958; *B,* from Esau, *Hilgardia,* 1957.)

In roots with secondary growth the endodermis is usually cast off together with the cortex, but in those remaining in primary state it often develops thick secondary walls (figs. 14.3 and 14.4). These walls typically consist of a suberin lamella covered by layers of lignified cellulose. The thickening is commonly uneven; it is rather thin or absent on the outer tangential walls (figs. 14.3 and 14.4). The formation of the secondary wall may be delayed in endodermal cells opposite the xylem. Such thin-walled cells (with Casparian strips) in an otherwise thick-walled endodermis are called passage cells. The endodermis is usually uniseriate, but in many Compositae some endodermal cells divide tangentially, and schizogenous secretory canals develop in the two-layered zones of the endodermis (Williams, 1954). In some species—pear and apple are good examples—the cortical layers outside the endodermis develop prominent wall thickenings, often restricted to the radial walls (e.g., Riedhart and Guard, 1957).

Exodermis. The exodermis occurs beneath the epidermis. Its cells may have Casparian strips, but more commonly they are described as having a suberin lamella covered by a thick cellulose wall. The exodermis contains either one kind of cells or short and long cells. While the root is young the walls of the short cells contain no suberin. Later they may become suberized. The exodermis may be several layers in thickness (fig. 14.3, *A*).

Vascular cylinder

The vascular cylinder comprises one or more layers of nonvascular cells—the pericycle—and the vascular tissues. The inclusion of the pericycle with the vascular tissues is done in part on developmental grounds—the pericycle arises from the same part of the apical meristem as the vascular tissues—and in part on historical grounds—the concept of the stele (chapter 16) defines the pericycle as the limiting layer of the stele. The pericycle may be composed entirely of parenchyma (figs. 14.1, *A*, and 14.2), or it may contain sclerenchyma (fig. 14.4) or elements of protoxylem (fig. 14.3, *B*); it is commonly one layer of cells in thickness but may be multiseriate (fig. 14.4, *B*). Schizogenous secretory canals occur in the pericycle of the Umbelliferae (Bruch, 1955). Lateral roots, part of the vascular cambium, and, in many roots, the phellogen arise in the pericycle.

The xylem frequently forms a solid core with ridge-like projections extending toward the pericycle (fig. 14.5). The strands of phloem alternate with the xylem ridges. If the xylem does not differentiate in the center of the root, a pith consisting of parenchyma or sclerenchyma is present (figs. 14.1, *A*, and 14.5, *D*).

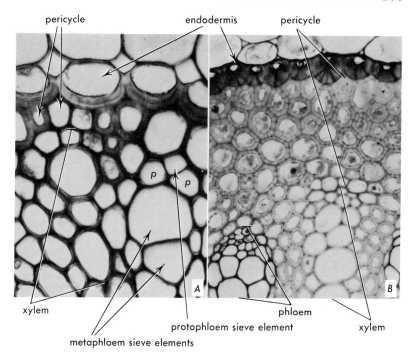

pericycle   endodermis   pericycle

xylem

metaphloem sieve elements

protophloem sieve element

phloem

xylem

Fig. 14.4. Structure of endodermis and pericycle in monocotyledon roots as seen in cross sections. *A, Zea mays* (corn), *B, Smilax herbacea* (carrion-flower). In both, endodermis with secondary wall thickenings. Uniseriate pericycle in *A*, multiseriate in *B*. The triangular protophloem group—one sieve element and two parenchyma cells (*p*)—in *A* is characteristic of grasses. (*A*, ×750; *B*, ×280.)

The number of xylem ridges varies, and in relation to this variation roots are referred to as being *diarch, triarch, tetrarch,* etc., or *polyarch* (fig. 14.5). The tracheary cells in the outermost position in each xylem are the narrowest and are the earliest to mature. They constitute the protoxylem (fig. 14.5) and have helical or scalariform-reticulate or sometimes annular secondary thickenings. Closer to the center are the increasingly wider metaxylem elements, most of which, especially the latest, commonly have secondary walls with bordered pits. As was mentioned in chapter 3, the xylem with the centripetal sequence of maturation as seen in roots is termed exarch.

The primary vascular structure of monocotyledon roots—they rarely have secondary growth—is highly variable and often complex. In some the center is occupied by a single metaxylem vessel; in others a circle of such vessels surrounds a pith (figs. 14.3 and 14.6, *A*). A more complicated pattern is illustrated by *Monstera* in which vessels

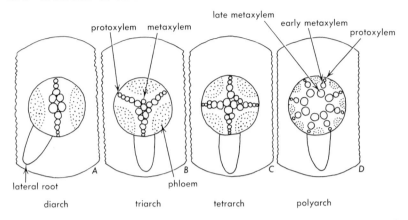

Fig. 14.5. Different patterns formed by primary xylem in cross sections of roots and position of lateral root with regard to xylem and phloem of main root. The patterns *A–C* are characteristic of dicotyledons; *D* is found in many monocotyledons. (After Esau, *Plant Anatomy*, John Wiley and Sons, 1953.)

and phloem strands are scattered throughout the central part of the vascular cylinder (fig. 14.6, *B*).

The first mature sieve elements, the protophloem elements, also occupy a peripheral position in the vascular cylinder; the metaphloem occurs farther inward. In other words, the primary phloem also shows a centripetal order of differentiation (fig. 14.4, *A*). Parenchyma cells occur among the sieve elements in the primary phloem. In the dicotyledons companion cells are characteristic of the metaphloem but are commonly absent in the protophloem. Information is deficient on this matter with regard to the monocotyledons. In grasses, each of the protophloem sieve elements is associated with two parenchyma cells (fig. 14.4, *A*, at *p*) that are derived from the same mother cell as the sieve element and, therefore, resemble companion cells. In roots having secondary growth the cells located between the xylem and the phloem eventually function as a vascular cambium (fig. 14.1, *B*). In permanently primary roots they mature as parenchyma or as sclerenchyma cells (fig. 14.3).

Rootcap

The rootcap (figs. 3.2 and 14.7) is interpreted as a structure protecting the apical meristem and assisting the growing root in penetrating the soil. The cells are living and often contain starch. The walls on the periphery of the cap and those next to the main body of the root often appear to have mucilaginous consistency, probably because of a peculiar condition of the pectic substances. This feature is assumed to facilitate the sloughing of the peripheral cells and the separation of the rootcap from the flanks of the growing root. Environmental conditions affect the structure of the rootcap. Roots normally growing in soil may lack rootcaps when grown in water cultures (Richardson, 1955). On the other hand, water plants often have particularly extensive rootcaps (e.g., *Eichhornia,* water hyacinth).

In ectotrophic mycorrhizae—mycorrhizae are said to be symbiotic associations of roots and fungi—a fungal mantle covers the root tip. Apparently the rootcap formation continues within such a mantle, but the outermost cells, which would be sloughed off in the soil, become decomposed within the fungal mantle (Clowes, 1954).

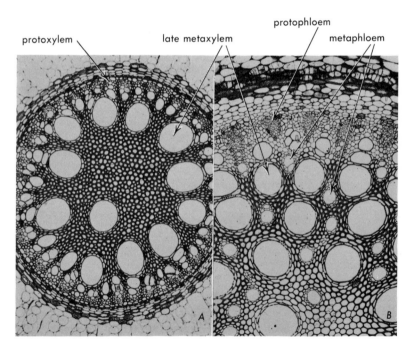

Fig. 14.6. Variations in structure of monocotyledon roots as seen in cross sections. *A,* African oil palm (*Elaeis guineensis*). Late metaxylem elements occur in a circle, except one which appears close to center. *B, Monstera deliciosa.* Numerous late metaxylem elements spread throughout center together with metaphloem groups. (*A,* ×76; *B,* ×50. Courtesy of V. I. Cheadle.)

DEVELOPMENT

Apical meristem

As was mentioned in chapter 2, the main phenomenon in the origin of the root in the embryo is the organization of the apical meristem, or promeristem, of the root (cf. Guttenberg et al., 1954, 1955) at the lower end of the hypocotyl; sometimes an embryonic root, the radicle, is present in the embryo also. After germination of the seed, the root promeristem of the embryo forms the taproot (primary root). As the latter grows, the promeristem assumes a definite organization, variable in different plant groups. Branch roots and adventitious roots, if present, also show characteristic arrangements of cells in the promeristem, more or less similar to that in the taproot. The architecture of apical meristems of roots has been studied most often for the purpose of revealing the origin of the tissue systems. When this architecture was found to have characteristic differences, investigators began relating the differences to the taxonomic groupings of plants; only a few attempts were made to discover the evolutionary lines in the apical organization (Voronin, 1956).

The differences in organization of the promeristem result from differences in spatial relation of the promeristem cells to the tissue regions in the root, a relation commonly taken as evidence of ontogenetic interrelation. Two main types of organization have been recognized. In one the vascular cylinder, the cortex, and the rootcap are traceable to independent layers of cells in the apical meristem, with the epidermis differentiating from the outermost layer of the cortex (fig. 14.7, *A, B*) or from cells having common origin with the rootcap cells (fig. 14.7, *C, D*). In the other type all regions, or at least the cortex and the rootcap, converge in one group of cells oriented transversely (fig. 14.7, *E, F*). These patterns were interpreted to mean that in the first type the three regions, vascular cylinder, cortex, and rootcap each has its own initials; in the second, all regions have common initials. The type of apical meristem with the common initials may be phylogenetically primitive (Voronin, 1956). In many lower vascular plants only one cell, the apical cell, appears in the focal position and is the common initial for all parts of the root.

With regard to a meristem the term initial implies that the cell divides repeatedly and thus adds cells to the body of the plant but that the cell itself remains in the meristem. Many studies on promeristems of roots, however, indicate a relative inactivity of the cells commonly referred to as the initials. Mitotic activity is more intense

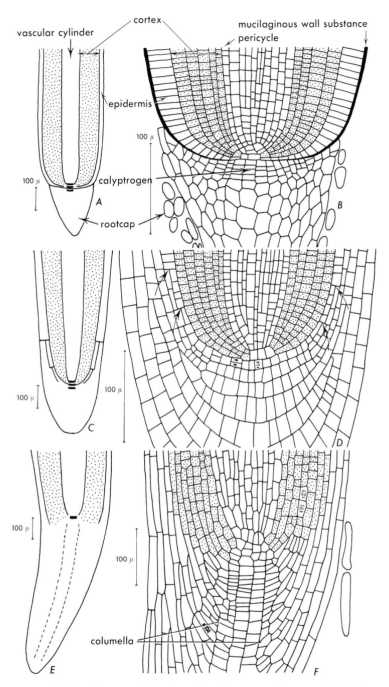

Fig. 14.7. Apical meristem and derivative regions in roots. *A, B,* grass (*Stipa*). Three tiers of initials, those of rootcap forming a calyptrogen. Epidermis has common origin with cortex. *C, D,* radish (*Raphanus*). Three tiers of initials. Epidermis has common origin with rootcap and becomes delimited on the flanks of root by periclinal walls (arrows in *D*). *E, F,* Spruce (*Picea*). All regions of root arise from one group of initials. Rootcap has a central columella of transversely dividing cells. It also gives off derivatives laterally.

a short distance from these cells (Clowes, 1958; Jensen and Kavaljian, 1958) so that the promeristem appears as a body of cells with the central "initials" relatively quiescent and the peripheral layers of cells dividing actively. Studies involving use of radioactive tracers support this concept of a quiescent center in the promeristem since they indicate that little or no protein and ribonucleic acid synthesis occurs in the apical initials (e.g., Jensen, 1957).

Growth of the root tip

The zone of actively dividing cells extends for a considerable distance from the apex. These divisions are combined with cell enlargement. Eventually divisions cease, and further growth results from an enlargement of cells. Because of the relatively simple organization of the root tip, it constitutes a favorite object for study of growth in a multicellular organism (cf. Torrey, 1956). In these studies physical, mathematical, and biochemical aspects of growth are considered. Attempts are made to distinguish the components of growth, cell division and cell enlargement, and to relate the biochemical changes at increasing distances from the promeristem to the distribution of the two components of growth. The results of these studies vary in detail, but they all indicate that maximum mitotic activity occurs not in the promeristem—not even in the most active peripheral part of it —but somewhat beyond it (fig. 14.8) and that the maximal increase in length of the root results from cell elongation. In a study of *Zea* roots (Erickson and Sax, 1956) elongation was recorded through 8–10 millimeters of root with a maximum rate at the 4-millimeter level where only elongation and no division of cells was taking place.

Primary differentiation

At various distances from the promeristem, cells enlarge and develop their specific characteristics in relation to their position in the root; that is, the cells become differentiated. The usual three regions of the root, the epidermis, the cortex, and the vascular cylinder, become delimited close to the promeristem (figs. 14.7 and 14.8). Differences in distribution of mitoses and in the degree of the early cell enlargement contribute to the initial differentiation of the tissue regions (Jensen and Kavaljian, 1958). In their meristematic state, the epidermis, cortex, and vascular cylinder may be called protoderm, cortical ground meristem, and procambium, respectively. (The use of procambium for the meristematic vascular cylinder is not entirely

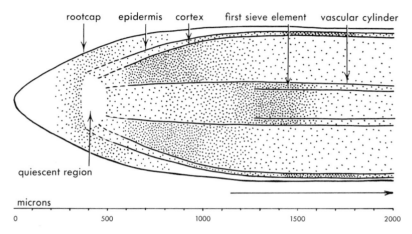

Fig. 14.8. Diagram of root tip of onion in longitudinal view illustrating distribution of meristematic activity. Frequency of mitosis is indicated by density of stippling. (After Jensen and Kavaljian, *Amer. Jour. Bot.,* 1958.)

satisfactory because at maturity this cylinder contains such nonvascular tissues as the pericycle and sometimes a pith.)

One of the most conspicuous features of epidermal differentiation is the appearance of root hairs. As was mentioned earlier, root hairs come to full development beyond the region of elongation, approximately at the level where the xylem begins to mature.

The cortex increases in width by periclinal divisions and radial enlargement of cells. In many roots the divisions occur entirely or partly in the innermost layer, that is, the layer located next to the pericycle (figs. 14.7, *B, D,* and 14.9, *B*). A repetition of the divisions in this layer often makes the resulting tissue appear orderly, arranged like a tissue derived from a vascular cambium (fig. 14.3, *A*). However, the number of divisions in the cortex is limited, and after the final division the innermost layer becomes the endodermis (fig. 14.9, *B*) when Casparian strips are differentiated. The intercellular spaces, so prominent in the root cortex, appear close to the promeristem.

In the differentiation of the vascular cylinder the pericycle is usually the first identifiable region. The vascular differentiation begins with the enlargement and increasing vacuolation of the tracheary elements of the metaxylem (figs. 14.9 and 14.10; Popham, 1955*b*). The maturation of the first phloem elements follows (figs. 14.9, *B,* and 14.10); then the first protoxylem elements next to the pericycle develop secondary walls and mature. Thus, the first phloem elements mature before the first xylem elements, and, in the xylem, the differentiation of the tracheary elements begins in the centrifugal direction but is completed in the centripetal direction. The exact distance between the apical meristem and the first mature vascular elements, especially xylem elements, varies in relation to the rate of growth of roots. In general slowly growing roots have mature vascular elements closer to the promeristem than fast growing roots (cf. Esau, 1953).

The development of the root is not a uniform continuous process. In noble fir (Wilcox, 1954), for example, the growth of roots slows

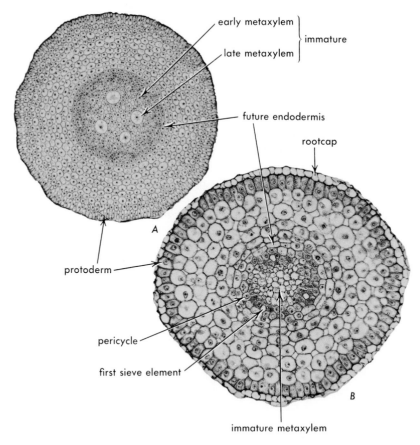

Fig. 14.9. Early stages of vascular differentiation in roots as seen in cross sections. *A*, *Zea mays* (corn). Metaxylem elements conspicuously enlarged and vacuolated before any other vascular elements are distinguishable. *B, Melilotus alba.* Immature triarch xylem is vacuolated; three phloem strands, densely cytoplasmic, each with one mature sieve element. (*A*, ×120; *B*, ×230. From J. E. Sass, *Botanical Microtechnique,* 3rd ed., The Iowa State College Press, 1958.)

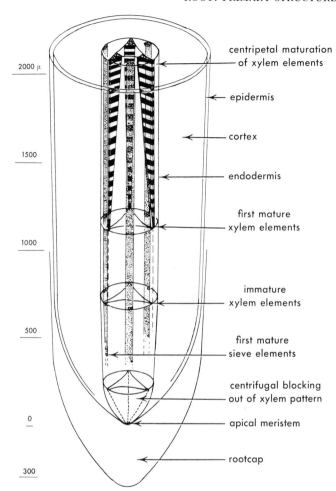

Fig. 14.10. Diagram of primary vascular differentiation in a root of pea (*Pisum sativum*). Root was grown isolated from the plant in a nutrient medium, but sequence and pattern of differentiation correspond with those observed in roots attached to plants. (From Torrey, *Amer. Jour. Bot.*, 1953.)

down periodically; maturation of cells progresses close to the apex; and, with the onset of dormancy, fatty materials, probably suberin, are deposited in the cortex and the rootcap. This deposition occurs through a layer of cells continuous with the endodermis and covering the promeristem. The promeristem thus becomes sealed off by a layer of suberized cells on its sides and toward the rootcap. Externally such root tips appear brown. When growth is resumed, the brown cap is broken, and the root tip pushes beyond it. Studies on excised roots indicate that roots may have a growth rhythm not dependent on seasonal changes but determined by internal factors (Street and Roberts, 1952).

Lateral roots

The lateral roots arise at some distance from the promeristem at the periphery of the vascular cylinder. Because of their deep-seated origin the lateral roots are spoken of as being endogenous. The lateral roots of gymnosperms and angiosperms, whether they arise on taproots or their branches or on adventitious roots, originate most commonly in the pericycle (fig. 14.11, *A*). The endodermis may participate to various degrees in the formation of new root primordia (fig. 14.11; Popham, 1955*a;* Schade and Guttenberg, 1951). In lower vascular plants, lateral roots are said to arise in the endodermis.

The pericyclic origin of the lateral root places it in close juxtaposition with the vascular tissues of the parent root with which the vascular tissues of the lateral root eventually become connected. The position of the lateral root with regard to xylem ridges of the parent root varies in relation to the vascular pattern of the parent root but is stable in a root with a given pattern (fig. 14.5). In a diarch root, the lateral root arises between the phloem and the xylem, in a triarch, tetrarch, etc., root opposite the xylem; in a polyarch root, opposite the phloem.

When a lateral root is initiated, several contiguous pericyclic cells divide periclinally. The products of these divisions divide again, periclinally and anticlinally. The accumulating cells form a protrusion, the root primordium (fig. 14.11, *A*). As the primordium increases in length, it penetrates the cortex and emerges on the surface. The endodermis often divides only anticlinally and thus keeps pace with the growth of the primordium, but the other cortical cells are partly deformed, crushed, pushed aside, and probably partly dissolved in front of the advancing tip. During growth of the young primordium through the cortex, the promeristem and rootcap are initiated and the vascular cylinder and cortex blocked out behind the promeristem (fig. 14.11, *B*). The endodermis, if it does not participate in the organization of the root primordium, is crushed and shed.

When phloem and xylem elements differentiate in the lateral root, they become connected with the equivalent elements in the parent root. The connection is formed by differentiation of pericyclic cells into vascular elements at the proximal end of the primordium.

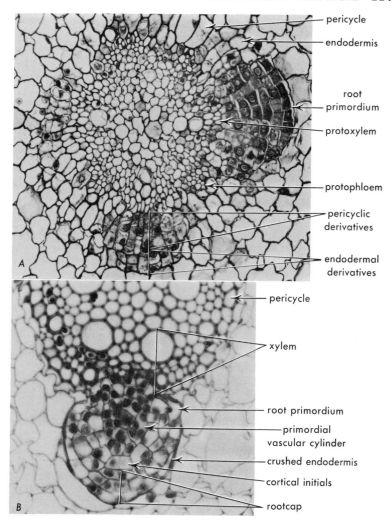

Fig. 14.11. Origin of lateral roots. *A, Helianthus annuus* (sunflower). Two root primordia are present. Metaxylem of parent root is still immature. Position of lateral root similar to that indicated in fig. 14.5, *C. B, Bromus mollis* (brome-grass). Position of lateral root similar to that indicated in fig. 14.5, *D.* (*A,* ×210; *B,* ×320. *B,* from Esau, *Hilgardia,* 1957.)

The development of new increments of growth by continued elongation of existing roots and by initiation of new lateral roots is considered to be an important feature with regard to absorption by roots. This growth creates new absorbing surfaces, and it brings these surfaces in contact with new areas of soil.

REFERENCES

Arnold, A. Über den Funktionsmechanismus der Endodermiszellen der Wurzeln. *Protoplasma* 41:189–211. 1952.

Beckel, D. K. B. Cortical disintegration in the roots of *Bouteloua gracilis*. *New Phytol.* 55:183–190. 1956.

Bouet, M. Études cytologiques sur le développement des poils absorbants. *Rev. Cyt. et Biol. Vég.* 15:261–305. 1954.

Bruch, H. Beiträge zur Morphologie und Entwicklungsgeschichte der Fenchelwurzel (*Foeniculum vulgare* Mill.). *Beitr. zur. Biol. der Pflanz.* 32:1–26. 1955.

Chan, T.-T. The development of the narcissus plant. *Daffodil and Tulip Yr. Bk.* 17:72–100. 1952.

Clowes, F. A. L. The root cap of ectotrophic mycorrhizas. *New Phytol.* 53:525–529. 1954.

Clowes, F. A. L. Protein synthesis in root meristems. *Jour. Exp. Bot.* 9:229–238. 1958.

Cormack, R. G. H. The development of the root hairs in angiosperms. *Bot. Rev.* 15:583–612. 1949.

Currier, H. B. Callose substance in plant cells. *Amer. Jour. Bot.* 44:478–488. 1957.

Dycus, A. M., and L. Knudson. The role of the velamen of the aerial roots of orchids. *Bot. Gaz.* 119:78–87. 1957.

Ekdahl, I. Studies on the growth and the osmotic conditions of root hairs. *Symb. Bot. Upsaliensis,* 11(6):5–83. 1953.

Erickson, R. O., and K. B. Sax. Rates of cell division and cell elongation in the growth of the primary root of *Zea mays*. *Amer. Phil. Soc. Proc.* 100:499–514. 1956.

Esau, K. Anatomical differentiation in root and shoot axes. In: *Growth and Differentiation in Plants*. W. E. Loomis, ed. Ames, Iowa State College Press. 1953.

Guttenberg, H. v., J. Burmeister, and H. J. Brosell. Studien über die Entwicklung des Wurzelvegetationspunktes der Dikotyledonen. II. *Planta* 46:179–222. 1955.

Guttenberg, H. v., H.-R. Heydel, and H. Pankow. Embryologische Studien an Monokotyledonen. I. Die Entstehung der Primärwurzel bei *Poa annua* L. *Flora* 141:298–311. 1954.

Jensen, W. A. The incorporation of C^{14}-adenine and C^{14}-phenylalanine by developing root-tip cells. *Natl. Acad. Sci. Proc.* 43:1039–1046. 1957.

Jensen, W. A., and L. G. Kavaljian. An analysis of cell morphology and the periodicity of division in the root tip of *Allium cepa*. *Amer. Jour. Bot.* 45:365–372. 1958.

Luhan, M. Das Abschlussgewebe der Wurzeln unserer Alpenpflanzen. *Deut. Bot. Gesell. Ber.* 68:87–92. 1955.

Popham, R. A. Zonation of primary and lateral root apices of *Pisum sativum*. *Amer. Jour. Bot.* 42:267–273. 1955a.

Popham, R. A. Levels of tissue differentiation in primary roots of *Pisum sativum*. *Amer. Jour. Bot.* 42:529–540. 1955b.

Richardson, S. D. The influence of rooting medium on the structure and development of the root cap in seedlings of *Acer saccharinum* L. *New Phytol.* 54:336–337. 1955.

Riedhart, J. M., and A. T. Guard. On the anatomy of the roots of apple seedlings. *Bot. Gaz.* 118:191–194. 1957.

Rosene, H. F. The water absorptive capacity of winter rye root-hairs. *New Phytol.* 54:95–97. 1955.

Row, H. C., anw J. R. Reeder. Root hair development as evidence of relationships among genera of Gramineae. *Amer. Jour. Bot.* 44:596–601. 1957.

Schade, C., and H. v. Guttenberg. Über die Entwicklung des Wurzelvegetationspunktes der Monokotyledonen. *Planta* 40:170–198. 1951.

Street, H. E., and E. H. Roberts. Factors controlling meristematic activity in excised roots. I. Experiments showing the operation of internal factors. *Physiol. Plantarum* 5:498–509. 1952.

Torrey, J. G. Physiology of root elongation. *Annu. Rev. Plant Physiol.* 7:237–266. 1956.

Voronin, N. S. *Ob evoliutsii korneĭ rasteniĭ.* [On evolution of roots of plants.] *Moskov. Obshch. Isp. Prirody, Otd. Biol., Biul.* 61:47–58. 1956.

Wilcox, H. Primary organization of active and dormant roots of noble fir, *Abies procera. Amer. Jour. Bot.* 41:812–821. 1954.

Williams, B. C. Observation on intercellular canals in root tips with reference to the Compositae. *Amer. Jour. Bot.* 41:104–106. 1954.

15. THE ROOT:
SECONDARY STATE OF GROWTH
AND ADVENTITIOUS ROOTS

THE SECONDARY GROWTH IN ROOTS, AS IN STEMS, CONSISTS OF THE formation of secondary vascular tissues from a vascular cambium and of periderm from a phellogen. Secondary growth is characteristic of roots of dicotyledons and gymnosperms. It occurs in various amounts in the dicotyledons although in some of them the roots remain in primary state throughout their life. As was mentioned in chapter 14 the roots of monocotyledons commonly lack secondary growth. The secondary growth of roots may show peculiarities in relation to functional specialization.

COMMON TYPE OF SECONDARY GROWTH

The vascular cambium is initiated by divisions of the procambial cells that remained undifferentiated between the primary phloem and the primary xylem (fig. 15.1, *A–D*). Thus, in the beginning, the cambium has the form of strips whose number depends on the type of root (fig. 15.1, *C*). There are two strips in a diarch root, three in a triarch root, and so forth. Subsequently, the pericyclic cells located outside the xylem ridges also become active as a cambium, and then the cambium completely encircles the xylem core. This early cambium has the same outline as the xylem; in cross sections it is oval in diarch roots, triangular in triarch roots, and so forth. However, the cambium located on the inner face of the phloem begins to function earlier than

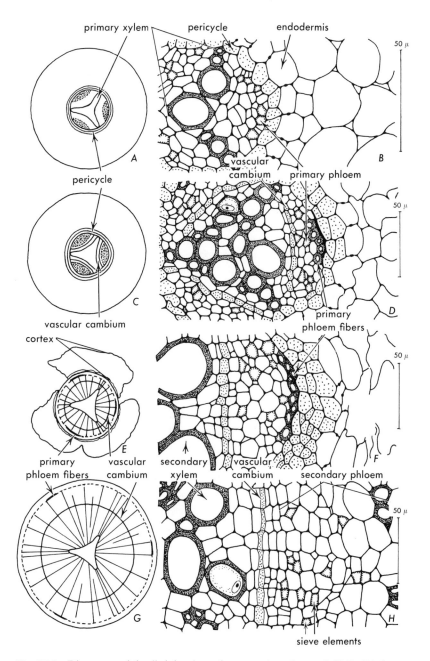

Fig. 15.1. Diagrams and detailed drawings of cross section of root of alfalfa (*Medicago sativa*) in different stages of development. *A, B,* primary stage of growth. *C, D,* initiation of vascular cambium. *E, F,* secondary growth of vascular cylinder, cell division in pericycle, and rupture of cortex. *G, H,* secondary growth is established.

the pericyclic part of the cambium. By formation of secondary xylem opposite the phloem the cambium is moved outward, and eventually its circumference becomes circular in cross section (fig. 15.1, *E*).

The cambium produces phloem and xylem cells (fig. 15.2) by periclinal divisions and increases in circumference by anticlinal divisions. The cambium arising on the inner face of the phloem produces conducting elements together with the other cells usually accompanying the conducting elements in the xylem and phloem (fig. 15.2, *A*). In some roots the cambium originating in the pericycle produces ray parenchyma (figs. 15.2, *B,* and 15.3, *B, C*). Rays also appear in the other parts of the secondary tissues (fig. 15.2, *C*), but those originating in the pericycle opposite the xylem ridges are frequently the widest. In some roots, however, no wide rays are formed, and the xylem appears rather homogeneous (fig. 15.3, *A, D*).

The formation of periderm follows the initiation of secondary vascular growth. The pericyclic cells undergo periclinal and anticlinal divisions (fig. 15.1, *D, F,* and 15.2, *A*). The periclinal divisions cause an increase of the number of pericyclic layers in radial extent. The combined increase in thickness of the vascular tissues and of the pericycle forces the cortex outward. The cortex does not undergo an increase in circumference but becomes ruptured and sloughed off together with the epidermis and endodermis (fig. 15.1, *E*). A phellogen arises in the outer part of the pericycle and forms phellem toward the outside. It may produce phelloderm toward the inside, but, if present, the phelloderm is difficult to distinguish from the pericycle which proliferated before the phellogen was initiated.

In perennial roots the activity of the vascular cambium continues through many years. The phellogen also continues its activity but may become replaced by phellogens arising at greater depths in the root. If such development occurs, the root, like a stem, has a rhytidome.

Herbaceous dicotyledon. Secondary growth in a herbaceous dicotyledon may be exemplified by the root of *Medicago sativa,* alfalfa (cf. Hayward, 1938; Simonds, 1935). The secondary xylem contains vessels of various diameters, mostly with scalariformly and reticulately pitted secondary walls. The vessels are accompanied by fibers and parenchyma cells. Wide rays of parenchyma divide the axial xylem into sectors (fig. 15.2, *C*). During the secondary growth the primary xylem becomes considerably modified by dilatation growth of the primary xylem parenchyma. The strands of primary conducting xylem cells are broken and partly crushed.

The phloem contains sieve tubes with companion cells, fibers, and parenchyma cells (fig. 15.1, *H*). The wide rays of the xylem are con-

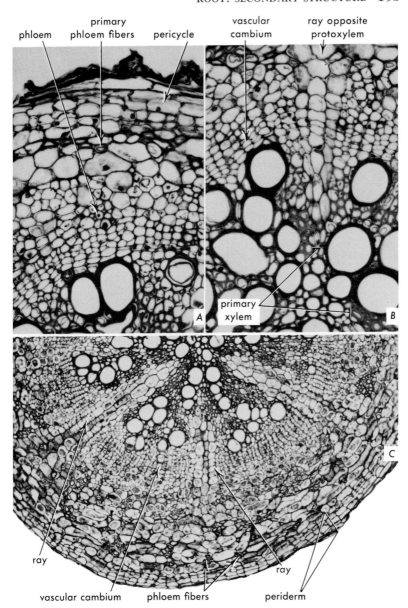

Fig. 15.2. Cross sections of root of alfalfa (*Medicago sativa*) with details of secondary growth. *A, B,* early stage of secondary growth similar to that in fig. 15.1, *E, F. A,* arrangement of tissues along radius through phloem; cortex (outside pericycle) is collapsed. *B,* region opposite a protoxylem pole. *C,* root in advanced stage of secondary growth similar to that in fig. 15.1, *G, H.* (*A, B,* ×250; *C,* ×90.)

tinuous through the cambium with similar rays of the phloem (fig. 15.2, *C*). The outer phloem contains only fibers and storage parenchyma; the old sieve tubes are crushed. The phloem merges imperceptibly with the pericyclic parenchyma beneath the periderm. Cork derived from the phellogen forms the protective tissue.

The amount of secondary growth varies in different herbaceous dicotyledons as do also the histology of the tissues and the distinctness of periderm (fig. 15.3).

Woody species. The organization of the secondary vascular tissues in roots of woody species resembles that just outlined for the root of alfalfa (e.g., Esau, 1943). Usually the roots of trees have a larger proportion of elements with lignified secondary walls (fig. 15.4), but roots of herbaceous plants also may become strongly sclerified (fig. 15.3, *A*). The roots of gymnosperms (fig. 15.4, *A*) have a type of secondary growth similar to that in trees of dicotyledons (fig. 15.4, *B*), but, of course, the basic elements of xylem and phloem are different in the two groups of plants.

When old roots and stems of trees are compared, they are found to resemble each other closely. There are, however, quantitative differences, mainly expressed in larger proportion of elements with lignified secondary walls in bark and wood of the stem, or, in other words, the roots have a greater relative volume of parenchyma tissue. The histologic differences between the secondary tissues of stems and roots are determined to a large extent by differences in the environment in which the two parts of the plant body develop. If roots of dicotyledon or gymnosperm trees are exposed to light and air, the wood that develops after the exposure assumes most of the characteristics of the wood in stems (Bannan, 1941; Morrison, 1953).

An important horticultural and ecological aspect of secondary growth in roots is the occurrence of natural root grafting between different trees, presumably of the same species, a phenomenon which is widespread in the tropics and in the temperate zone among dicotyledons and gymnosperms (LaRue, 1952). Where the roots come in contact with each other they become united through secondary growth. The details of this phenomenon are apparently not known, but it has been pointed out that removal of bark by friction at the point of contact is not necessary to produce root grafts. The grafting of roots is one of the effective means of transmission of infectious diseases from one tree to another.

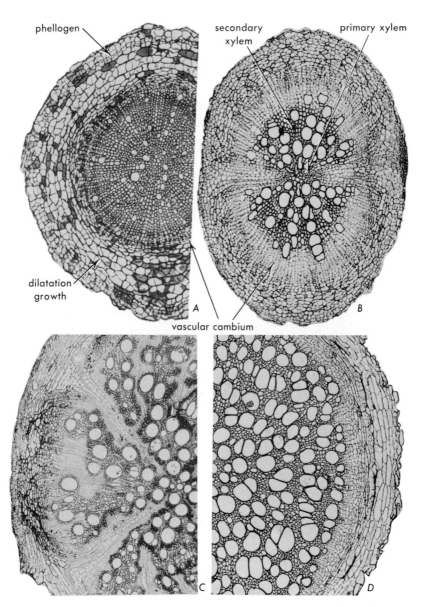

Fig. 15.3. Cross sections of roots of herbaceous plants in secondary state of growth. *A*, tomato (*Lycopersicon esculentum*). *B*, cabbage (*Brassica oleracea*). *C*, pumpkin (*Cucurbita* sp.). *D*, potato (*Solanum tuberosum*). (*A, B,* ×55; *C,* ×23; *D,* ×45.)

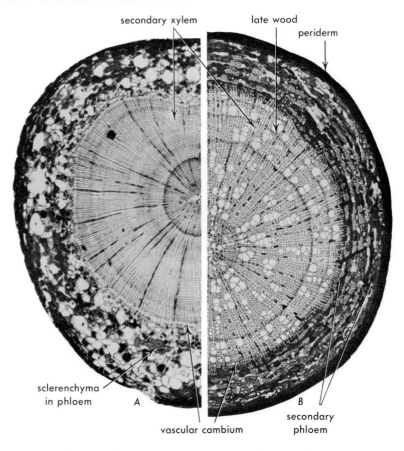

Fig. 15.4. Cross sections of roots of woody species in secondary state of growth. *A, Abies,* fir (×24). *B, Tilia* (×33). (After Esau, *Plant Anatomy,* John Wiley and Sons, 1953.)

VARIATIONS IN SECONDARY GROWTH

Herbaceous dicotyledons often have a limited amount of secondary growth, which may be associated with characteristic features. In *Actaea,* for example, the conducting part of the secondary vascular tissues appears in discrete strands separated from each other by wide rays of large-celled parenchyma. These rays originate in the pericycle opposite the protoxylem ridges. The divisions extending these rays radially occur in line with the vascular cambium between the xylem and the phloem. These dividing cells in the rays may be interpreted accordingly as part of the cambium, although they are not readily identified in sections.

Actaea, Convolvulus, and some other herbaceous plants with roots having a limited amount of secondary growth have a superficial periderm and therefore retain their cortex. The endodermis may increase in circumference by radial cell division and tangential cell enlargement, like the rest of the cortex (*Actaea*), or it may be crushed (*Convolvulus*). In roots of *Citrus sinensis* the periderm is first formed beneath the epidermis; later a deeper periderm arises in the pericycle (Hayward and Long, 1942). In certain families (Rosaceae, Myrtaceae, Onagraceae, Hypericaceae) a special kind of protective tissue called polyderm is formed by the pericycle (chapter 12). The products of the initiating layer of polyderm are not ordinary cork cells but consist of rows of nonsuberized parenchyma cells alternating with rows of cells containing suberin (Casparian strips) and generally resembling endodermal cells.

Storage roots. Several variations in secondary structure occur in connection with the development of storage roots (usually a combination of root and hypocotyl). In roots of Umbelliferae as, for example, fennel (*Foeniculum,* Bruch, 1955) or carrot (*Daucus,* Esau, 1940), the secondary growth is of the ordinary kind, but parenchyma predominates in the xylem and phloem. In the beet (*Beta,* cf. Hayward, 1938), however, the main increase in thickness results from the so-called anomalous type of growth (fig. 15.5). A series of supernumerary cambia arranged approximately concentrically, like growth rings in a tree, arise outside the normal vascular core. The cambial cells are derived from cells of pericycle and phloem and produce several increments of vascular tissues, each composed of storage parenchyma and strands of xylem and phloem separated from one another by wide radial panels of parenchyma (fig. 15.5, *B*).

Another complex type of anomalous growth occurs in the fleshy adventitious roots of *Ipomoea batatas,* the sweet potato (cf. Hayward, 1938). The xylem contains a large proportion of parenchyma and arises in the usual manner. Cambia develop in the parenchyma around individual vessels or vessel groups and produce a few tracheary elements toward the vessels, a few sieve tubes and laticifers away from the vessels, and a considerable number of storage parenchyma cells in both directions (fig. 15.6, *A, B*). Thus, phloem elements appear within the part of the root that originally differentiated as xylem. A cambium in normal position separates the xylem from the phloem in normal position, and a periderm of pericyclic origin occurs on the periphery (fig. 15.6, *B*). In the fleshy taproots and stems of some

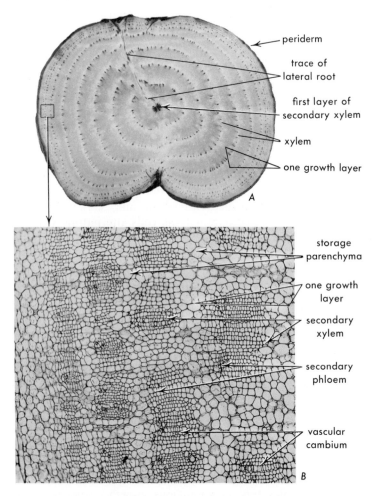

Fig. 15.5. Cross sections of root of sugar beet (*Beta vulgaris*) illustrating anomalous secondary growth resulting from formation of many cambial layers outside the ordinary cambium, each of which gives rise to xylem and phloem cells together with storage parenchyma. *A,* ×0.66; *B,* ×60. (From Esau, *Plant Anatomy,* John Wiley and Sons, 1953; from Artschwager, *Jour. Agr. Res.,* 1926.)

Cruciferae (turnip, radish, kohlrabi, rutabaga, and others) parenchyma of the xylem and pith (if present) proliferates, and subsequently cambia and vascular tissues arise in this parenchyma (fig. 15.6, *C, D;* cf. Hayward, 1938).

The common character of all the fleshy storage organs derived from

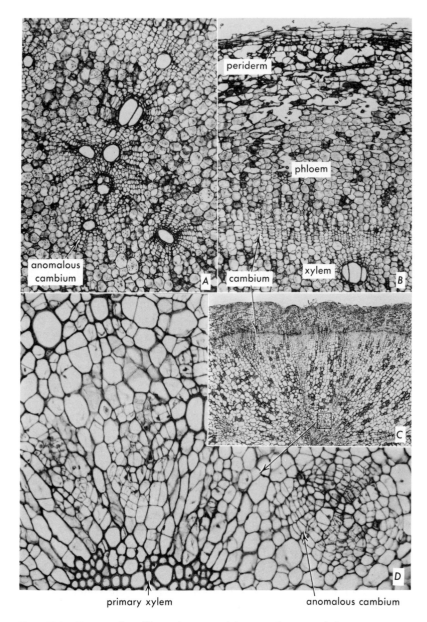

Fig. 15.6. Cross sections illustrating anomalous secondary growth in storage roots. *A, B,* sweet potato (*Ipomoea batatas*). Anomalous cambium around vessels of secondary xylem in *A;* normal vascular cambium in *B. C, D,* radish (*Raphanus sativus*) with anomalous cambium in secondary xylem. (*A,* ×54; *B,* ×43; *C,* ×16; *D,* ×90.)

hypocotyls, roots, and sometimes stems is the possession of abundant storage parenchyma permeated by vascular tissues. This close association between conducting and storage tissues is attained by various types of secondary growth.

ADVENTITIOUS ROOTS

The term adventitious root has somewhat varied meanings. In this book it is used broadly to designate roots that arise on aerial plant parts, on underground stems, and on more or less old root parts. In contrast, lateral roots arise in typical acropetal succession in the young parts of the taproot and its branches or of adventitious roots and their branches. Adventitious roots are widely distributed in all vascular plants and are formed in many locations on the plant (Baranova, 1951; Hayward, 1938, p. 54). They may occur at nodes in association with axillary shoots (e.g., tillering in grasses) but may also be independent of axillary buds and may develop on internodes. Leaves sometimes form adventitious roots. The development of adventitious roots is important in the propagation of plants by means of stem cuttings, and the phenomenon has been widely explored in connection with research on growth regulating substances. Adventitious roots constitute the main root system in lower vascular plants, monocotyledons, dicotyledons that are propagated by means of rhizomes and runners and in water plants, saprophytes, and parasites.

The origin and development of adventitious roots resemble those of lateral roots: they usually have an endogenous origin and arise close to the vascular tissues (fig. 15.7), and they grow through tissues located outside the point of origin. During this growth the adventitious root originating in a relatively old stem may find an obstacle in the form of a sheath of perivascular sclerenchyma which deflects the root from its normally radial course (e.g., Stangler, 1956).

In young stems of dicotyledons and gymnosperms the adventitious roots commonly arise in the interfascicular parenchyma and in older stems in the vascular ray near the cambium. Thus the new root appears close to both xylem and phloem (Satoo, 1955; Stangler, 1956). Commonly the region of origin of adventitious roots in young stems is referred to as pericycle. In gymnosperms and dicotyledons the region called pericycle in stems is most commonly the outermost part of the primary phloem in which the sieve elements are no longer functioning. When adventitious roots are formed in cuttings they may originate in the callus tissue that is often formed at the base of cuttings.

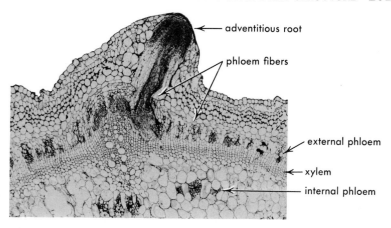

adventitious root

phloem fibers

external phloem

xylem

internal phloem

Fig. 15.7. Cross section of tomato stem infected with aster-yellows virus. Adventitious root originated in parenchyma of external phloem. External and internal phloem is degenerating in response to virus infection. (×43. Slide, courtesy of E. A. Rasa.)

The formation of adventitious roots ("replacement roots") during regeneration of injured roots in noble fir (Wilcox, 1955) illustrates the variability in the origin of adventitious roots. Wound healing and callus formation (chapter 17) occur on the pruned surfaces of roots. Subsequently adventitious roots arise in part from the undisturbed tissues beneath the wound surface, in part from derivatives of the callus. Younger pruned roots produce replacement roots in the same positions where lateral roots arise, that is, opposite the protoxylem ridges, and they are initiated in the pericycle or in the pericyclic callus. If the root is pruned after some secondary tissues have been formed, the adventitious roots arise in various positions around the circumference of the vascular cylinder and are more numerous than in younger roots. In these roots the replacement roots are initiated in the vascular cambium or in cambium that has been regenerated by callus growth.

The adventitious root primordia are initiated by divisions of parenchyma cells—callus cells or other parenchyma cells—resembling the divisions initiating the lateral roots in the pericycle of young roots. Before the adventitious root emerges from the stem or root it differentiates a promeristem, a rootcap, and the beginning of the vascular cylinder and cortex. When vascular elements differentiate in the adventitious root the callus cells or other parenchyma cells located at the proximal end of the primordium differentiate into vascular elements and provide a connection with corresponding elements of the initiating organ.

REFERENCES

Bannan, M. W.   Variability in wood structure of native Ontario conifers.   *Torrey Bot. Club. Bul.* 68:173–194.   1941.

Baranova, E. A.   *Zakonomernosti obrazovaniĭa pridatochnykh korneĭ u rasteniĭ* [Laws of formation of adventitious roots in plants.]   *Trudy Glav. Bot. Sada* 2:168–193.   1951.

Bruch, H.   Beiträge zur Morphologie und Entwicklungsgeschichte der Fenchelwurzel (*Foeniculum vulgare* Mill.).   *Beitr. zur Biol. der Pflanz.* 32:1–26.   1955.

Esau, K.   Developmental anatomy of the fleshy storage organ of *Daucus carota. Hilgardia* 13:175–226.   1940.

Esau, K.   Vascular differentiation in the pear root.   *Hilgardia* 15:299–311.   1943.

Hayward, H. E.   *The structure of economic plants.*   New York, The Macmillan Company.   1938.

Hayward, H. E., and E. M. Long.   The anatomy of the seedling and roots of the Valencia orange.   *U. S. Dept. Agr. Tech. Bul.* 786.   1942.

LaRue, C. D.   Root-grafting in tropical trees.   *Science* 115:296.   1952.

Morrison, T. M.   Comparative histology of secondary xylem in buried and exposed roots of dicotyledonous trees.   *Phytomorphology* 3:427–430.   1953.

Satoo, S.   Origin and development of adventitious roots in layered branches of 4 species of conifers.   *Japanese Forest. Soc. Jour.* 37:314–316.   1955.

Simonds, A. O.   Histological studies of the development of the root and crown of alfalfa.   *Iowa State Col. Jour. Sci.* 9:641–659.   1935.

Stangler, B. B.   Origin and development of adventitious roots in stem cuttings of chrysanthemum, carnation, and rose.   *N. Y. Agr. Expt. Sta. Mem.* 342.   1956.

Wilcox, H.   Regeneration of injured root systems in noble fir.   *Bot. Gaz.* 116:221–234.   1955.

16. THE STEM:
PRIMARY STATE OF GROWTH

THE STEM AS PART OF THE SHOOT

The close association of the stem with the leaves makes this part of the plant axis more complex than the root. Because of this association one often uses the term shoot, which refers to the stem and leaves as one system.

In contrast to the root, the stem is divided into nodes and internodes, with one or more leaves attached at each node. Depending on the degree of development of the internodes, the shoot assumes different aspects. It may be an elongated structure with easily recognizable nodes and internodes, or it may be a condensed structure without discernible internodes and with leaves crowded in a so-called rosette. The other main features that contribute to the variation in the external aspect of the shoot are the arrangement of leaves, their manner of insertion, development or nondevelopment of axillary buds into lateral shoots, and the level at which branching, if present, occurs. Differences are also associated with the type of growth and habitat of the stem, that is, whether the stem grows in air or underground, in water or on land, and whether it is upright, or climbing, or creeping.

PRIMARY STRUCTURE

The stem, like the root, consists of three tissue systems, the dermal, the fundamental, and the fascicular or vascular (fig. 16.1). The variations in the primary structure in stems of different species and the larger groups of plants are based chiefly on differences in the relative distribution of the fundamental and vascular tissues. In the conifers and dicotyledons the vascular system of the internode commonly appears as a hollow cylinder delimiting an outer and an inner region of fundamental tissue, the cortex and the pith, respectively. The subdivisions of the vascular system, the vascular bundles or larger complexes, may appear close together, or they may be separated from each other by more or less wide panels of parenchyma tissue—the interfascicular parenchyma, also a part of the ground tissue. It is called interfascicular because it occurs between the fascicles, a name of Latin origin for bundles. A panel of interfascicular parenchyma is often called the medullary or pith ray.

Stems of many ferns, some herbaceous dicotyledons, and most monocotyledons have a more complex arrangement of tissues in that the vascular tissues, as seen in transverse section, do not appear as a single ring of bundles between the cortex and the pith. The bundles may occur in more than one ring (fig. 17.7) or may appear scattered throughout the cross section (fig. 17.8). The delimitation of the ground tissue into cortex and pith is less precise or does not exist when the vascular bundles do not form one ring in cross sections of internodes.

Epidermis

The principal features of the epidermis of aerial plant parts are described in chapter 7. Stomata constitute a less prominent component of the stem epidermis than of the leaf epidermis. The tissue commonly consists of one layer of cells and has a cuticle and cutinized walls. It is a living tissue capable of mitotic activity—an important characteristic in view of the stresses to which the tissue is subjected during the primary and secondary increase in thickness of the stem. The epidermal cells respond to these stresses by tangential enlargement and radial divisions. The persistence of mitotic activity in the stem epidermis is particularly impressive in species with long delayed periderm formation (chapter 12).

Cortex and pith

The cortex of stems contains parenchyma, usually with chloroplasts (fig. 16.1, B). Intercellular spaces are prominent but sometimes are largely restricted to the median part of the cortex. The peripheral part of the cortex frequently contains collenchyma (fig. 16.1, *A*), in strands or in a more or less continuous layer.   In some plants, notably grasses, sclerenchyma rather than collenchyma develops as the primary supporting tissue in the peripheral parts of the stem.   The conifers typically have no special strengthening tissue in the cortex.

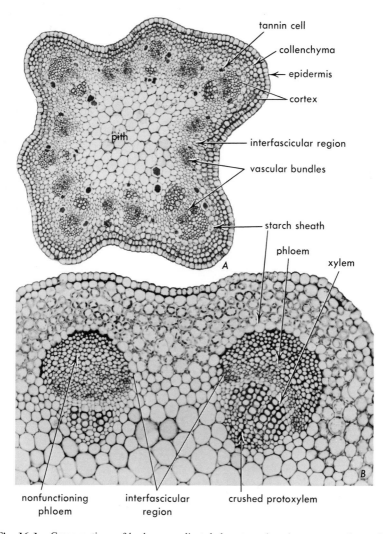

tannin cell

collenchyma

epidermis

cortex

pith

interfascicular region

vascular bundles

starch sheath

phloem

xylem

A

nonfunctioning
phloem

interfascicular
region

crushed protoxylem

B

Fig. 16.1.   Cross sections of herbaceous dicotyledon stems in primary state of growth. *A, Lotus corniculatus,* complete section (×96).   *B, Trifolium hybridum,* clover, partial section (×184).   (*A,* from Hansen, *Iowa State Coll. Jour. Sci.,* 1953; *B,* from J. E. Sass, *Botanical Microtechnique,* 3rd ed.   The Iowa State College Press, 1958.)

The delimitation of the cortex from the vascular region is an important morphological aspect that is related to the concept of stele reviewed below. As was shown in chapter 14 the innermost layer of the cortex of roots of vascular plants has special wall characteristics so that the layer merits a separate designation as endodermis. The stems of conifers and angiosperms commonly lack a morphologically differentiated endodermis. In young stems the innermost layer or layers may contain abundant starch and thus be recognizable as a starch sheath (fig. 16.1). Some dicotyledons, however, develop Casparian strips in the innermost cortical layer (fig. 16.2; cf. Esau, 1953, p. 362; Ziegenspeck, 1952), and many lower vascular plants have a clearly differentiated endodermis in stems.

When no starch accumulates and no special wall characteristics develop in the innermost cortical layer, the delimitation of the cortex from the vascular region may be difficult or impossible. Nevertheless, in the region where the fundamental tissue of the peripheral part of the stem joins the vascular tissue, certain reactions take place between materials derived from the vascular tissue and those present in the cortical parenchyma (Van Fleet, 1950), so that the boundary between the two kinds of tissues has distinct chemical and physiological char-

Casparian strip

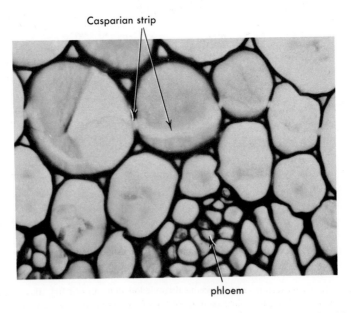

phloem

Fig. 16.2. Cross section of ragweed (*Ambrosia*) stem showing endodermis with Casparian strip (pale band crossing the radial walls), an unusual type of differentiation for stems of seed plants. (×800. Slide, courtesy of N. H. Boke.)

acteristics similar to those present in the morphologically differentiated endodermis (Ziegenspeck, 1952). In other words, there is a physiological boundary between the cortex and the vascular system that may or may not be associated with a morphological specialization. In this book the term endodermis is used only when the layer has Casparian strips or other wall characteristics found in root endodermis. Layers in similar position—and probably differing physiologically from adjacent layers—but without distinctive wall characteristics are called endodermoid.

The consistent occurrence of a visibly differentiated endodermis in roots and its common absence in stems of higher vascular plants led to the thought that environment influences the development or nondevelopment of the endodermis. Accordingly, some investigators reported a development of Casparian strips in the innermost cortical layer in stems that were grown in darkness; newer work (Venning, 1954), however, failed to confirm these findings.

The inner part of the ground tissue, the pith, is commonly composed of parenchyma, which may contain chloroplasts. In many stems the central part of the pith is destroyed during growth. Frequently this destruction is limited to the internodes, whereas the nodes retain their pith (nodal diaphragms). Sometimes series of horizontal plates of pith tissue remain in the internodes also (*Juglans, Pterocarya*). The pith has prominent intercellular spaces, at least in the central part. The peripheral part may be distinct from the inner in having compactly arranged small cells and greater longevity. Since the pith is also referred to as the medulla, the distinct peripheral zone in the pith is called perimedullary zone or medullary sheath (fig. 17.2).

Both cortex and pith may contain various idioblasts, including cells with crystals and other ergastic contents, and sclereids. If the plant has laticifers, they may be present in pith and cortex.

Vascular system

As was mentioned earlier, the vascular tissues commonly appear as a cylinder between the cortex and the pith or assume one of the more complex patterns. Sometimes the cylinder appears to be continuous, or almost so, around the circumference of an internode, but more frequently it is separated by interfascicular regions into smaller and larger units. These units are commonly referred to as vascular bundles (fig. 16.1). If the cylinder appears continuous, a detailed analysis will show that it, too, consists of units but that they occur close together (fig. 17.5).

The two kinds of vascular tissues, the phloem and the xylem, commonly occur in a *collateral* arrangement, as seen in transverse sections, with the phloem located outside the xylem (fig. 16.1). The collateral arrangement contrasts with the so-called *radial*, or *alternate*, arrangement in the root in which the phloem strands alternate with the peripheral parts of the xylem around the circumference of the vascular cylinder.

In certain ferns and in many families of dicotyledons (for example, Apocynaceae, Asclepiadaceae, Convolvulaceae, Cucurbitaceae, Solanaceae, and certain tribes of Compositae) one part of the phloem occurs outside, another on the inner side of the xylem. Thus one can distinguish between the *external* and *internal phloem*. An individual vascular bundle containing phloem on two sides of the xylem is *bicollateral* (fig. 17.5).

A further modification of the relative arrangement of vascular tissues is represented by the *concentric* vascular bundles in which either the phloem surrounds the xylem (amphicribral bundles) or the xylem surrounds the phloem (amphivasal bundles). The amphivasal bundles appear to be phylogenetically rather specialized and are found in certain dicotyledons (e.g., medullary bundles in *Rheum, Rumex, Mesembryanthemum crystallinum,* and *Begonia*) and monocotyledons (e.g., Araceae, Liliaceae, Juncaceae, and Cyperaceae). Amphicribral bundles are most common in ferns but may be found in angiosperms. Small bundles in flowers, fruits, and ovules may be amphicribral.

As was indicated in the discussion regarding the endodermis, in stems of higher vascular plants the vascular tissues are usually not sharply delimited from the nonvascular ground tissue. If no endodermis is present, there is a complete continuity between the interfascicular region and the cortex. The outer part of the phloem, the protophloem, may contain fibers at the end of primary growth. These fibers help to identify the outer limit of the phloem. If no protophloem fibers are present, as in the conifers, the cortex merges with the parenchyma of the phloem. The sieve elements of the protophloem can serve for the identification of the outer part of the phloem only in the youngest parts of the shoot, because later these cells are destroyed (chapter 11).

In the root, nonvascular tissue of variable width, the pericycle, occurs between the endodermis and the outermost vascular elements. In the stems of the higher vascular plants usually no separate layer is present between the phloem and the cortex, and the term pericycle is then applied, in the literature, to the outermost part of the phloem in which the sieve elements are no longer functional. This terminology is not used in this book but is modified as outlined below.

In some stems of the dicotyledons a continuous or nearly continuous cylinder of fibers occurs on the periphery of the vascular cylinder (for example, *Aristolochia,* before secondary growth occurs; *Pelargonium;* and *Cucurbita*). These fibers may arise from the same meristem as the phloem (*Pelargonium*) or may originate outside the phloem (*Aristolochia, Cucurbita*), but inside of the starch sheath, that is, the innermost layer of the cortex (Blyth, 1958). Thus, in some instances, a tissue of nonphloic origin occurs between the vascular tissue and the cortex. It was precisely such tissue for which the term pericycle was coined originally, then extended to all stems, despite the fact that in most stems of seed plants the phloem is in contact with the cortex. In view of this dubious meaning of pericycle with reference to the stems of seed plants, the fibers on the periphery of the vascular tissue in such stems are called, in this book, *primary phloem fibers* when they are obviously of phloic origin and *perivascular fibers* if they arise outside the phloem. Both kinds of fibers may be combined under the term *primary extraxylary fibers.*

Leaf traces and leaf gaps. Since the stem and leaves are structurally continuous, the vascular system of the stem, to be properly understood, must be studied with reference to its connection with the vascular system of the leaves. The vascular connection of leaf and stem

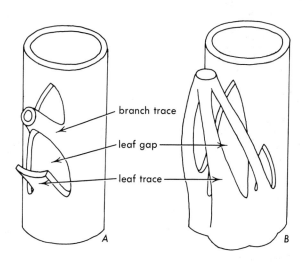

Fig. 16.3. Diagrams of vascular system from stems with a unilacunar node (*A*) and trilacunar node (*B*). In *A*, a single leaf trace diverges into a leaf and two branch traces into a branch; in *B*, the leaf has three traces, each confronted by a separate gap. (After Esau, *Plant Anatomy*, John Wiley and Sons, 1953.)

is seen in the nodal region where one or more strands in the stem bend away from the stem and toward the leaf (fig. 16.3). If a bundle connecting the leaf and the stem is followed through successive levels in the stem below the leaf insertion, it may be recognized as an independent bundle through one or more nodes and internodes (fig. 16.4). At some level this bundle is attached to another part of the vascular system in the stem. The bundle extending from the base of the leaf to the point of its junction with another strand in the stem is called the leaf trace (fig. 16.4, *B*). Thus, the leaf trace may be defined as the cauline (that is, occurring within the caulis, stem) part of the vascular supply of the leaf. The foliar part of this supply begins at the base of the petiole and extends into the blade where it may be extensively branched.

In the nodal region, where a leaf trace is bent away from the center of the stem toward the leaf base, a region of parenchyma occurs in the vascular cylinder of the stem (figs. 16.3 and 16.4). This region, which in transverse sections appears like a rather wide interfascicular

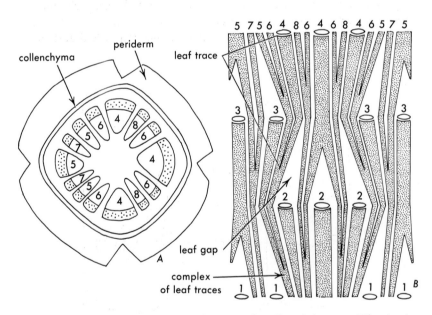

Fig. 16.4. Diagrams illustrating vascular system in a dicotyledon stem (*Ulmus*). *A*, cross section. *B*, longitudinal view showing the vascular cylinder as though cut through median leaf trace 5 and spread out in one plane. The numbers in both views indicate leaf traces. The cross section in *A* corresponds with the topmost view in *B*. (After Smithson, *Leeds Phil. Soc. Proc.*, 1954.)

region, is called leaf gap. A leaf gap thus is a parenchymatous region in the vascular cylinder of the stem located opposite (adaxially to) the upper part of a leaf trace, that is, approximately at the level of insertion of the leaf. If this trace did not diverge into the leaf, it would have occupied the gap region. This consideration explains the meaning of the term gap.

Leaf gaps vary in extent, laterally and vertically. If the vascular cylinder has many axially elongate interfascicular regions, the gaps may be confluent with these (fig. 16.4, *B*). In such instances the delimitation of the gap can be made only arbitrarily. The gap may be easily recognized after some secondary growth has occurred, however, because a gap is wider than an interfascicular region and the vascular cambium takes more time to form a layer outside the gap than outside an ordinary interfascicular region. Thus, with regard to stems with discrete vascular bundles and no secondary growth and particularly to those with "scattered" vascular bundles, the concept of leaf gap may be applied only on theoretical grounds.

The number of leaf traces and the number of leaf gaps vary in different plants (fig. 16.5) and may vary in the same plant at different levels. The variations are of phylogenetic significance and are, therefore, an object of comparative studies. The various forms of nodal anatomy are designated by special terms. The term gap is replaced by that of lacuna, and the nodes are said to be unilacunar, trilacunar, and multilacunar depending upon whether one leaf is associated with one, three, or several lacunae at the node (fig. 16.5). If more than one leaf is inserted at a node, the node is characterized with reference to one leaf. That is, if each leaf is associated with one gap, the nodal form is identified as unilacunar, although actually there may be two or more gaps at the node (fig. 16.5, *D*). If a leaf is confronted by three or more gaps, the gap (and the associated trace) that occurs in the median position with reference to the leaf is referred to as median and the others as lateral (figs. 16.3, *B*, and 16.5, *B, C*).

In many plants only one trace is related to each gap. In other words, if the node is unilacunar, the leaf has one trace; if it is trilacunar, each leaf has three traces; and at a multilacunar node many traces diverge into a single leaf. A more primitive condition, however, is that of two leaf traces confronting a single gap, that is, a two-trace unilacunar condition (fig. 16.5, *F*). When more than two traces confront one gap, the form may be interpreted as a modification of the two-trace unilacunar (Bailey, 1956).

The evolution of the nodal structure is presently interpreted as having followed the sequences of (1) two-trace unilacunar, trilacunar,

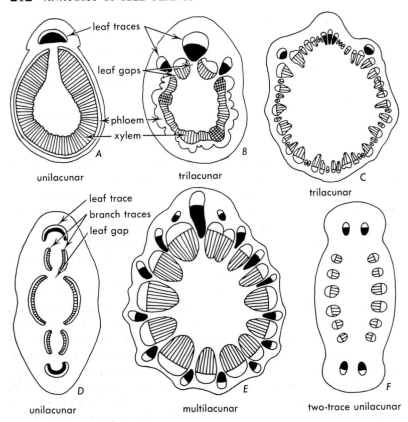

Fig. 16.5. Cross sections of stems with different types of nodal structure. Leaf traces are indicated by blackened xylem regions. *A, Spiraea. B, Salix. C, Brassica. D, Veronica. E, Rumex. F, Clerodendron.* (*A–E,* after Esau, *Plant Anatomy,* John Wiley and Sons, 1953.)

and multilacunar or (2) two-trace unilacunar, one-trace unilacunar, trilacunar, and multilacunar. The one-trace unilacunar could have been derived also from a trilacunar. The various changes have involved fusions, losses, and additions of traces.

As was mentioned above, a leaf trace, when followed through the stem, is found connected with a part of the vascular tissue of the stem at some level below the insertion of the leaf. In the Pteropsida this part of the vascular tissue in the stem is also usually related to one or more leaves at some other level of the axis; that is, it also represents leaf-trace material. One can state, descriptively, therefore, that the primary vascular system in Pteropsida stems is composed of leaf traces and their complexes (figs. 16.4, *B*), that the vascular system of the stem is organized in relation to leaves, that leaf bundles and leaf traces are parts of a vascular system of a unit structure, the shoot.

Surgical experiments with vascular plants clearly illustrate that leaves can have a developmental effect upon the structure of the vascular system of the stem: if leaf primordia are destroyed by puncturing in their earliest stages of development, the vascular system of the stem develops into a relatively simple structure without gaps and without projections into lateral organs (Wetmore and Wardlaw, 1951),

The complex vascular system of the shoots of Pteridopsida is regarded as more highly advanced phylogenetically than the relatively simple system of roots. Stems of some of the lower vascular plants, e.g., *Selaginella* and *Lycopodium,* have relatively simple vascular systems resembling that of roots of the higher plants. Their leaves are small, and the thin traces are not confronted by gaps and have a shallow insertion on the vascular cylinder of the axis. Here the vascular system of the axis is relatively independent of the leaves in its organization and development.

The organization of the vascular system of the stem of a pteropsid into leaf traces and their complexes explains the variable appearance of the vascular units in a given cross section of a stem (fig. 16.1, *A*). The clearly circumscribed units with largest amounts of xylem are leaf traces of leaves attached at some not too distant nodes above; the units that are less well defined are complexes of traces. Some of the smallest bundles may be either lateral traces, which are usually smaller than the median, or the lower ends of median (or of single) traces of some younger leaves inserted at levels far above the section.

Leaf gaps may be identifiable not only in nodes but also in internodes because leaf traces often have an oblique course through part of the internode; that is, if followed level by level, the leaf traces appear to bend away from the central cylinder of bundles gradually. Where a leaf trace is partly bent away, ground tissue, continuous with the pith, fills an indentation (part of leaf gap) in the vascular cylinder. Thus the shape of the vascular cylinder in the internode may clearly reflect the nodal structure of the plant.

Leaf arrangement and vascular organization. The pattern formed by the vascular system is more or less regular and shows a relation to the arrangement of leaves. As is well known, the leaf arrangement, or phyllotaxy, shows many variations, but basically there are three main types: the whorled, with several leaves at each node (the opposite-leaved decussate type with two opposite leaves at each node may be regarded as a subtype of the whorled); the distichous or two ranked, with leaves single at each node but disposed in two opposite ranks (e.g., grasses); and the alternate or helical.

Students of phyllotaxy are concerned with the meaning of the specific arrangements and use various approaches to designate the patterns by numerical values and formulas. The classical method is to use the angle of divergence of two successive leaves. Examples of divergence are ½, ⅓, ⅖, ⅜, ⁵⁄₁₃, and still smaller fractions of the circumference of the axis. Thus one speaks of the ½, ⅓, ⅖, etc. phyllotaxes. If one draws a line from leaf to leaf connecting the leaves in the order of their appearance, one obtains a helix—the "genetic spiral" of the classical botanists—which follows the leaves in the order of their origin at the shoot tip. The fractional expression of the angle of divergence tells something about the distribution of the leaves along the genetic spiral. In ⅖ phyllotaxy, for example, two windings about the axis will include 5 leaves, with leaves *n* and *n* plus 5 located one above the other. A vertical series of leaves *n, n* plus 5, *n* plus 5 plus 5, etc. (for example, leaves 1, 6, 11, 16, etc.) is referred to as orthostichy. The term implies arrangement along a straight line. In a plant with a helical phyllotaxy the orthostichy in a mature shoot may appear to be a straight line, but at origin of leaves it forms a steep helix. It is thus not a true orthostichy but a parastichy. (True orthostichies occur in decussate and distichous leaf arrangements.) Many other parastichies may be projected on a shoot, some flatter, some steeper, some curving in a clockwise direction, others in counterclockwise.

The interconnections of leaf traces in the vascular system of the stem may be analyzed in terms of leaf arrangement. The trace connections may be restricted to leaves belonging to the same parastichy, or there may be connections within and between parastichies. The arrangement of the vascular bundles in a cross section of the stem may show groupings according to closely connected parastichies. Sometimes the interconnections are less regular and may change, like the phyllotaxy itself, during the development of the plant.

Branch traces and branch gaps. Buds commonly develop in the axils of leaves so that, in addition to the leaf traces, the vascular bundles that connect the main stem with the branch may be recognized in the nodal region. These strands are called branch traces (fig. 16.5, *D*). Actually the branch traces are also leaf traces, namely, leaf traces of the first leaves, the prophylls, of the branch (fig. 16.6). In the conifers and the dicotyledons two prophylls occur opposite one another oriented so that a plane bisecting both prophylls would be parallel with the plane of the subtending leaf. Because of this arrangement, two traces, each composed of one or more bundles, connect the bud with the main axis. In the monocotyledons the side shoot also has two traces, although there is only one prophyll (sometimes interpreted as a double structure) at the base of the axillary shoot.

The branch traces usually diverge from the main stem right and left of the median leaf trace (or the single trace if no laterals are present) so that they have a common gap with the subtending leaf (fig. 16.3, *A*). The branch traces extend through variable distances in the main axis and at some level connect with the vascular system of this axis.

The concept of the stele. A well-known concept pertaining to the phylogeny of the form of the vascular system in the axis is that of the stele. The stelar concept was introduced by Van Tieghem and Douliot (1886) in their attempt to express the unity of structure of the plant axis. The stele, meaning a column, was defined as the core of the plant axis (stem and root) including the vascular system with all its inter-fascicular regions, gaps, pith (if present), and some fundamental tissue on the periphery of the vascular system, the pericycle. The plant axis was thus envisioned as consisting of a central column covered by the cortex, with the epidermis on the surface of the latter.

The stelar concept was received favorably and has proved useful in comparative and phylogenetic studies of vascular plants. As is true of all much used concepts, the stelar concept underwent many changes, as did the classification and nomenclature of the steles. One of the results of these changes is that many modern workers, speaking of the stele, refer only to the vascular system and not to the column consist-

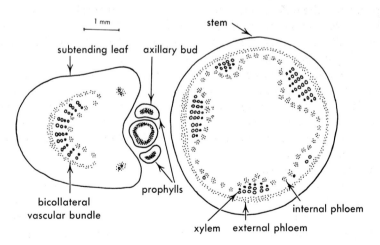

Fig. 16.6. Cross section of a shoot of potato showing spatial relation between main stem (right), axillary bud, and leaf subtending the bud.

ing of vascular tissue and the associated ground tissue. In fact, it is difficult to apply the original concept of the stele meaningfully to the stem of seed plants because of the usual absence of a clear morphological delimitation between the cortex and the stele. A delimitation of the stele by an endodermis and a pericycle was considered an important evidence of the reality of the stele. As was stressed previously, the stems of seed plants usually have no morphologically differentiated endodermis and no pericycle in the sense of a separate region between the cortex and the phloem. Thus the concept of the stele needs a complete discussion giving a rational explanation of the differences between the lower plants with their clearly circumscribed central cylinders and the seed plants with their more diffuse structure. In the absence of such a discussion, the structure of the stem of a seed plant may be made more comprehensible without reference to the stele. It may be said to be composed in its primary state of a system of vascular tissues, on the one hand, and of systems of nonvascular tissues, ground tissues and epidermis, on the other. If it is necessary to refer to the vascular region as a three-dimensional unit distinguished from the cortex and pith, the term vascular cylinder may be employed (cf. Esau, 1953, p. 360).

The classification of steles into types is based mainly on the relative distribution of the vascular and nonvascular tissues as seen in the primary state of development of the axis. In the simplest type of stele—which is also considered to be the most primitive phylogenetically—the vascular tissue forms a solid column. This is the protostele. In a protostele the phloem may surround the xylem in a relatively uniform layer, or the two vascular tissues may intermingle in the form of strands or plates. Protosteles are most common in plants below Pteropsida, but they occur also in the lower parts of ferns and in the stems of some water plants in the angiosperms. The vascular cylinder of roots of seed plants is classified as a protostele.

The second form of stele is the siphonostele or tubular stele in which the tubular refers to the arrangement of the vascular tissue around a nonvascular core, the pith (fig. 16.3). The siphonostele and its variations are most characteristic of Pteropsida. Some of these variations are the ectophloic siphonostele (fig. 16.5), with the phloem tissue appearing only on the outside of the xylem, and the amphiphloic siphonostele (fig. 16.6), with the phloem present on both the outer and the inner sides of the xylem cylinder. In its simplest form the siphonostele has no leaf gaps. In some ferns the leaf gaps are relatively short vertically, and since interfascicular regions are absent the vascular cylinder appears as a continuous ring in an internode. In other ferns the leaf gaps are vertically elongate and overlap in the internodes so that the vascular cylinder appears dissected into strands, each with the phloem surrounding the xylem (amphicribral concentric vascular bundles). Such a modification is called a dictyostele.

The stele of the gymnosperms, dicotyledons, and some monocotyledons, which is interpreted as being dissected not only by leaf gaps but also by other interfascicular regions, is called the eustele (figs. 16.4 and 16.5). When the vascular system consists of a widely dispersed network of bundles, as in some monocotyledons, the stele is called an atactostele (fig. 17.8).

As was mentioned earlier, the stelar types refer to the primary structure of the axis. In the gymnosperms and dicotyledons, secondary growth occurs in the stele, and the deposition of the secondary xylem usually obscures the original form of the stele. The interfascicular regions and later also the gaps are "closed" by the layer of xylem. The identification of the type of stele at this stage of growth requires study of the pattern formed by the primary xylem on the periphery of the pith.

DEVELOPMENT

Apical meristem

In chapter 2 the origin of the apical meristem of the shoot was traced to the early stages of embryo development. This meristem becomes identifiable as such after the appearance of the cotyledons—or single cotyledon in the monocotyledons—although some consider that the cotyledons themselves arise from the apical meristem.

Research on apical meristems of shoots is concerned with such problems as architecture of the meristem, that is, form, size, and arrangement of cells; cytologic characteristics of the meristematic cells; mitotic activity, its distribution, frequency, planes of division and, accordingly, direction of growth; and relation between the organization of the meristem and its mitotic activity to the origin of tissues and organs. The basic aim of studies on apical meristems, however, is to increase our understanding of plant growth and differentiation.

The apical meristem of the shoot is more complex than that of the root because its activity involves the formation of leaf primordia, and even the origin of the lateral branches is commonly traced to the apical meristem. Thus, the discussion of the structure and activity of the apical meristem of the shoot must include reference to the origin of the lateral organs. Another difference between the apical meristems of shoot and root is the absence on the shoot of any structure comparable to the rootcap.

The apical meristem of the shoot, like that of the root, merges with the differentiating tissue. If one agrees that the course of development of cells and tissues becomes more and more determined as they pass from the meristematic to the mature state, the most distal part of the apical meristem may be regarded as the least determined and may be referred to as promeristem. Beneath the promeristem, a gradual determination of tissue regions occurs (fig. 16.7). The peripheral region, from which originate the foliar primordia, the epidermis, the cortex, and the vascular tissue, becomes distinguishable from the future pith. In the peripheral region, or peripheral meristem, the cells remain meristematic in appearance — relatively unvacuolated and small — longer than in pith region. Within the peripheral meristem the vascular tissue is initiated in the form of procambium, the cells of which become rather narrow and elongated because of the predominance of longitudinal divisions. They thus become distinct from the less elongated and wider cells of the ground meristem, the precursor of the ground tissues (chapter 3). In the meantime, the meristematic epider-

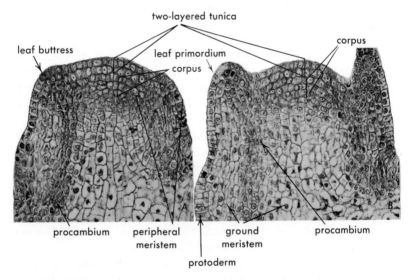

Fig. 16.7. Shoot apices of potato showing tunica-corpus organization of apical meristem and two stages in the initiation of a leaf primordium; leaf buttress stage in *A,* and beginning of upward growth in *B.* Longitudinal sections. (Both, ×170. From Sussex, *Phytomorphology,* 1955.)

mis, or protoderm, also assumes more and more of the specific characteristics of the mature epidermis. Thus, increasingly farther from the promeristem the three tissue systems, the epidermal, the vascular, and the fundamental, become differentiated, first in the form of their meristematic precursors, the protoderm, the procambium, and the ground meristem, later as mature tissues.

According to a common concept, an apical meristem of a vegetative shoot consists (*a*) of certain cells, the initials, that are the source of all body cells and (*b*) of the derivatives of these initials that are still actively dividing but by differentiation in size of cells, degree of vacuolation, and rate and orientation of cell divisions foreshadow the future body organization. Thus, the apical meristem is regarded not as an unorganized mass of embryonic cells but as a structure with distinguishable areas and a certain degree of cytologic differentiation related to the organization of the mature shoot. The promeristem includes the initials and their most recent derivatives that together may be regarded as the least determined part of the apical meristem. In the following, variations in the organization of the apical meristem of the shoot seen in different groups of plants are reviewed.

Apical meristems with apical cells. Structurally the simplest apical meristems are those that have a single large initial cell at the apex of the promeristem. It varies in shape but is often pyramidal (fig. 16.8, *A, B*). Its immediate derivatives show a rather orderly arrangement indicating a regular alternation of divisions along the faces of the initial—three in a pyramidal cell. Promeristems with single initials, the apical cells, occur among the lower vascular plants. A closely related organizational pattern found in the same group of plants is characterized by more than one initial at the apex of the promeristem. The solitary apical cells are not only relatively large but they are also conspicuously vacuolated. The nearest derivatives are also strongly vacuolated, but, as they divide, smaller cells with denser protoplasts are eventually produced. This change in cytological appearance is noted mainly along the periphery of the shoot tip where leaf primordia originate; the pith meristem (if present) is more highly vacuolated.

Tunica-corpus organization. Because the first detailed studies on apical meristems were carried out on lower vascular plants, the idea developed that the presence of apical cells was a basic structural feature of apical meristems. With the extension of research to other vascular plants the concept was found to be erroneous. In the gymnosperms and angiosperms groups of apical initials occur. Moreover, in the angiosperms there is typically a stratification in the promeristem suggesting that two or more layers grow independently from one

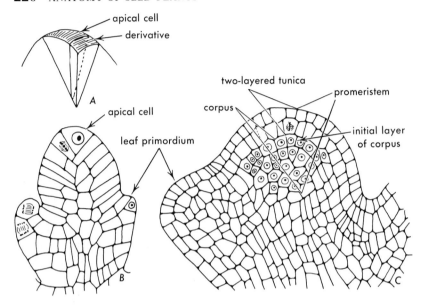

Fig. 16.8. Shoot apices in longitudinal sections. *A, B,* apical meristem with a single initial cell—apical cell—of *Equisetum* (horsetail). *C,* apical meristem with a tunica-corpus organization of *Pisum* (pea). (*A, B,* after Esau, *Plant Anatomy,* John Wiley and Sons, 1953.)

another, that, in other words, two or more superimposed tiers of initials are present. The initials are usually not distinguishable morphologically from their immediate derivatives.

The modern formulation of the idea of separate tiers of initials is embodied in the tunica-corpus concept introduced by Schmidt (cf. Esau, 1953, p. 97). It states that the initial region of the apical meristem consists of (*a*) the tunica, one or more surface layers of cells which divide in planes perpendicular to the surface of the meristem (anticlinal divisions), and (*b*) the corpus, a body of cells several layers deep in which the cells divide in various planes (figs. 16.7 and 16.8, *C*). Thus, the corpus is a core of cells that adds bulk to the apical portion of the shoot by increase in volume, and the tunica is a covering of one or more cell layers that maintains its continuity over the enlarging body by surface growth. Of course, the tunica and the corpus do not increase in extent and volume indefinitely, for, as they form new cells, the older cells become incorporated in the various regions below the promeristem.

The corpus and each layer of tunica are visualized as having their own initials. In the tunica the initials are disposed in the median position. By anticlinal divisions these cells form derivatives which, by subsequent divisions, contribute cells to the peripheral part of the shoot (fig. 16.9, *A*). The initials of the corpus appear beneath those of the tunica. By periclinal divisions (transverse with regard to the apical surface of the shoot) these initials contribute derivatives to the corpus below, the cells of which divide in various planes. Divisions near the periphery of the corpus add cells to the center of the axis, that is, to the pith meristem, and commonly also to the peripheral region (fig. 16.9, *A*).

The initials of the corpus may form an orderly layer in contrast to the rather haphazardly arranged cells in the mass of the corpus. When the corpus initials form a layer, the delimitation between the tunica and the corpus is difficult (fig. 16.7) and requires the study of many shoot tips collected in different stages of development of the plant. Then the uppermost layer of the corpus will be found undergoing periodic periclinal divisions. After such a division a second orderly layer may appear temporarily in the corpus. The corpus then may be characterized as being stratified (fig. 16.7, *B*).

The concept of the tunica-corpus organization outlined above agrees with its original formulation (cf. Jentsch, 1957). It restricts the tunica

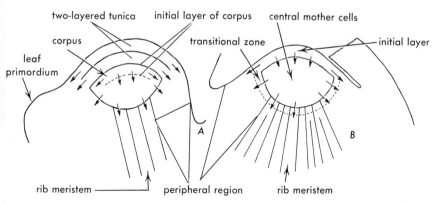

Fig. 16.9. Diagrams of shoot apices in longitudinal views. *A*, dicotyledon (*Pisum*) with tunica-corpus organization of apical meristem. *B*, gymnosperm (*Pinus*) with periclinally dividing initial layer and central mother-cell zone. Both illustrate also relation of promeristem to derivative regions below it. Arrows indicate direction in which cells are given off from layers in apical meristems.

to layers that with rare exceptions show no periclinal divisions in the median (or distal) apical position in the promeristem; layers that show such divisions in the distal position in the meristem pertain to the corpus. Some of the subsequent investigators have modified the concept to include in the tunica all parallel layers and have interpreted the tunica as showing a fluctuation in the number of layers.

The number of tunica layers varies in angiosperms (Gifford, 1954). More than half of the species studied among the dicotyledons have a two-layered tunica. The higher numbers reported, four and five, are subject to the qualification that some workers include the innermost layer or layers in the tunica, others in the corpus. One and two are common numbers of tunica layers in the monocotyledons. The specific variation in the number of tunica layers in grasses is reported to be of taxonomic significance (Brown et al., 1957).

The majority of gymnosperms do not show a tunica-corpus organization of the apical meristem (Johnson, 1951); that is, they do not have stable surface layers dividing only anticlinally. The outermost layer of the promeristem undergoes periclinal divisions and contributes to the interior tissues, and by anticlinal divisions it adds to the peripheral zone (figs. 16.9, *B*, and 16.10). The surface cells located in the median position in the promeristem are interpreted as initials. Several conifers, such as for example *Araucaria* (Griffith, 1952) and *Ephedra* (Seeliger, 1954), have a single stable surface layer and are therefore regarded as having a tunica-corpus apical pattern. The occurrence of this pattern in these gymnosperms is regarded as an evolutionary advance.

The view that the layers in the promeristem of plants with the tunica-corpus organization are relatively independent is strongly supported by observations on periclinal cytochimeras. Such chimeras are plants in which one or more layers of the plant body show a chromosome number different from that in the other layers. If such differences are present, they may be traced through continuous cell lineages to similar differences in the promeristem.

Development of cytochimeras may be induced by treatment with colchicine, and, thus, it is possible to tag experimentally one or another layer in the promeristem and relate this layer to its progeny in the mature plant body (Dermen, 1953). The plants studied by this method thus far are dicotyledons with a two-layered tunica. In such plants, studies of periclinal cytochimeras have clearly revealed the existence of three independent layers (two layers of tunica and one of corpus initials) in the promeristem. The outermost layer of tunica gave rise only to the epidermis. The derivatives of the second layer of tunica divided anticlinally and periclinally and contributed various amounts to the subepidermal part of the peripheral region. The derivatives of the third layer also divided variously and gave rise to the central part and variable amounts of peripheral tissues.

The results obtained with cytochimeras clearly support the basic premise of the tunica-corpus concept that the differentiation of the various regions of the plant is not predetermined in the organization of the promeristem despite the independence of its layers. True, the epidermis arises consistently from the outermost layer of the promeristem, but the destinies of the derivatives of the deeper layers of the promeristem are not fixed. Moreover, in apices without a tunica and corpus combination the epidermis and the subepidermal tissue are derived from the same surface layer.

The premise about the lack of predestination of tissue regions in the promeristem distinguishes the tunica-corpus concept from the classic histogen concept of Hanstein (cf. Esau, 1953, pp. 95 and 96), according to which the epidermis, the cortex, and the vascular cylinder have their own precursory meristems, the dermatogen, the periblem, and the plerome, respectively, each with its own initials in the promeristem.

Cytohistological zonation. Whereas the tunica-corpus concept helps our understanding of the structure and growth of the apical meristem, the recognition of cytohistological zonation reveals the connection between the apical meristem and the differentiation in the shoot (Kondrat'eva, 1955). The term cytohistological zonation refers to the differentiation of regions with distinctive cytological characteristics in the apical meristem. Such zonation was first described in detail for *Ginkgo* by Foster (1938) and has since been recognized in other gymnosperms (cf. Johnson, 1951) and in many angiosperms (cf. Gifford, 1954). Because of the research on zonation, emphasis did not remain centered on the least determined part of the apical meristem (the promeristem) but was extended to the derivative regions, and, unavoidably, the concept of apical meristem of the shoot became broader. Formerly it was possible to treat promeristem and apical meristem as synonyms (Esau, 1953, pp. 92, 114); now it is more appropriate, with regard to the shoot, to consider the promeristem as the most distal part of the apical meristem. (In the root, apical meristem usually has a narrower meaning than in the shoot and may be treated synonymously with promeristem; chapter 14.)

The basic features of zonation in an apical meristem were given previously when the promeristem was distinguished from the peripheral zone and the pith meristem. The description of the zonation in *Ginkgo* or *Pinus* (figs. 16.9, *B,* and 16.10, *A*) may serve to complement this information. The promeristem of these two genera includes a superficial group of initials, their lateral derivatives, and the subjacent zone of enlarged cells—the central mother cells—also derivatives of the apical initials. The lateral derivatives are produced by anticlinal divisions of the initials; the additions to the subsurface promeristem group are made by periclinal divisions. The initials and their immediate derivatives, especially the central mother cells, are conspicuously vacuolated. Moreover, the central mother cells often have relatively thick and distinctly pitted primary walls. The vacuolated appearance of the promeristem cells—of the surface and the subsurface cells—is associated with a relatively low rate of mitotic activity.

The promeristem is flanked by the peripheral region, or peripheral meristem, and beneath the central mother cells is the pith meristem. The outermost part of the peripheral meristem originates from the lateral derivatives of the apical initials; the deeper layers, from the central mother cells. The pith meristem is derived from the central mother cells. The derivation of cells from the central mother cells occurs in such a way that those mother cells that are displaced toward the periphery of the group by additions from the apical initials and by divisions within the group are considerably activated and undergo frequent mitoses. The divisions may be orderly, with the dividing cells arranged in anticlinal files with reference to the periphery of the central group. This active layer of cells is sometimes so conspicuous that it has re-

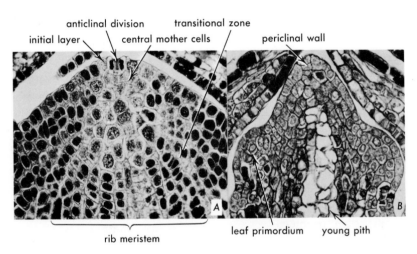

Fig. 16.10. Shoot apices of conifers in longitudinal sections. *A, Pinus strobus,* pine. *B, Cupressus macrocarpa,* Monterey cypress. In the cypress, the mother cells are not well differentiated, and the pith is vacuolated close to the apex. (Both, ×220. Slides, courtesy of A. R. Spurr, *A;* A. S. Foster, *B.*)

ceived a special designation, the transitional zone (it is often compared with cambium because of a superficial resemblance to a cambium in transverse section; Popham, 1951). The degree of discreteness of the transition zone apparently depends on the vigor of growth since it may vary in the same plant in relation to seasonal periodicity or stages in leaf formation.

The peripheral zone is densely cytoplasmic and is active mitotically. A particularly intense activity in localized positions results in the formation of protrusions, the leaf primordia (fig. 16.10, *B*). The peripheral zone is also concerned with the elongation of the shoot (anticlinal divisions) and increase in width (periclinal divisions). The pith meristem is commonly more highly vacuolated than the peripheral meristem, but it is also active mitotically. In some gymnosperms the central group of vacuolated cells may be poorly differentiated, and the pith meristem appears only a few cell layers beneath the apex of the promeristem (fig. 16.10, *B*). The divisions in the pith meristem are often regularly oriented in the transverse plane so that derivatives of individual cells soon form vertical files of cells. The series of files gives the meristem a characteristic appearance, and accordingly it has been named *rib meristem* or *file meristem* (fig. 16.10, *A*). Some vertical divisions occur also and result in an increase in the number of files.

The cytohistological zonation of different plants varies in details, and plants with rather homogeneous apical meristems may not show it at all. Some investigators, who emphasize differences in zonation and architecture of shoot tips, have established "types" of apical meristems (Popham, 1951), but these are differences of detail in an organization that is fundamentally uniform (Millington and Fisk, 1956).

Concept of quiescent promeristem. The relatively low rate of mitotic activity in the promeristem, especially as compared with the peripheral zone below it, has led to a development of a new concept of apical growth (Bersillon, 1956; Buvat, 1955; Camefort, 1956; and other French workers). The vacuolated distal part of the apical meristem is regarded as being of small significance in the vegetative growth and organogenesis of the shoot. The peripheral meristem, with its actively dividing cells and production of foliar primordia, and the pith meristem, with its contributions to the elongation of the axis, play the major role in building the shoot. The common interpretation of the cells in apical and subapical positions as initials is rejected by the French workers. As was reviewed in chapter 14, the notion of a quiescent state of cells in the distal part of the apical meristem has been expressed with reference to the roots also; in both, low tests for ribonucleic acids were taken as supporting evidence of low mitotic activity.

The views presented above have not been left uncontested. Some workers find evidence of considerable mitotic activity in the promeristem (e.g., Popham, 1958); others stress ontogenetic continuity between the promeristem and the mature tissues, as revealed by the work on cytochimeras (e.g., Clowes, 1957). Moreover, surgical and tissue culture techniques suggest that presence of the promeristem may be necessary for the restoration of an entire shoot (cf. Wetmore, 1956).

Origin of leaves

The preceding discussion has referred to the origin of leaf primordia from the peripheral meristem of the shoot tip. The histologic details of this process have been obtained for many plants in different taxonomic groups. In the gymnosperms and dicotyledons the divisions initiating leaf primordia commonly occur in the second or third layer from the surface (fig. 16.7, *A*). Periclinal divisions add cells toward the periphery and are largely responsible for the lateral protrusion of the primordium. Anticlinal divisions occur in the surface layer and also accompany the periclinal divisions in the deeper layers. In the monocotyledons, leaf primordia are frequently initiated by periclinal divisions in the surface layer.

Derivatives of either tunica or corpus may initiate divisions leading to the formation of a primordium. If the tunica is deep, the entire primordium may originate from its derivatives. Otherwise the origin of at least some leaf tissue may be traced to the corpus.

The initial lateral protrusion is commonly referred to as *leaf buttress* (fig. 16.7, *A*), although obviously it is not a part of the leaf but the entire leaf at that moment. The subsequent growth of the primordium occurs in the upward direction from the buttress (fig. 16.7, *B*).

The leaf primordia arise in positions around the circumference of the apical meristem that are in accordance with the phyllotaxy of the shoot. The causal relations of the orderly initiation of leaves have interested botanists for a long time. One prevalent view is that the new leaf is initiated in a locus that is removed from inhibitions exercised by the distal part of the apical meristem and by the most recently formed leaf primordia (cf. Wetmore, 1956). Thus, the physiological environment determines the position of a new leaf. The opposite view is that the leaves arise in contact with one another along two or more helices, each of which terminates in the peripheral meristem with its own generative center (cf. Buvat, 1955). As a primordium is initiated, the generative center moves upward and sideways in continuation of the helix.

The peripheral meristem bearing the generative centers is referred to as the initial ring. Both of these hypotheses attempt to explain the causal relations in the existing patterns, but neither has approached the explanation of how the pattern—leaf arrangement in this instance —is established in the embryo.

In descriptions of leaf origin a convenient term, *plastochron,* has been adopted to designate the time interval between the origin of two successive leaves, or whorls of leaves, at the apex. A precise technique is necessary to determine the duration of a plastochron. Available information indicates that plastochrons vary in length during different stages of development of a plant and may be a fraction of a day to several days long (e.g., Abbe and Phinney, 1951).

Origin of buds

Axillary buds. The axillary buds originate at somewhat variable plastochronic distances from the apical meristem. It is not uncommon for the first divisions initiating an axillary bud to occur in relation to the second leaf from the apex (e.g., Seeliger, 1954; Sussex, 1955). At this level the tissue in the axil of the leaf may have the same appearance as the peripheral meristem above (fig. 16.11, *A*), and, therefore, the axillary bud is considered to arise in continuity with the apical meristem (Garrison, 1955). With further growth of the parent shoot, vacuolation above the axillary meristem (fig. 16.11, *B*) makes it appear isolated from the apical meristem. To express this developmental sequence the term detached meristem is being used for the axillary meristem. Axillary buds may arise somewhat distant from the apical meristem, in partly vacuolated tissue. Because the bud eventually becomes more densely cytoplasmic than the tissue of origin, the change is often interpreted as a dedifferentiation; in other words, a more mature tissue becomes less mature.

If axillary buds arise close to the apex they may be connected by discernible bud traces (branch traces) to the main axis from the earliest stages of vacuolation in the peripheral meristem. The explanation of this phenomenon is that the bud traces do not undergo the early vacuolation characteristic of the ground meristem and thus appear as meristematic strands between the bud and the vascular tissue of the axis. If the buds arise late in the growth of the shoot, the vascular connection differentiates through a relatively strongly vacuolated ground tissue and is likely to progress from the bud toward the main axis. When an axillary bud remains dormant, secondary growth removes the bud

subtending leaves

axillary bud   axillary bud

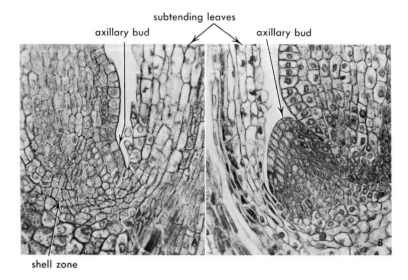

shell zone

Fig. 16.11. Origin of axillary buds in potato. Longitudinal sections of nodes showing an earlier (*A*) and a later (*B*) stage of bud development. (Both, ×225. From Sussex, *Phytomorphology*, 1955.)

farther and farther from its place of origin (MacDaniels, 1953). Its traces then elongate, probably by addition of vascular elements by the vascular cambium.

The divisions initiating a bud often occur on the axis slightly above the insertion of the subtending leaf but during subsequent growth become displaced toward the axil. Periclinal and anticlinal divisions occur as at the origin of leaves, and the bud comes to protrude above the surface. During these initial divisions the cells become arranged as they are in the apical meristem of the parent shoot; that is, the apical meristem of the bud is organized (fig. 16.11, *B*). In many plants, in early stages of bud development, orderly divisions occur along the basal and lateral limits of the bud and form a zone of parallel curving layers reminiscent of the transition zone found in some shoot apices at the inner limits of the mother-cell zone (fig. 16.10, *A*). This zone is referred to as the shell zone (fig. 16.11, *A*) because of its shell-like form. It has been suggested that the shell zone has to do with the upward growth of the bud, just as the transition zone appears to be associated with an active upward thrust of the promeristem. If the bud is not dormant, its upward growth is followed by the initiation of the first leaves, the prophylls, and in time the subsequent foliar structures.

Adventitious buds. In contrast to the axillary buds, the adventitious buds arise with no connection with the apical meristem. They originate in callus tissue of cuttings or near wounds, vascular cambium, or on the periphery of the vascular cylinder (MacDaniels, 1953), and even epidermis. They may develop on roots, stems, hypocotyls, and leaves. If the adventitious buds arise in mature tissues, they involve the phenomenon previously characterized as dedifferentiation. Depending on their place of initiation they may be exogenous (arising near the surface) or endogenous (arising in deeply seated tissues) in origin (Priestley and Swingle, 1929). They are initiated by divisions that become so oriented as to organize an apical meristem. When this meristem gives rise to the first leaves, a vascular connection is established between the bud and the parent structure, usually by differentiation from the bud toward the existing vascular system.

Primary growth of the stem

When leaves are initiated at the apex of the shoot, they appear at close levels to one another so that nodes and internodes do not exist as separate units. Later, meristematic activity between the leaf insertions elongate the parts of the stem that thus become recognizable as internodes. Transverse divisions in the peripheral region and in the pith are the basis of the earlier stage of elongation; later, cell enlargement predominates. Growth proceeds in the pith, as earlier in the pith meristem, according to a rib-meristem pattern. During the rosette stage of growth characteristic of some plants, internodal elongation does not occur and the leaves remain crowded.

The meristematic activity causing the elongation of an internode may be rather uniform throughout the growing internode, but more frequently it is more intense at the base of the arising internode than elsewhere. If internodes have a prolonged period of growth in length, the meristem at the base of the internode is rather pronounced and is called an *intercalary meristem*—intercalated between regions of more mature tissues. Some vascular elements differentiate within the intercalary meristem and connect the more mature regions of the stem above and below the meristematic region. Intercalary meristems have been studied particularly intensively in grasses and horsetails (*Equisetum*). The long retention of a meristematic region at the base of the internode (and of the leaf sheath) in grasses makes it possible for lodged plants to lift their culms from the ground by growth of the intercalary meristem on the side turned toward the ground.

The growth in thickness of the axis involves periclinal divisions and cell enlargement in both the pith and the peripheral region. The primary thickening varies in detail in different plants. It is usually moderate in species in which secondary growth is prominent. Herbaceous dicotyledons of specialized types, such as rosette types and many succulents (Rauh and Rappert, 1954), and many monocotyledons (cf. Esau, 1953, pp. 386 and 387) may have intense primary thickening growth; and it may occur so close to the apical meristem that the latter appears inserted on a flat plateau or even in a depression. In some dicotyledons the primary thickening may be emphasized in the pith (medullary thickening) or in the peripheral region (cortical thickening) or be dispersed throughout the axis. In the monocotyledons the thickening may be restricted to a relatively narrow region near the periphery, the so-called primary thickening meristem (fig. 16.12). Its cells are orderly arranged in anticlinal series.

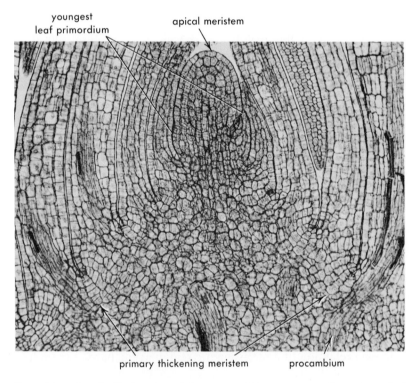

Fig. 16.12. Longitudinal section of shoot apex of *Zea mays,* corn, illustrating apical meristem, leaf primordia, and primary thickening meristem. (×90.)

The primary thickening varies in degree at different levels of the same plant. It usually increases through several internodes and then again diminishes through the higher internodes (cf. Esau, 1954). Such distribution of growth gives the lower part of the primary body of the shoot the shape of an inverted cone. However, secondary growth, when present, commonly obscures the effect of the upward increase in thickness of the primary body.

Vascular differentiation

The origin of the vascular tissue in the shoot of a seed plant is comprehensible only against the background of leaf and stem interrelationship: the vascular system of the stem differentiates in relation to the initiation of the leaves.

Origin of procambium. The sequence of differentiation of procambium is best explained by reference to the changes in the aspect of the axis from the apical meristem downward in a plant having a single cylinder of vascular bundles in the mature state (gymnosperm or dicotyledon). No vascular differentiation is detectable above the level of leaf initiation, that is, in the promeristem (figs. 16.7 and 16.13, *A*). Farther below, where leaf primordia appear, the peripheral meristem shows increased vacuolation in its outer layers. This is the beginning of cortical development (fig. 16.13, *B*). The vacuolation in the cortex and pith delimits a more or less irregular circular zone, the future vascular region (fig. 16.13, *C*). The zone is not homogeneous in structure, for beneath the insertion of leaf primordia longitudinal divisions, without a concomitant increase in width of cells, produce somewhat elongated cells, the first procambial cells (figs. 16.7 and 16.13, *C*). In transverse sections the procambial cells may be difficult to recognize at this level because they still differ too little from the adjacent meristematic cells. Later, or farther below, the distinction is intensified as the procambial cells become relatively narrower after additional longitudinal divisions occur.

The ring-like (as seen in transverse sections) highly meristematic zone containing the procambial strands has been the subject of much controversy regarding its interpretation and appropriate terminology (cf. Esau, 1954). In this book this zone is interpreted as composed of procambial strands (traces of nearest leaves) and of less differentiated meristem, the residual meristem (residuum of the peripheral meristem).

At successively later stages of development more and more procambial strands may be recognized in the vascular region (fig. 16.13, *D*).

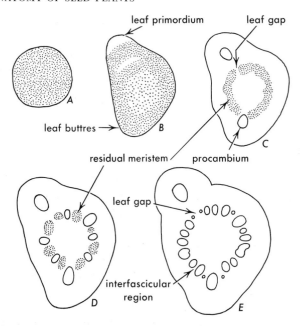

Fig. 16.13. Cross sections of flax stem taken at different levels beginning with apical meristem (*A*) and ending with node in which all primary tissues are differentiated (*E*). Illustrate stages in differentiation of vascular system: vacuolation of cortex (starts in leaf primordium in *B*) and pith, the resulting blocking out of prospective vascular region as residual meristem associated with procambial strands, and completion of procambial differentiation within the residual meristem.

These are the leaf traces—and their complexes—of newly developing leaves. Since the traces of the younger leaves are inserted among the older in a given section, evidently additional residual meristem has become procambium. In other words, with increasing age of an internode, more and more of the residual meristem differentiates into procambium. Finally, after all traces (and trace complexes) characteristic of the given level have differentiated, the remaining residual meristem differentiates into interfascicular parenchyma (fig. 16.13, *E*). At the nodes some of the residual meristem becomes leaf-gap parenchyma. The gaps are usually discernible in earlier stages of stem development, that is, at higher levels of the shoot, than are the interfascicular regions (fig. 16.13, *C*).

The sequence described above implies an initial continuity between the sites of leaf primordia and the vascular system of the axis. As the leaf primordia are elevated upon their buttresses, procambium differentiates within them also, in continuity with the procambium of the leaf traces (fig. 16.7). The differentiation of procambium in the successively younger internodes in the order of their appearance, and its continuation into the leaf primordia is usually referred to as *acropetal* differentiation of procambium. It is a continuous acropetal differentiation, for the procambium of the younger internodes is connected with that of the older internodes from its inception.

The opposite view that procambium differentiates first at the bases of leaves and then *basipetally* toward the older part of the axis of gymnosperms and dicotyledons is not well substantiated (cf. Esau, 1954). There is, however, some evidence that in the complex systems in monocotyledons at least some of the numerous vascular bundles differentiate downward in the axis.

Origin of phloem and xylem. As seen in a cross section, the first phloem appears in the outer part and the xylem in the inner part of a procambial strand (chapter 11). The subsequent differentiation of the phloem is centripetal; that is, new phloem elements appear closer to the center of the stem. The xylem differentiates in the opposite direction, that is, centrifugally. These sequences were reviewed previously with reference to the differences between root and stem structure, in particular, to the exarch xylem of the root and the endarch xylem of the shoot (chapter 3).

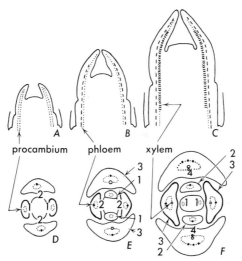

Fig. 16.14. Diagrams illustrating sequence in vascular differentiation, as seen in longitudinal (*A–C*) and cross (*D–F*) sections of a shoot with decussate leaf arrangement (*Lonicera,* honeysuckle): acropetal and continuous differentiation of procambium and phloem; discontinuous initiation and bidirectional differentiation of xylem. The first phloem elements mature before the first xylem elements.

In the vertical direction the first phloem and xylem show contrasting developmental sequences (Jacobs and Morrow, 1957, 1958; Sloover, 1958). The phloem follows the procambium more or less closely in its acropetal course of differentiation (fig. 16.14, *A, B*). The xylem, however, commonly differentiates first in the basal part of the leaf or below it somewhere in the leaf trace and then progresses acropetally in the leaf and basipetally in the axis (figs. 8.9, 16.14, 16.15) until a connection with older xylem is formed. The first phloem appears before the xylem in a given leaf trace.

Information is scanty regarding how much of the primary xylem has the discontinuous initiation outlined above, but it is not a matter of one or few series of tracheary elements (Sloover, 1958). A considerable amount of xylem may be present in the leaf itself and in the leaf trace before a connection with the older xylem in the axis is established, and there may be several leaves with discontinuous trace xylem in a given shoot. The significance of this method of development is obviously of much interest from the physiological standpoint but is at present completely obscure.

The sequence of development of xylem and phloem just described explains in part the heterogeneous appearance of vascular bundles in a cross section of a young internode (fig. 16.15, *B*). Some bundles are in procambial state; these are the youngest traces. Somewhat older ones have the first phloem; still older ones, xylem and phloem.

In lower vascular plants with small leaves and correspondingly small leaf traces the procambium of the vascular system may be recognizable above the youngest leaf primordia. It develops relatively independently of the leaves and their traces, somewhat like the vascular system of roots.

REFERENCES

Abbe, E. C., and B. O. Phinney. The growth of the shoot apex in maize: external features. *Amer. Jour. Bot.* 38:737–743. 1951.

Bailey, I. W. Nodal anatomy in retrospect. *Arnold Arboretum Jour.* 37:269–287. 1956.

Bersillon, G. Recherches sur les Papavéracées. Contribution à l'étude du développement des dicotylédones herbacées. *Ann. des Sci. Nat., Bot. Ser.* 11. 16:225–447. 1956.

Blyth, A. Origin of primary extraxylary stem fibers in dicotyledons. *Calif. Univ., Publs., Bot.* 30:145–232. 1958.

Brown, W. V., C. Heimsch, and W. H. P. Emery. The organization of the grass shoot apex and systematics. *Amer. Jour. Bot.* 44:590—595. 1957.

Buvat, R. Le méristème apical de la tige. *Ann. Biol.* 31:595–656. 1955.

Camefort, H. Étude de la structure du point végétatif et des variations phyllotaxiques chez quelques gymnospermes. *Ann. des Sci. Nat., Bot. Ser.* 11. 17:1–185. 1956.

Clowes, F. A. L. Chimeras and meristems. *Heredity* 11 (Pt. 1):141–148. 1957.

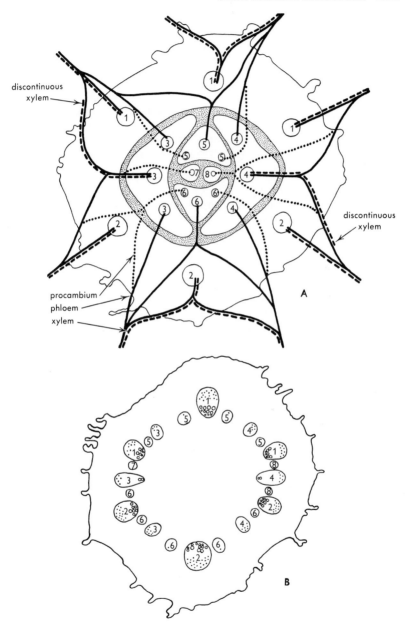

Fig. 16.15. The vascular system of sunflower (*Helianthus*) seedling. *A*, cross section of leaves above apical meristem. Illustrates: vascular bundles of leaves of various ages; interconnections of leaf traces of these leaves; and sequence of differentiation of procambium, phloem, and xylem. *B*, cross section of stem below node of leaf pair 1–2. Shows distribution of leaf traces 1–8. (After Esau, *Biol. Reviews,* 1954.)

Dermen, H. Periclinal cytochimeras and origin of tissues in stem and leaf of peach. *Amer. Jour. Bot.* 40:154–168. 1953.

Esau, K. *Plant anatomy.* New York, John Wiley and Sons. 1953.

Esau, K. Primary vascular differentiation in plants. *Biol. Revs. Cambridge Philos, Soc.* 29:46–86. 1954.

Foster, A. S. Structure and growth of the shoot apex in *Ginkgo biloba. Torrey Bot. Club Bul.* 65:531–556. 1938.

Garrison, R. Studies in the development of axillary buds. *Amer. Jour. Bot.* 42:257–266. 1955.

Gifford, E. M., Jr. The shoot apex in angiosperms. *Bot. Rev.* 20:477–529. 1954.

Griffith, M. M. The structure and growth of the shoot apex in *Araucaria. Amer. Jour. Bot.* 39:253–263. 1952.

Jacobs, W. P., and I. B. Morrow. A quantitative study of xylem development in the vegetative shoot apex of *Coleus. Amer. Jour. Bot.* 44:823–842. 1957.

Jacobs, W. P., and I. B. Morrow. Quantitative relation between stages of leaf development and differentiation of sieve tubes. *Science* 128:1084–1085. 1958.

Jentsch, R. Untersuchungen an den Sprossvegetationspunkten einiger Saxifragaceen. *Flora* 144:251–289. 1957.

Johnson. M. A. The shoot apex in gymnosperms. *Phytomorphology* 1:188–204. 1951.

Kondrat'eva, E. A. O stroenii verkhushki vegetativnogo pobega pokrytosemennykh. [Concerning structure of vegetative shoot apex in angiosperms.] *Leningrad Univ. Vest. Ser. Biol., Geog. i Geolog.* 10:3–15. 1955.

MacDaniels, L. H. Anatomical basis of so-called adventitious buds in apple. *New York Agric. Exp. Sta. Mem.* 325. 1953.

Millington, W. F., and E. L. Fisk. Shoot development in *Xanthium pennsylvanicum.* I. The vegetative plant. *Amer. Jour. Bot.* 43:655–665. 1956.

Popham, R. A. Principal types of vegetative shoot apex organization in vascular plants. *Ohio Jour. Sci.* 51:249–270. 1951.

Popham, R. A. Cytogenesis and zonation in the shoot apex of *Chrysanthemum morifolium. Amer. Jour. Bot.* 45:198–206. 1958.

Priestley, J. H., and C. F. Swingle. Vegetative propagation from the standpoint of plant anatomy. *U. S. Dept. Agr. Tech. Bul.* 151. 1929.

Rauh, W., and F. Rappert. Über das Vorkommen und die Histogenese von Scheitelgruben bei krautigen Dikotylen, mit besonderer Berücksichtigung der Ganz- und Halbrosettenpflanzen. *Planta* 43:325–360. 1954.

Seeliger, I. Studien am Sprossvegetationskegel von *Ephedra fragilis* var. *campylopoda* (C.A.Mey.) Stapf. *Flora* 141:114–162. 1954.

Sloover, J. De. Le sens longitudinal de la différenciation du procambium, du xylème et du phloème chez *Coleus, Ligustrum, Anagallis* et *Taxus. Cellule* 59:55–202. 1958.

Sussex, I. M. Morphogenesis in *Solanum tuberosum* L.: Apical structure and developmental pattern of the juvenile shoot. *Phytomorphology* 5:253–273. 1955.

Van Fleet, D. S. The cell forms, and their common substance reactions, in the parenchyma-vascular boundary. *Torrey Bot. Club Bul.* 77:340–353. 1950.

Van Tieghem, P., and H. Douliot. Sur la polystélie. *Ann. des Sci. Nat., Bot. Ser.* 7. 3:275–322. 1886.

Venning, F. D. The relationship of illumination to the differentiation of a morphologically specialized endodermis in the axis of potato, *Solanum tuberosum* L. *Phytomorphology* 4:132–139. 1954.

Wetmore, R. H. Growth and development in the shoot system of plants. In: *Cellular Mechanisms in Differentiation and Growth.* Princeton Univ. Press. 1956.

Wetmore, R. H., and C. W. Wardlaw.   Experimental morphogenesis in plants.   *Annu. Rev. Plant Physiol.* 2:269–292.   1951.

Ziegenspeck, H.   Vorkommen und Bedeutung von Endodermen und Endodermoiden bei oberirdischen Organen der Phanerogamen im Lichte der Fluoroskopie.   *Mikroskopie* 7:202–208.   1952.

17. THE STEM:
SECONDARY STATE OF GROWTH
AND STRUCTURAL TYPES

SECONDARY GROWTH

The secondary growth increases the amount of vascular tissues in stems, beginning with the part of the shoot or seedling axis that has ceased to elongate. It contributes only to the thickness of the axis. Secondary growth occurs mainly in the axis but may be observed in limited amounts in the leaves, particularly the petiole and midribs. This growth is characteristic of the gymnosperms and of woody dicotyledons but is found in variable amounts in the herbaceous dicotyledons also. Some herbaceous dicotyledons and most monocotyledons have no secondary thickening. The secondary growth that does occur in monocotyledons is of a special type.

The term secondary growth includes the formation of periderm, as well as that of the secondary vascular tissues. Chapter 12 deals in detail with the structure and development of the periderm, the protective tissue that replaces the epidermis in many trees and herbaceous plants.

Establishment and extent of vascular cambium

The vascular cambium, the structure and function of which are described in chapter 10, arises in part from the procambium within the vascular bundles and in part from the interfascicular parenchyma (fig. 17.1). The parts of the vascular cambium arising in the two positions are called *fascicular* and *interfascicular cambium*, respectively. As was stated in chapter 16, in some stems the leaf traces and their complexes are laterally almost contiguous (fig. 17.2). In such stems it is difficult to recognize interfascicular cambium, except in the leaf gaps. The origin of the cambium in the leaf gaps is similar to that in the interfascicular regions—that is, it is derived from parenchyma—so that it may be included under the designation of interfascicular.

In the most common type of secondary growth the vascular cambium becomes a complete cylinder and produces continuous cylinders of secondary vascular tissues (figs. 17.2 and 17.3), phloem and xylem, each with its axial and radial systems of cells (chapters 8–11), the axial derived from the fusiform cambial initials, the radial from the ray initials (chapter 10).

In the branches developing from axillary buds the vascular cambium appears in the same manner as in the main axis, and the cambia of the two are a continuous sheath. The stem cambium is also continuous with that of the root. The cambium appears before the end of the first year of growth of the stem or its branches, but this occurs in the part of the axis that has completed elongation. The stem and the older branches obviously develop cambium before the younger branches, so that the secondary tissues formed, for example, during the seventh year of growth of the trunk of a tree are continuous, through successively younger branches, with the first-year secondary growth in a young branch formed during the current year, and furthermore the secondary growth at the base of the young branch is continuous with the primary vascular tissues of its upper part. Thus, the latest increment of vascular tissues, partly secondary and partly primary, is continuous in any given year. The connection of the leaf traces with the youngest secondary wood can be readily demonstrated by introducing an eosin solution through a cut in a leaf (Wareing and Roberts, 1956). The continuity of the latest increment of secondary tissues makes possible an uninterrupted movement of water and materials through the most active parts of the tissues.

Effect of secondary growth on the primary body

The interpolation of secondary vascular tissues between the primary phloem and the primary xylem creates considerable stress in the interior of the stem, especially with regard to tissues located outside the cambium. The pith and the primary xylem—the latter now isolated from the primary phloem—become covered by the secondary xylem. Usually they remain in their original position and form, and their

nonfunctioning phloem

phloem   tannin cell

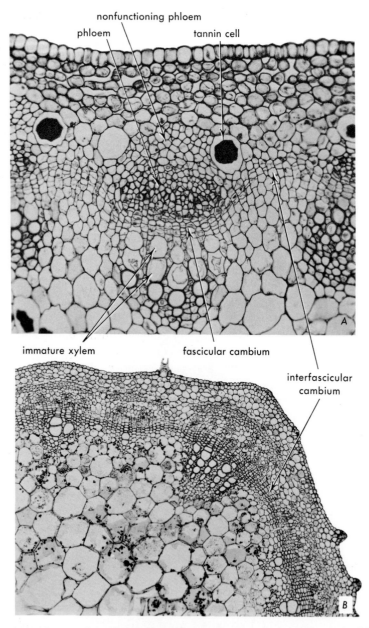

immature xylem   fascicular cambium

interfascicular
cambium

Fig. 17.1. Cross sections of stems showing an earlier (*A*) and a later (*B*) stage in differentiation of fascicular and interfascicular cambium. *A, Lotus corniculatus. B, Medicago sativa,* alfalfa. (*A,* ×184; *B,* ×60. Courtesy of J. E. Sass. *B,* from J. E. Sass, *Botanical Microtechnique,* 3rd ed. The Iowa State College Press, 1958.)

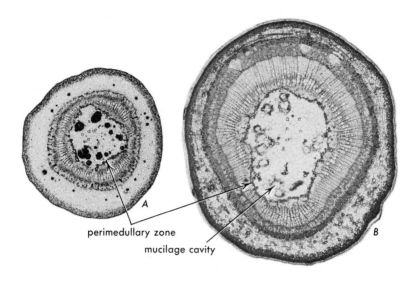

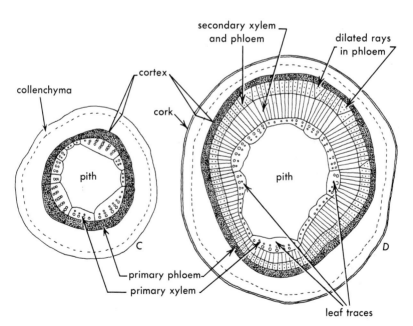

Fig. 17.2. Cross sections of *Tilia* stem made before (*A, C*) and after *(B, D)* secondary growth was initiated. (All, ×23; from *Plant Anatomy*, John Wiley and Sons, 1953.)

parenchyma may remain alive for many years. Sometimes, however, the pith is deformed by an inward pressure of the enlarging secondary body. The conducting elements of the primary xylem become non-functioning, not as a result of secondary growth, however (chapters 8 and 11).

The primary phloem is pushed outward (fig. 17.2). It becomes non-conducting and in many dicotyledons contains fibers. The thin-walled components, however, are more or less thoroughly crushed. The cortex may persist for many years. It increases in circumference through expansion in cell size in periclinal direction and divisions in the anticlinal plane. The epidermis also may persist by growing—cell enlargement followed by cell division—in accordance with the increase in circumference of the stem.

With continued accumulation of secondary tissues, the secondary phloem also is subjected to pressure from the inside because of the enlarging woody cylinder. Depending on the structure of the phloem it is more or less changed in aspect with increase in stem circumference. When fibers are abundant and occur in tangential bands (*Tilia*) the old nonconducting phloem may not be crushed. In other species with or without fibers, large masses of cells—sieve elements and associated parenchyma cells—are completely crushed. The accommodation to the increasing circumference occurs by cell division in phloem parenchyma and rays. Sometimes this growth is restricted to certain rays, and they become especially conspicuous by their increased width, "flaring" or dilatation, toward the periphery (figs. 17.2 and 17.3, *B*).

In many species periderm is formed during secondary growth. As was described in chapter 12, the periderm may be superficial and remain in the same position for many years. The cortex and phloem underneath then undergo changes described above. On the other hand, rhytidome may be formed by development of a series of successively deeper periderms. The primary tissues are gradually cut off, then also successive layers of secondary phloem. This elimination of peripheral layers periodically relieves the stress of increasing circumference.

Effect on leaf gaps and leaf traces. As was mentioned above, vascular cambium eventually appears in the gap parenchyma. The divisions initiating this meristem occur first along the margins of the gap; then they progress toward its middle. This cambium functions like that of an interfascicular region in producing xylem toward the inside and phloem toward the outside (Fig. 17.4). If the gap is wide, it may take two or more years for the cambium to form through the entire

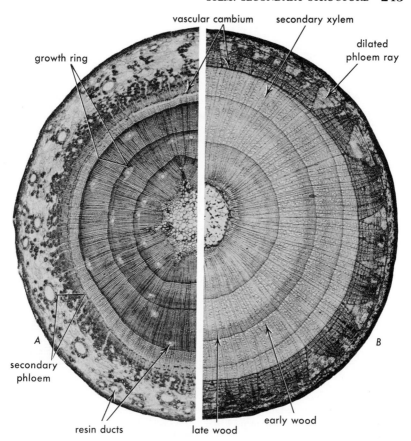

growth ring

vascular cambium   secondary xylem

dilated
phloem ray

A

B

secondary
phloem

resin ducts   late wood   early wood

Fig. 17.3. Cross sections of stems of *Pinus* (*A*) and *Tilia* (*B*), each with periderm and several increments of secondary vascular tissues. (*A,* ×21; *B,* ×12. After Esau, *Plant Anatomy*, John Wiley and Sons, 1953.)

expanse of the gap. The gap, therefore, would appear narrower and narrower in the successive increments of the xylem. When the process of xylem formation in front of the gap is completed, the gap is said to be closed.

The leaf trace confronting the gap undergoes complex changes (fig. 17.4). In deciduous species it becomes severed from the leaf bundle at the end of the first season. The lower part of the trace is within the cylinder of bundles and, before the leaf falls, develops fascicular cambium in line with that of the other vascular bundles in the stem.

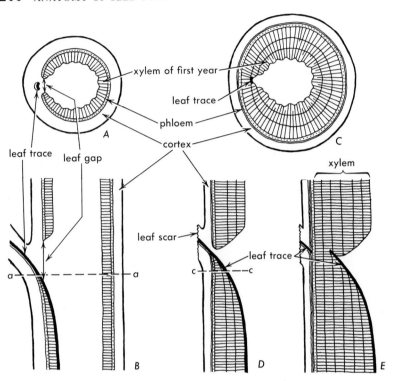

Fig. 17.4. Diagrams illustrating effect of secondary growth on leaf traces and leaf gaps in cross (*A, C*) and longitudinal (*B, D, E*) sections of stems made through the nodal region. The sequence is: gap open (*A, B*); gap somewhat restricted by secondary vascular tissues (*D*); gap closed and leaf trace ruptured (*C, E*). (From Esau, *Plant Anatomy,* John Wiley and Sons, 1953.)

The upper part, however, is directed outward and ends at the leaf scar. This stub-like end has little or no cambial activity. If this end of the trace makes a small angle with the vascular cylinder of the stem, the vascular tissues formed above it in the gap region eventually cause a rupture of the trace and the severed end is carried outward. If the end has a nearly horizontal orientation in the cortex, it becomes imbedded in the secondary tissues. In evergreen species the leaves and their traces are retained for variable lengths of time. During secondary growth the continuity of the traces of such leaves is reported to be maintained by complex adjustments between the trace and the vascular cambium (Eames and MacDaniels, 1947, p. 185).

Seasonal activity of the vascular cambium

In temperate zones the vascular cambium is inactive during winter dormancy. It is reactivated in the spring and produces a new increment of xylem and phloem during a certain period of growth, the duration and fluctuations of which are determined by seasonal and other environmental conditions and by specific characteristics of the plants. Studies on the nature of factors determining the seasonal activity of the cambium have revealed a close correlation between the initiation of cambial activity and the resumption of growth of the buds in the spring. This relation became comprehensible when it was shown that auxin, abundantly produced in the growing buds, was transmitted basipetally in the stem and there induced cambial divisions. The degree of development of the bud associated with the inception of cambial activity varies somewhat. The bud may still be closed, or just opening, or be in some later stage of obvious growth (Ladefoged, 1952). The basipetal progress of cambial activity may be underway for weeks before the cambium becomes reactivated in the root.

The cessation of cambial activity is not necessarily dependent on the state of shoot growth. In fact, there is experimental evidence that duration of cambial activity is affected by day-length conditions (Wareing and Roberts, 1956). In *Robinia pseudoacacia,* for example, the cambium is dormant under short-day conditions, and this effect is apparently photoperiodic rather than resulting from differences in photosynthesis.

During dormancy, walls of the cells in the vascular cambium may become considerably thickened. The cambial zone is narrower than during the active season, and little or no undifferentiated xylem and phloem are present next to it. Reactivation of the cambium occurs in two steps. The cells extend radially, and their radial walls become thinner. The cambium is sensitive to frost at this stage and is also easily broken; that is, the bark begins to "slip" at this stage. Cell division is the second stage, and it may occur one to several weeks later than the first stage.

The seasonal growth of the xylem (Ladefoged, 1952) is much better known than that of the phloem because of technical difficulties involved in studies of the phloem. In young branches the main part of the annual growth increment of the xylem is formed during the time when the shoot as a whole is undergoing the most intensive elongation. The duration of wood formation increases from the younger to the older branches and down to the trunk of the tree. The greater part of the annual increment is formed in late summer (Ladefoged, 1952). After the addition of cells ceases, wall thickening in the new xylem continues for several weeks longer. The xylem increment produced during one season is usually wider than that of the phloem.

Wound healing and grafting

Secondary growth and cambial activity are often involved in wound-healing phenomena and formation of the union during grafting. Injuries in leaves, young twigs, and other plant parts not having secondary growth usually result in the formation of a cicatrice—deposition of substances that appear to protect the surface from desiccation and from outside injuries—and development of a periderm from living cells underlying the cicatrice. When branches or trunks having secondary growth are injured, the formation of periderm is preceded by the development of callus, a parenchyma tissue resulting from a proliferation of various cells near the surface of the wound. Callus also provides the tissue through which cambial continuity is restored if it was severed by the wound.

The establishment of the graft unions involves phenomena similar to those associated with wound healing. Callus tissue forms from the stock and the scion and fills the space between the two and the cambia of the two become continuous by differentiation of connecting cambium from the callus cells. The matching of the cambia when the stock and scion are put together facilitates the establishment of cambial connection.

The question regarding the type of cells giving rise to callus in wound repair and grafting is often raised in the literature. The xylem rays or the newly formed derivatives of the cambium (Barker, 1954; Buck, 1954) are mentioned as the principal sources of callus.

The establishment of graft unions involves many problems, and the causes of successes or failures of grafts are incompletely known. Sometimes the cause of failure is purely technical as, for example, a poor matching of cambia of stock and scion. The beneficial effect of the removal of wood from a bud scion (Scaramuzzi, 1952) is another example of the importance of technique. In many instances, however, the exact cause of poor unions is not understood. Sometimes the phloem specifically fails to unite (McClintock, 1948), or, under certain conditions, the phloem degenerates and the graft does not survive (Stigter, 1956). Peculiar structure is sometimes regarded as an obstacle to a successful union, but apparently refinement of methods may overcome this difficulty. Thus, successful grafts have been obtained even in monocotyledons that have no cambial activity at all (Krenke, 1933; Muzik and LaRue, 1954). The largest problem, however, is not the nature of the union but incompatibility, a phenomenon that involves interaction between the stock and the scion (Roberts, 1949).

TYPES OF STEMS

Stems differ in primary and secondary structure so that it is sometimes convenient to distinguish between types of stems. It is customary to speak of woody and herbaceous stems, of vines, of monocotyledonous stems, and of stems with anomalous secondary growth.

Conifer

The stem of pine is used here as an example of a conifer stem. In the primary state the stem has discrete vascular bundles—the leaf traces and their complexes—separated from one another by relatively narrow interfascicular regions. The vascular cambium composed of fascicular and interfascicular parts forms a continuous cylinder of secondary xylem and secondary phloem (fig. 17.3, *A*). (The structure of these tissues is dealt with in chapters 9 and 11.) Opposite the gaps secondary tissues are formed gradually so that the gap parenchyma projects into the earlier part of the secondary wood. The primary xylem of the original vascular bundles may be recognized next to the pith, but the primary phloem is completely obliterated. When the crushed primary phloem is still evident, a demarkation between the phloem and the cortex may be made. Otherwise the boundary is obscure because the primary phloem forms no fibers. The cortex contains resin ducts which become considerably enlarged tangentially as the stem increases in circumference. The initial periderm arises beneath the epidermis and is not replaced by deeper periderms for many years.

Woody dicotyledon

As was mentioned in chapter 16 the stems of dicotyledons vary in discreteness of vascular bundles and interfascicular regions. In most of the arborescent dicotyledons the interfascicular regions are narrow (*Salix, Quercus, Prunus*) to very narrow (*Tilia*). In all these species the secondary tissues form a continuous cylinder.

Tilia (figs. 17.2 and 17.3, *B*) or *Liriodendron* illustrates some of the common features of the stem of a woody dicotyledon. On the inner edge of the continuous secondary xylem, the primary xylem has a slightly uneven outline around the pith and can be delimited from the secondary only approximately. The secondary xylem has a somewhat denser appearance than the primary and contains, in the axial system, vessel elements, tracheids, fibers, and xylem parenchyma cells in banded paratracheal arrangement. Wide and narrow rays are present. The secondary phloem has a characteristic appearance because of the dilatation of some of the rays and because of the alternation of bands of fibers and bands containing sieve tubes, companion cells, and parenchyma cells. The initial periderm arises beneath the epidermis and persists for many years. The cortex is retained during this time. The cortex may be easily delimited from the phloem because the latter contains fibers in its peripheral part (primary phloem fibers), as well as in deeper layers (secondary phloem fibers). The pith is parenchymatous and contains mucilage cells or cavities. The outer part of the pith long remains active as a storage tissue. This is the medullary sheath or perimedullary region.

Herbaceous dicotyledon

Many herbaceous dicotyledons have secondary growth of the ordinary type and therefore resemble young woody dicotyledons of comparable age. Stems of *Helianthus* and *Ricinus* have distinct interfascicular regions in which the interfascicular parts of the cambium arise. A continuous cylinder of secondary vascular tissues is formed. Both *Helianthus* and *Ricinus* have primary phloem fibers, and the cortex and phloem can thus be delimited. Some Compositae have an endodermis with Casparian strips in the stem (fig. 16.2).

In contrast to the two genera just mentioned, *Pelargonium* has closely approximated vascular bundles so that the interfascicular regions are difficult to distinguish (fig. 17.5, *A*). The secondary tissues form a complete cylinder. The vascular region is surrounded by several rows of primary fibers with lignified secondary walls. These fibers arise from procambium, most of them in association with sieve tubes, some between phloem strands. They may be called primary-phloem fibers. In an older stem the epidermis is replaced by periderm of subepidermal origin. The cortex and pith are parenchymatous.

In the stem of *Medicago*, alfalfa, the vascular bundles are clearly separated from one another by relatively wide interfascicular regions (fig. 17.1, *B*). Some secondary growth occurs at the base of the stem, but the interfascicular cambium produces cells mostly on the xylem

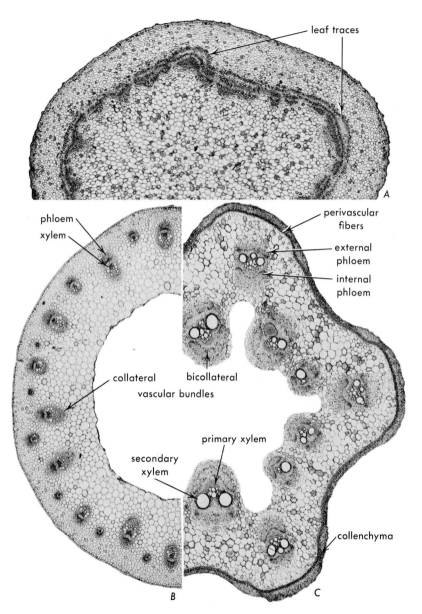

Fig. 17.5. Cross sections of herbaceous dicotyledon stems. *A, Pelargonium* (×10). *B, Ranunculus* (×19). *C, Cucurbita* (×12). (After Esau, *Plant Anatomy,* John Wiley and Sons, 1953.)

side, and these are mainly sclerenchyma cells. Stems of *Ranunculus* species may be mentioned as examples of those without secondary growth (fig. 17.5, *B*).

All the stems described thus far have collateral arrangement of xylem and phloem; that is, phloem appears only on the outer side of the xylem. In the Solanaceae, for example, tomato, potato, and tobacco, internal phloem is present. When secondary growth occurs in these stems, the cambium appears only between the outer phloem and the xylem.

Herbaceous stems may become more or less modified with reference to their principal function. An example of conspicuous modification is found in the tuber of the potato (cf. Hayward, 1938). The primary vascular system is more homogeneous in appearance than that of the aerial stem because the leaves are scale-like and their traces are small. The internodes remain short. A small amount of secondary growth occurs. The main mass of the storage tissue is derived from the peri-medullary region containing the internal phloem. The pith, including the perimedullary region, is very thick, and the internal phloem is dispersed more widely than in the aerial stem or in the rhizome.

Dicotyledon vine

A common characteristic ascribed to vine types of stems is the presence in the secondary body of rays so wide that the vascular system in both primary and secondary states appears to consist of strands. The stem of the grapevine, *Vitis* (fig. 12.6), may be considered as an example (Esau, 1948). In the primary state the vascular system consists of strands of various sizes. Originating in the fascicular and interfascicular regions, the vascular cambium becomes continuous. However, the interfascicular cambium forms parenchyma so that wide rays are formed in continuity with the interfascicular regions. New wide rays arise from time to time within the strands of the vertical system. These rays are, of course, not continuous with the interfascicular regions, but they maintain the dissected appearance of the secondary body. The primary phloem develops fibers, which clearly delimit the periphery of the vascular system. The secondary phloem also contains fibers, arranged in tangential bands. The cortex is composed of collenchyma and parenchyma, both with chloroplasts. The innermost layer of the cortex is a starch sheath. The pith is composed of parenchyma.

As was mentioned in chapter 13, *Vitis* exemplifies a stem in which the initial periderm arises not immediately beneath the epidermis but deeper (fig. 12.6). It first appears within the primary phloem beneath the primary phloem fibers and is derived from phloem parenchyma cells. The phellogen becomes continuous between the vascular bundles by differentiating also from the interfascicular parenchyma. The outer part of the phloem and the cortex are sloughed off in one continuous piece ("ring bark"). The sequent periderms arise in progressively deeper layers of the secondary phloem.

The separation of the vascular tissues into strands is even more pronounced in *Aristolochia* (fig. 17.6). This plant shows some additional peculiarities so that it is sometimes classified as having anomalous secondary growth. The widely spaced collateral vascular bundles

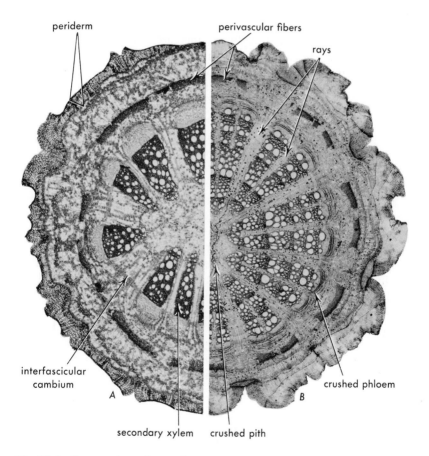

Fig. 17.6. Cross sections of stem of *Aristolochia*, a vine, in an earlier (*A*) and a later (*B*) stage of secondary growth. (*A*, ×19; *B*, ×10. After Esau, *Plant Anatomy*, John Wiley and Sons, 1953.)

surround a parenchymatous pith. In the primary state of growth the vascular system together with the ground parenchyma in which it is imbedded is surrounded by a cylinder of sclerenchyma, the perivascular fibers (cf. Esau, 1953, p. 666). The phloem, however, contains no fibers. The cortex is composed of parenchyma and collenchyma. Its innermost layer, that is, the layer located next to the perivascular fibers, is differentiated as a starch sheath in young stems. During secondary growth the individual strands are extended radially without becoming confluent with adjacent strands. Cell division occurs in the interfascicular regions in line with the cambium arising in the fascicular areas. The zone of divisions may be called interfascicular cambium although it is not sharply delimited. The interfascicular cambium produces rays composed of parenchyma similar to that of the primary interfascicular regions. As the vascular cylinder increases in circumference, the sclerenchyma cylinder is ruptured, mostly in front of the rays (fig. 17.6). Adjacent parenchyma invades the breaks (intrusive growth) and may differentiate into sclereids. The pith and the interfascicular regions are partly crushed. The periderm develops in the collenchyma beneath the epidermis. It is initiated as isolated vertical strips extending from node to node, and several years pass before it spreads over the entire surface. The cork is layered because of an alternation of radially unextended cells with cells that are larger in radial dimension. A considerable amount of phelloderm is present.

Cucurbita (fig. 17.5, *C*) resembles *Aristolochia* in general structure but has much less secondary growth. The bicollateral vascular bundles appear in two series, an outer and an inner. They are imbedded in ground parenchyma. The inner part of the pith breaks down in early stages of primary growth. The vascular system and the associated ground parenchyma are enclosed in a cylinder of perivascular fibers with lignified secondary walls and living contents. The cortex is composed of parenchyma and collenchyma. A starch sheath occurs immediately outside the perivascular fibers. Secondary growth is limited to the vascular bundles, specifically to the region between the outer phloem and the xylem. Some Cucurbitaceae have more extensive secondary growth, including addition of cells to the interfascicular regions and formation of periderm.

Dicotyledons with anomalous secondary growth

The forms of anomalous secondary growth vary considerably and intergrade with normal forms. They are called anomalous mainly because they are less familiar in the flora of temperate regions where most of the earlier anatomic investigations have been made. In some plants with anomalous growth the vascular cambium occurs in normal position, but the secondary body shows an unusual distribution of xylem and phloem. In *Leptadenia* (Asclepiadaceae), *Strychnos* (Loganiaceae), *Thunbergia* (Acanthaceae; Mullenders, 1947) phloem is formed not only toward the outside but from time to time also toward the inside. Thus strands of secondary phloem become imbedded in the secondary xylem. In the Amaranthaceae, Chenopodiaceae, Menispermaceae, and Nyctaginaceae secondary growth begins from a vascular cambium in normal position; then a series of other cambia arise successively farther outward from the first, each producing xylem toward the inside and phloem toward the outside. Thus, several alternating layers of xylem and phloem are formed more or less according to the pattern described in chapter 15 for the root of the beet. Anomalous structure of secondary xylem results sometimes from growth of parenchyma of pith and xylem (*Bauhinia*). At the end of this growth the xylem appears split into units of various sizes.

Monocotyledons

Grass stem. Like most of the monocotyledons, the Gramineae have widely spaced vascular bundles not restricted to one circle in transection. The bundles either are in two circles (fig. 17.7; *Avena, Hordeum, Secale, Triticum, Oryza*) or are scattered throughout the section (fig. 17.8; *Bambusa, Saccharum, Sorghum, Zea*). In grasses with a circular arrangement of bundles a continuous cylinder of sclerenchyma occurs close to the periphery. The outer smaller bundles are imbedded in this sclerenchyma. Fiber strands occur between these bundles and the epidermis, and strands of chlorenchyma alternate with the fiber strands. Stomata occur in the epidermis adjoining the chlorenchyma. The pith often breaks down—except near and at the nodes—in grass stems that have the bundles arranged in circles. In stems with scattered bundles no cylinder of sclerenchyma develops, but the subepidermal parenchyma may be sclerified (fig. 17.8). In both kinds of stems the vascular bundles are entirely primary and are enclosed in sheaths of sclerenchyma.

Secondary growth. Monocotyledons usually lack secondary growth but may develop thick stems (e.g., palms) by primary growth alone. Secondary growth occurs in herbaceous and woody Liliflorae (*Agave, Aloe, Cordyline, Dracaena, Sansevieria, Yucca*) and other groups (Cheadle, 1937). The cambium that occurs in these plants is continuous with the primary thickening meristèm (chapter 16) if the latter is

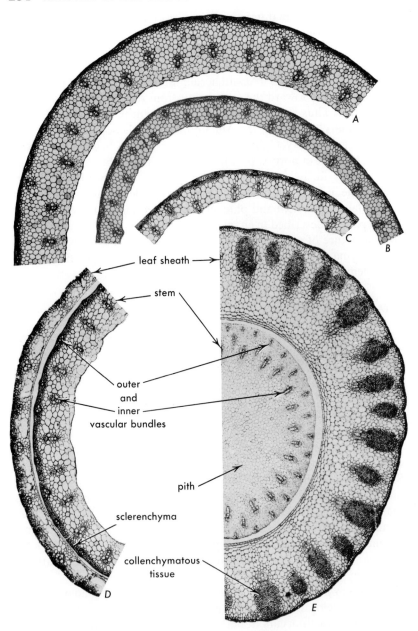

leaf sheath

stem

outer
and
inner
vascular bundles

pith

sclerenchyma

collenchymatous
tissue

Fig. 17.7. Stems of grasses in cross sections. *A, Avena,* oat (×16). *B, Hordeum,* barley (×21). *C, Secale,* rye (×23). *D, E, Triticum,* wheat (×21), both showing stem and leaf sheath as they appear in middle of internode (*D*) and near node (*E*).

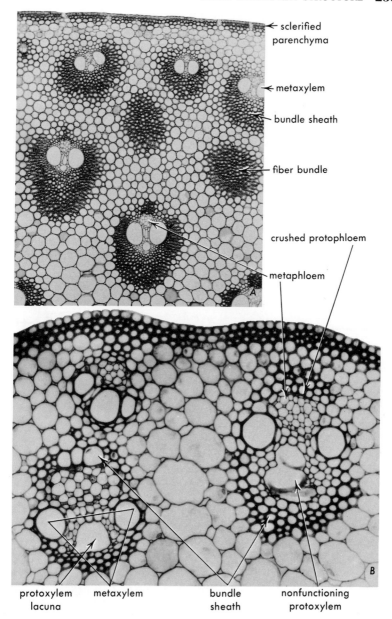

Fig. 17.8. Stem of *Zea* (corn) in cross sections. *A*, vascular bundles with thick strongly lignified sheaths of sclerenchyma ($\times 60$). *B*, bundle sheaths are not massive and their cell walls relatively thin ($\times 280$). (*A*, from J. E. Sass, *Botanical Microtechnique*, 3rd ed. The Iowa State College Press, 1958.)

discernible, but the cambium functions in the part of the stem that has completed elongation. The cambium arises in the parenchyma outside the vascular bundles and produces vascular bundles and ground parenchyma toward the inside and a small amount of parenchyma toward the outside (fig. 17.9). In the development of the vascular bundles, individual cells derived from the cambium divide longitudinally; then two or three of the resulting cells divide also by longitudinal walls. The products of the final divisions differentiate into vascular elements and associated sclerenchyma cells. Vertically, many tiers of cells combine to form a vascular bundle. These vascular bundles may be collateral or amphivasal. Their arrangement is more orderly than that of the primary bundles; they occur in more or less definite radial files.

As was mentioned in chapter 12 some monocotyledons form the type of periderm found in dicotyledons (e.g., *Aloe, Cocos, Roystonia*); others have the so-called storied cork as a protective tissue (fig. 12.8).

cambium   young bundle   phloem   xylem

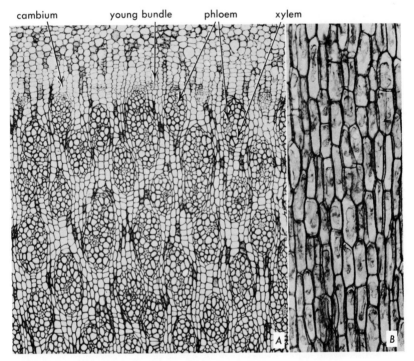

Fig. 17.9. Secondary growth in a stem of a monocotyledon, *Cordyline*. *A*, cross section of secondary vascular tissues (×50). *B*, tangential section of cambium (×90). (From Cheadle, *A*, *Amer. Jour. Bot.*, 1943; *B*, *Bot. Gaz.*, 1937.)

REFERENCES

Barker, W. G. A contribution to the concept of wound repair in woody stems. *Canad. Jour. Bot.* 32:486–490. 1954.

Buck, G. J. The histology of the bud graft union in roses. *Iowa State Coll. Jour. Sci.* 28:587–602. 1954.

Cheadle, V. I. Secondary growth by means of a thickening ring in certain monocotyledons. *Bot. Gaz.* 98:535–555. 1937.

Eames, A. J., and L. H. MacDaniels. *Introduction to plant anatomy.* 2nd ed. New York, McGraw-Hill Book Company. 1947.

Esau, K. Phloem structure in the grapevine, and its seasonal changes. *Hilgardia* 18:217–296. 1948.

Esau, K. *Plant anatomy.* New York, John Wiley and Sons. 1953.

Hayward, H. E. *The structure of economic plants.* New York, The Macmillan Company. 1938.

Krenke, N. P. *Wundkompensation, Transplantation und Chimären bei Pflanzen.* Berlin, Julius Springer. 1933.

Ladefoged, K. The periodicity of wood formation. *Danske Biol. Skr.* 7:1–98. 1952.

McClintock, J. A. A study of uncongeniality between peaches as scions and the Marianna plum as stock. *Jour. Agr. Res.* 77:253–260. 1948.

Mullenders, W. L'origine du phloème interxylémien chez *Stylidium* et *Thunbergia.* Étude anatomique. *Cellule* 51:5–48. 1947.

Muzik, T. J., and C. D. LaRue. Further studies on the grafting of monocotyledonous plants. *Amer. Jour. Bot.* 41:448–455. 1954.

Roberts, R. H. Theoretical aspects of graftage. *Bot. Rev.* 15:423–463. 1949.

Scaramuzzi, F. Le basi istogenetiche dell'innesto "ad occhio." Ricerche sul pesco. [Researches on the histogenetic process in bud-union of peach trees.] *Ann. della Sper. Agr.* 6:517–537. 1952.

Stigter, H. C. M. De. Studies on the nature of the incompatibility in a cucurbitaceous graft. *Lanbouwhogesch. Wageningen Meded.* 56:1–51. 1956.

Wareing, P. F., and D. L. Roberts. Photoperiodic control of cambial activity in *Robinia pseudoacacia* L. *New Phytol.* 55:356–366. 1956.

18. THE LEAF:
BASIC STRUCTURE AND DEVELOPMENT

EXTERNAL MORPHOLOGY

The leaf, in the wide sense of the term, is highly variable in both structure and function. The foliage leaf usually clearly shows its specialization as a photosynthetic structure in the expanded flat form of its blade, the lamina. The blade may be attached to the stem by means of a narrow stalk-like part, the petiole, or be without such structure (sessile leaf). If the base of either a sessile or a petiolate leaf encircles the stem, it is referred to as a sheathing base. Sometimes the sheathing part of the leaf is developed into a conspicuous structure, the leaf sheath. Plants with multilacunar nodes characteristically have sheathing bases. Outgrowths at the base of the leaf, the stipules, are often present in leaves associated with trilacunar nodes. In a simple leaf one blade is present; in a compound leaf two or more leaflet blades are attached to a common axis.

In the classification of leaf forms, the foliage leaf may be distinguished from the cotyledons, the first leaves on the plant, and the cataphylls. The cataphylls are represented by various protective and storage bracts or scales. They are simpler than the foliage leaves in shape and histology. The first bracts on a lateral shoot are called prophylls (chapter 16). The prophylls may be followed by foliage leaves or by a succession of other bracts intergrading with the foliage leaves.

258

HISTOLOGY OF ANGIOSPERM LEAF

The variations in structure of leaves of angiosperms and gymnosperms are reviewed in chapter 19. In the present chapter the basic features of leaves of angiosperms are considered. Like the root and the stem the leaf consists of the dermal system, the vascular system, and the ground tissue system (fig. 18.1). Since the leaf commonly has no secondary growth—sometimes in a limited amount in petioles and large veins—the epidermis persists as the dermal system. Bud scales may develop some periderm, however.

Epidermis

The description in chapter 7 of the main features of the epidermis referred chiefly to the leaf epidermis. Compact arrangement of epidermal cells and presence of cuticle and stomata are the main features of the leaf epidermis; they are related to the function of the leaf as a transpiring and a photosynthesizing organ. The stomata may occur on both sides of the leaf (fig. 18.1) or only on one side, most commonly on the abaxial, or dorsal (lower), side. In the more or less broad leaves of dicotyledons the stomata are scattered (fig. 18.1, *F*), apparently at random. In the narrow elongated leaves characteristic of the monocotyledons and the conifers, the stomata occur in rows parallel with the long axis of the leaf. As was pointed out in chapter 7, the stomata may be on the same level as the other epidermal cells, or they may be located above the surface of the epidermis (raised stomata), or below it (sunken stomata). Several stomata may appear in a depression called stomatal crypt. The position of stomata is often analyzed with regard to the ecologic adaptations of plants (chapter 19). Raised stomata are associated with a habitat providing a large supply of available water (hydrophytic), sunken stomata with a habitat characterized by a low supply of available water (xerophytic). Although this relation is not strict, experiments show that a raised position of stomata may be induced by enclosing developing leaves in vapor-saturated atmosphere (Aykin, 1952). This observation is used to explain the frequently raised position of stomata that are located in stomatal crypts: the crypts, usually protected by trichomes, probably contain a moist atmosphere. Thus, characteristics ascribed to hydrophytes (raised stomata) and xerophytes (stomatal crypts) are combined in the same leaf.

Mesophyll

The main part of the ground tissue of a leaf blade is differentiated as mesophyll characterized by an abundance of chloroplasts and a

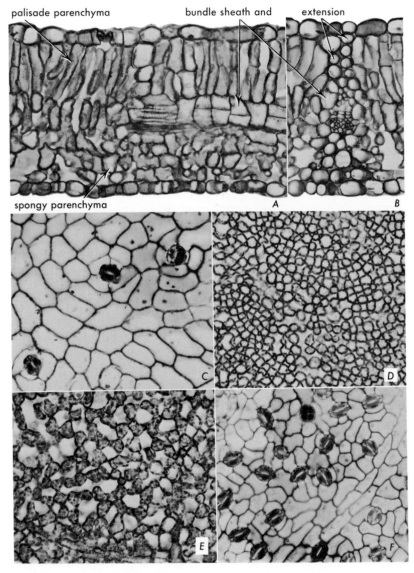

Fig. 18.1. Structure of lilac (*Syringa*) leaf. *A, B,* cross sections of mesophyll with a small vascular bundle imbedded in it. *C–F,* paradermal sections showing upper epidermis (*C*), palisade parenchyma (*D*), spongy parenchyma (*E*), and lower epidermis (*F*). (All, ×210.)

large intercellular space system. The mesophyll may be relatively homogeneous or may be differentiated into palisade parenchyma and spongy parenchyma (fig. 18.1). The palisade parenchyma consists of cells elongated perpendicularly to the surface of the blade. Although the palisade tissue appears more compact than the spongy tissue (fig. 18.1, A) a considerable part of the long sides of the palisade cells is exposed to the intercellular air (fig. 18.1, D). Leaves may have one or more rows of palisade parenchyma. In plants of the temperate regions characterized by abundant available water in the soil (mesophytic habitat) the palisade is usually located on the upper (adaxial, or ventral) side of the blade, the spongy parenchyma on the lower (abaxial, or dorsal). A leaf with such structure of mesophyll is called dorsiventral, or bifacial. If the palisade tissue occurs on both sides of the leaf, as often occurs in leaves of xerophytic habitats (fig. 19.1), the leaf is called isobilateral or isolateral.

The spongy parenchyma consists of cells of various shapes, frequently irregular, with branches (arms) extending from one cell to the other. Limitation of such connections to the ends of the branches gives the spongy parenchyma an appearance of a three-dimensional net, with the meshes enclosing the intercellular spaces (fig. 18.1, E). The spongy-parenchyma tissue has a dominantly horizontal continuity (that is, connections with other cells) parallel with the surface of the leaf. In contrast, the palisade parenchyma is continuous mainly in the direction perpendicular to the surface of the leaf.

The loose structure of the mesophyll is responsible for a large total surface area between the cells and the internal air in the leaf tissue, a surface that many times exceeds that between the epidermis and the external air; and, of the two kinds of mesophyll tissue, the palisade parenchyma has a larger internal free surface than the spongy parenchyma (cf. Esau, 1953, p. 419).

Vascular system

The principal characteristic of the vascular system of the leaf blade is the close spatial relation between the mesophyll and the vascular tissues. Vascular strands form an interconnected system in the median plane of the blade parallel with the surface of the leaf. The vascular bundles in the leaf are commonly called veins, and the pattern formed by these veins, venation. As seen with the unaided eye, the venation appears in two main patterns, the so-called reticulate, or netted, and parallel. Reticulate venation may be described as a branching pattern with successively thinner veins diverging as branches from the thicker (fig. 18.2). In the parallel-veined leaves, strands of relatively uniform size are oriented longitudinally side by side (fig. 18.3, *A, B*). The designation "parallel" is, of course, an approximation, since the bundles converge and fuse at the apex of the leaf and are not truly parallel even in the widest part of the leaf. Netted venation is most common in the dicotyledons, parallel venation in the monocotyledons.

Leaves with reticulate venation often have the largest veing along the median longitudinal axis of the leaf (fig. 18.2). This is the midvein. It is connected laterally with somewhat smaller lateral veins. Each of these veins is connected with still smaller veins, and, from them,

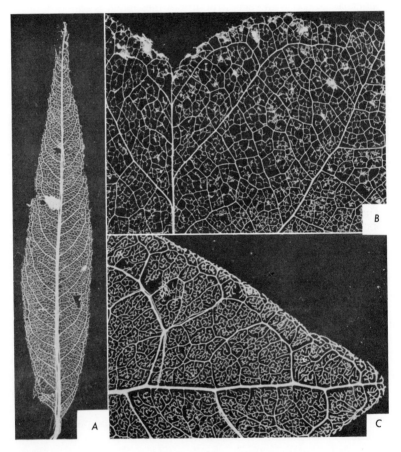

Fig. 18.2. Leaf "skeletons" showing reticulate type of venation. *A*, willow (*Salix*). *B*, tulip tree (*Liriodendron*). *C*, privet (*Ligustrum*). Natural size. (From Whittenberger and Naghski, *Amer. Jour. Bot.*, 1948.)

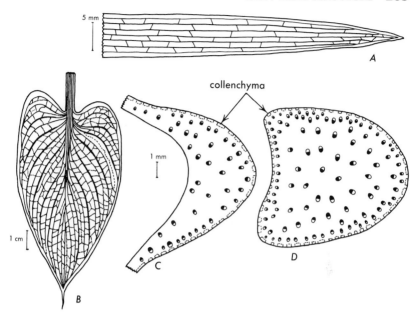

Fig. 18.3. Vascular system in monocotyledon leaves. *A,* venation in a grass leaf. *B,* venation in *Zantedeschia* leaf. *C, D,* arrangement of vascular bundles as seen in cross sections of midvein (*C*) and petiole (*D*) of *Zantedeschia* leaf. (After Esau, *Plant Anatomy,* John Wiley and Sons, 1953.)

other smaller veins diverge. The ultimate branchings form meshes delimiting small areas, or areoles, of mesophyll. Freely terminating vein endings may or may not extend into these areoles (cf. Foster, 1952; Pray, 1954). In the venation pattern designated as dendroid (Wylie, 1952) closed meshes are absent.

In the parallel-veined leaves the main veins may vary in size, with the smaller and larger veins alternating. These longitudinal veins are interconnected by considerably smaller veins, which may appear singly as simple cross connections or as a complex network (e.g., *Hosta,* Pray, 1955*b*). Thus, at the microscopic level, the parallel-veined leaves also have a reticulate arrangement of vascular bundles (fig. 18.3, *A, B*).

In petiolate leaves, the number and arrangement of the vascular bundles in the petiole and midvein vary greatly (figs. 18.3, *C, D,* and 18.4). The lateral veins usually consist of single strands and in these the vascular tissues diminish in amount from the first lateral veins toward the ultimate ramifications. In the bundle ends, xylem elements frequently extend farther than the phloem elements. In some plants

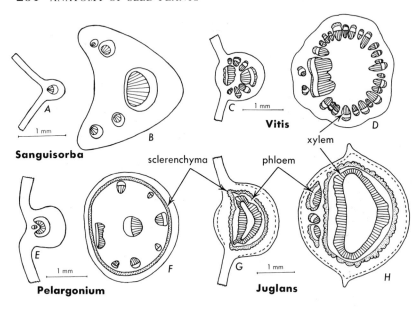

Fig. 18.4. Arrangement of vascular tissues in midveins (*A, C, E, G*) and petioles (*B, D, F,* and *H*) of dicotyledon leaves.

(e.g., *Syringa*, fig. 18.5) the phloem accompanies the xylem to the ends of veins. The xylem of vein endings usually consists of short tracheids (fig. 18.5, *A*), the phloem of short narrow sieve elements and enlarged companion cells (fig. 18.5, *B*). Depending on the plant group the vascular bundles are collateral or bicollateral, although in leaves having internal (adaxial) phloem this phloem does not extend into the smaller veins. When the bundles are collateral, the xylem occurs on the adaxial side of the leaf, the phloem on the abaxial.

In the dicotyledons the smaller veins are imbedded in the mesophyll (figs. 18.1, *B,* and 18.6, *B,* arrows) but the larger are enclosed in ground tissue that is not differentiated as mesophyll and has relatively few chloroplasts. The tissue associated with the larger veins rises above the surface of the leaf and thus forms ribs, most commonly on the dorsal side (fig. 18.6). Collenchyma is often present in the vein rib, on one or both sides of the vein beneath the epidermis.

The small vascular bundles located in the mesophyll are enclosed in one or more layers of compactly arranged cells forming the bundle sheath (fig. 18.1, *B*). The cells of a bundle sheath may resemble mesophyll cells in the development of chloroplasts, or they may have only few small chloroplasts. Bundle sheaths extend to the ends of the bundles so that no part of the vascular tissue in its course in the leaf is exposed to the intercellular air (fig. 18.5). An exception is found in hydathode regions where the xylem elements of the bundle ends are exposed to the intercellular spaces into which the water is released (chapter 13). In many dicotyledons the bundle sheaths are connected with the epidermis by panels of cells resembling the bundle-sheath cells (fig. 18.1, *B*). These panels are called bundle-sheath extensions (Wylie, 1952). The term bundle-sheath extension is applied only to the smaller vascular bundles imbedded in the mesophyll; it does not refer to the vein-rib tissue associated with the larger veins. Bundle sheaths and bundle-sheath extensions occur in the monocotyledons also. In some plants they contain sclerenchyma.

Extensive comparative studies on leaves of dicotyledons have re-

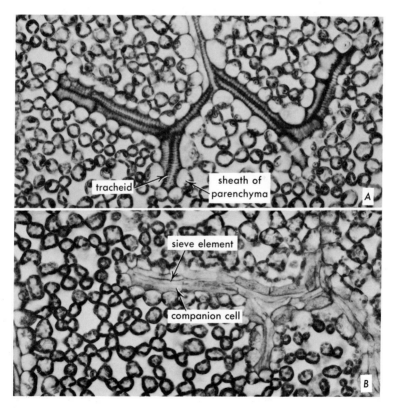

Fig. 18.5. Bundle ends in lilac (*Syringa*) leaf as seen in paradermal sections. *A*, section through terminal tracheids of xylem. *B*, section through terminal sieve elements and companion cells of phloem. (Both, ×210.)

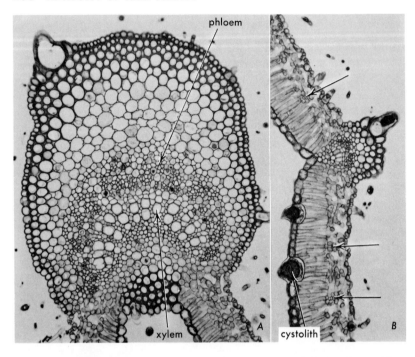

Fig. 18.6. Structure of midvein (*A*) and small vein and mesophyll (*B*) of hemp (*Cannabis*) leaf. Unlabeled arrows in *B* indicate minor veins in mesophyll. (*A*, ×130; *B*, ×150.)

vealed a significant correlation between the character of the vascular system and those structural features of the nonvascular tissues that may have an influence upon conduction (cf. Esau, 1953, pp. 429 and 430; Philpott, 1953; Wylie, 1952). Among the nonvascular tissues the epidermis and the spongy mesophyll, both tissues in which cells have extensive lateral contacts, may be assumed to be better adapted for lateral conduction than the palisade with its dominant cell connection in the abaxial-adaxial direction. In conformity with this concept, the ratio of palisade tissue to spongy tissue is closely related to vein spacing: the greater this ratio, the closer the vein spacing. Then, there is evidence that bundle-sheath extensions of the parenchymatous type conduct water toward the epidermis, where a lateral spread occurs. Accordingly, the presence of sheath extensions is negatively correlated with density of venation. When sheath extensions are absent, the distance between the nearest minor veins, that is, the diameters of the mesophyll areas delimited by the ultimate branchings of the veins, are smaller than when sheath extensions are present.

DEVELOPMENT

Initiation of leaf primordia

As was described in chapter 16, the leaf is initiated by periclinal divisions near the surface of the apical meristem below the promeristem region. Concomitantly with these divisions the surface layer or layers undergo anticlinal divisions. The localized meristematic activity produces a bulge on the side of the apical meristem—the leaf buttress. The growth of the leaf buttress occurs laterally from the point of origin so that the apical meristem is encircled to a lesser or greater extent depending on the degree to which the leaf of the given species ensheaths the stem. In the following stages the leaf grows upward from the leaf base and soon reveals the dorsiventral form characteristic of leaves.

In chapter 16 the views on the causal relations in the orderly initiation of foliar primordia, resulting in characteristic phyllotactic patterns, have been mentioned. Investigators are also inquiring into the possible causes of the development of the dorsiventral form in the emerging primordium. In surgical experiments with a juvenile shoot of potato, incisions isolating the site of a new primordium from the promeristem induced the development of a structure with a radial symmetry in place of a dorsiventral leaf (Sussex, 1955). Other kinds of incisions did not have this effect. These results suggest that the development of dorsiventrality is in some manner influenced by the apical meristem.

Apical and marginal growth

The upward growth of the primordium commonly begins with repeated divisions of cells at one locus, which becomes the apex of the primordium (figs. 18.7, *A, B,* and 18.8, *A–C*). If the repeatedly dividing cells are clearly distinguishable, they are referred to as initials (figs. 18.7, *E,* and 18.8, *A–C*). There may be an apical initial and a subapical initial. The apical initial divides anticlinally; the subapical, periclinally and anticlinally and thus produces subsurface layers as well as an inner layer (fig. 18.7, *E*). Periclinal (longitudinal) divisions in the derivatives of the subapical initial increase the thickness of the primordium and anticlinal (transverse) divisions in all layers increase its length. These divisions among the derivatives of the initials are referred to as intercalary divisions. The apical growth is of relatively short duration so that most of the increase in length occurs through intercalary growth.

In dicotyledons the primordium resulting from the initial upward

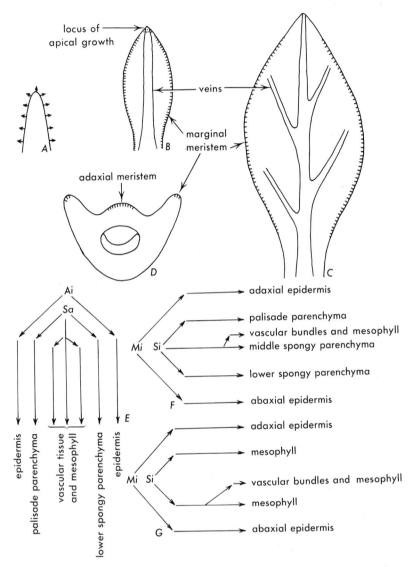

Fig. 18.7. Diagrams interpreting growth of a dicotyledon leaf. *A,* undifferentiated leaf primordium without a blade; arrows indicate direction of growth of prospective blade. *B, C,* two stages of growth of blade derived from marginal meristem; also differentiation of major veins. *D,* cross section of leaf showing position of adaxial and marginal meristems. *E,* ontogenetic relation of leaf tissues to the initiating cells at apex of leaf primordium (longitudinal view). *F, G,* two patterns of ontogenetic relation of leaf tissues to the initiating cells in marginal meristem (transverse views).

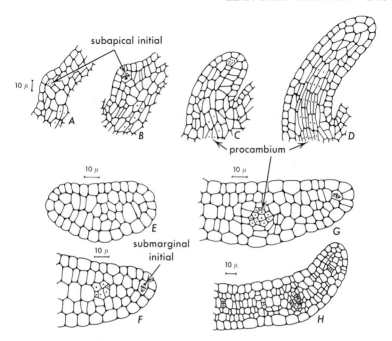

Fig. 18.8. Origin of leaf primordium and blade in flax (*Linum*). *A–D*, longitudinal sections; *E–H*, cross sections. *A, B*, emergence of leaf buttress through enlargement and division of subsurface cells; a subapical initial is defined. *C, D*, upward growth of primordium; the subapical initial is still discernible. *E*, primordium before initiation of blade. *F–H*, growth of blade. Derivatives of submarginal initials from a plate meristem in which intercalary anticlinal divisions predominate. (After Girolami, *Amer. Jour. Bot.*, 1954.)

growth is a peg-like structure having no blade (fig. 18.7, *A*). The flattened form characteristic of the blade is developed through a localized meristematic activity along two sides of the primordium beginning during the early stages of elongation of the primordium (fig. 18.7, *B, C*). The two meristematic bands are not exactly opposite each other, but both are turned inwardly (fig. 18.7, *D*), that is, toward the adaxial side of the leaf, so that the lamina develops toward the apical meristem and eventually encloses it.

The meristems producing the lamina are called marginal meristems. Each marginal meristem consists of a row of surface initials (marginal initials), a row of subsurface initials (submarginal initials), and the immediate derivatives of these initials (fig. 18.7, *F, G*). The marginal initials divide anticlinally and produce the protoderm. Leaves of some dicotyledons and monocotyledons, however, form at some later stage of ontogeny uniseriate membranous margins by periclinal divisions in the marginal initials (cf. Esau, 1953, p. 453; Schneider, 1952). The mesophyll and the vascular bundles with associated tissues originate from the derivatives of the submarginal initials.

The submarginal initials divide according to two types of sequences. One of these is based on an alternation of anticlinal and periclinal divisions (planes of divisions are here referred to the surface of the future blade). The anticlinal divisions result in the formation of adaxial and abaxial mesophyll layers, the periclinal in the origin of a middle layer (fig. 18.7, *F*). Subsequent periclinal divisions in the middle layer or all inner layers increase the thickness of the mesophyll. In the second type of sequence the submarginal initials divide only anticlinally, and the deeper inner layers are derived from the abaxial or the adaxial layer (figs. 18.7, *G*, and 18.8, *F–H*). Thus the initiation of the lamina follows two patterns, but the multiplication of mesophyll layers is variable (e.g., Girolami, 1954; Hara, 1957; Schneider, 1952). Moreover, variations may occur in different parts of the same leaf (e.g., Girolami, 1954).

The number of mesophyll layers is rather characteristic for leaves of a given species, and this number is established more or less close to the marginal meristem. After all layers are formed they divide only anticlinally. Thus the surface area of the leaf is increased, but in depth the number of cell layers does not change. A meristematic tissue consisting of parallel layers of cells dividing only anticlinally with reference to the wide surface of the tissue is called *plate meristem*. The plate-meristem divisions in the leaf constitute part of the intercalary growth by means of which the leaf attains its final size.

Plate-meristem growth is disturbed in the areas where the vascular bundles differentiate. Here, anticlinal and periclinal divisions give origin to the procambium and the tissues associated with the vascular bundles like bundle sheath or ground tissue of the vein rib (figs. 18.8, *F–H*, and 18.9, *A, B*).

The foregoing paragraphs outlined a common pattern of development in dicotyledons of a foliage leaf with an expanded lamina. The formation of a leaf buttress followed by apical and marginal growth of the primordium may be recognized, with modifications, in the cataphylls of dicotyledons and in the foliage leaves of monocotyledons and gymnosperms (cf. Esau, 1953, pp. 447–455). Limited duration of the marginal growth is responsible, for example, for the small width of the blades in many monocotyledons and most conifers. A more profound deviation is found in the so-called unifacial leaves common in monocotyledons. These leaves may be tubular (*Allium*) or flat (*Iris*) and involve changes in direction of apical growth and absence of marginal growth (Troll, 1955). In compound and variously dissected leaves of the dicotyledons the individual units originate from separate local-ized growth centers; that is, they originate as separate entities each passing through developmental stages similar to those of a single leaf. One might say, in general, that variations in histogenetic patterns result in variations in shapes of leaves; at the same time, similar structures may be produced by dissimilar methods of growth, and, contrariwise, similar methods may produce dissimilar structures by quantitative differences in growth (Roth, 1957). In addition to differ-ential growth, splitting or dissociation of tissues is reported to be in-volved in the development of palm leaves (Eames, 1953; Venkatana-rayana, 1957).

Intercalary growth

As was pointed out earlier, the derivatives of the meristems initiating the leaf primordium and its parts continue to divide and to enlarge until the final form and size of the leaf are attained. This growth shows quantitative and temporal differences in different parts of the leaf and thus brings about the development of the specific form of the leaf. In a dicotyledon leaf, for example, growth is less extensive and ceases earlier at the apex of the leaf than farther below where the leaf is wider at maturity. The earlier cessation of growth means that the apex ma-tures before the base, that, in other words, maturation progresses basipetally. The basipetal order of maturation is even more pro-nounced in the long narrow leaves of the monocotyledons. In such leaves growth is conspicuously localized at the bases of leaf blade and sheath. The basal growing zones in such leaves are referred to as intercalary meristems.

The intercalary growth of leaves of woody species continues for more than one season. According to a detailed study of overwintering buds (Artiushenko and Sokolov, 1952), a leaf is laid down as a small protuberance during the first year, enlarges within the bud during the second year, and emerges from the bud and attains the final size during the third year. Both division and enlargement of cells occur during the three growth periods, the two phenomena varying in relative degree in different species and in different tissues of the same leaf.

As was mentioned with reference to the marginal growth, the number of mesophyll layers present in the mature leaf is established more or less close to the marginal meristem, and the subsequent growth of the blade occurs through intercalary growth of the plate-meristem type, except where the vascular bundles develop. Some studies indicate a one hundred-fold increase in surface area of the leaf without an increase in number of layers in the thickness of leaf (Schneider, 1952), and leaves emerging from winter buds apparently may increase in thickness entirely by cell enlargement, although at this time divisions still occur in the anticlinal planes (Artiushenko and Sokolov, 1952).

Differentiation of mesophyll

Differential enlargement and division of cells bring about the development of the specific characteristics of the mesophyll and the establishment of differences between the palisade and the spongy parenchyma (cf. Esau, 1953, pp. 457 and 458; Artiushenko and Sokolov, 1952; Schneider, 1952). Usually the differentiation of mesophyll begins with an anticlinal elongation of the future palisade cells accompanied by anticlinal divisions (fig. 18.9, *A*). The spongy parenchyma cells also divide anticlinally, but less frequently than the palisade cells; and, commonly, they remain approximately isodiametric in form during these divisions. While the divisions are still in progress in the palisade tissue, the adjacent epidermal cells cease dividing and enlarge, particularly in the plane parallel with the surface of the leaf (paradermal plane). Thus, eventually, several palisade cells are found to be

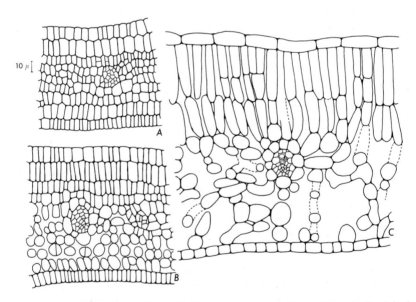

Fig. 18.9. Differentiation of mesophyll as seen in cross sections of *Pyrus* leaf. *A*, leaf still compact. *B*, spongy parenchyma with conspicuous intercellular spaces; palisade cells in characteristic orderly arrangement. *C*, mature leaf.

attached to one epidermal cell (fig. 18.9, *C*). Usually divisions continue longest in the palisade tissue. After the divisions are completed a separation of the palisade cells from one another occurs along the anticlinal walls (fig. 18.9, *C*). The partial separation of cells and formation of intercellular spaces in the spongy parenchyma precedes these phenomena in the palisade tissue (fig. 18.9, *B*). The separation of the spongy parenchyma cells is combined with localized growth of the cells resulting, in many species, in the development of branched, or armed, cells.

Development of vascular tissues

Vascular development in a leaf of a dicotyledon begins with the differentiation of procambium in the future midvein, a phenomenon that usually may be discerned while the leaf still has the peg-like form. This procambium differentiates in continuity with the leaf-trace procambium in the axis. The lateral veins of various orders originate among the derivatives of the marginal meristems. The larger lateral veins are initiated earlier and closer to the marginal meristems than the smaller ones. According to some observations, formation of new vascular bundles occurs during the entire stage of intercalary growth (Schneider, 1952); in other words, the ground tissue located between the earlier appearing veins may retain for a long time the capacity to produce new procambial strands (Foster, 1952).

Fewer cells are involved in the initiation of the smaller veins than in that of the larger. The smallest veins may be uniseriate in origin; that is, they may arise from cell series which are only one cell in diameter (Pray, 1955*a*). The differentiation of the procambium is usually a continuous process since the successively formed procambial strands arise in continuity with those formed earlier (Pray, 1955, *a*, *c*). The phloem differentiates in similar manner, but the first mature xylem appears in isolated areas, the eventual continuity resulting from subsequent differentiation of xylem in the intervening lengths of procambium.

The vertical course of differentiation of the midvein in a leaf of a dicotyledon is acropetal in the sense that it occurs first at the base of the leaf and later at the higher levels. The lateral veins of the first order develop from the midrib toward the margins (Pray, 1955*a*). In leaves with parallel veins the several veins of similar size develop acropetally (Pray, 1955*c*). The smaller veins in both dicotyledons and monocotyledons develop among the larger veins, usually first near the apex of the leaf, then successively farther down.

ABSCISSION

The separation of a leaf from a branch, without injury to the branch, is called leaf abscission. Seasonal defoliation of trees results from leaf abscission, but various injuries also may induce this phenomenon. In woody deciduous species abscission is commonly prepared, near the base of the petiole, by cytologic and chemical changes in cells along which the leaf will separate. This region is referred to as the abscission region or abscission zone (fig. 18.10). Two layers may be discernible in the abscission zone: an abscission, or separation, layer, in which structural changes facilitate the separation of the leaf, and a protective layer. The latter occurs beneath the separation layer and protects the surface that becomes exposed when the leaf falls from desiccation and invasion by parasites. The separation of the leaf along the abscission layer may be caused by three kinds of dissolution phenomena (cf. Addicott and Lynch, 1955). In the first, only the middle lamella becomes dissolved; in the second, part or all of the primary wall is dissolved in addition to the middle lamella; in the third, entire cells are dissolved. Some observations indicate that the intercellular spaces in the abscission zone become filled with liquid—perhaps a result of a loss of selective permeability on the part of the cells—and dissolution sets in subsequently (Sacher, 1957). In woody dicotyledons the separation layer is often prepared by cell division (fig. 18.10, *B*). The new cells become affected by dissolution phenomena at the time of leaf fall.

The phenomena described above occur in the ground tissue. The vascular bundles usually are broken mechanically at the end of the process of separation, although tyloses may appear in the tracheary elements before or soon after the break occurs.

The protective layer is formed through deposition of various substances in the cell walls and the intercellular spaces. Among these substances suberin and wound gum have been recognized. Lignification is also mentioned sometimes, but wound gum and lignin show certain similar staining reactions and may be confused. In woody species the protective layer is sooner or later replaced by periderm that develops beneath the protective layer in continuity with the periderm in the branch.

The abscission of leaves is not necessarily associated with dissolution phenomena. In most monocotyledons and in the herbaceous dicotyledons no dissolution occurs, and physical stresses appear to bring about the separation of the leaves. Mechanical breakage but no chemical changes have been observed in certain conifers (Facey, 1956).

Numerous factors affect abscission, growth regulating substances

abscission layer   protective layer

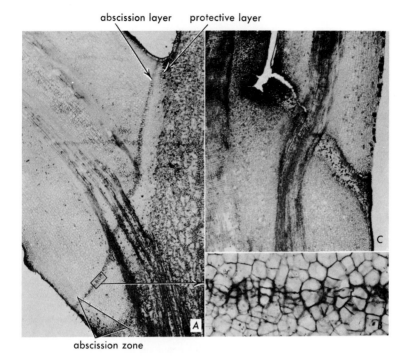

abscission zone

Fig. 18.10. Abscission zone in walnut, *Juglans* (*A, B*), and cherry *Prunus* (*C*), as seen in longitudinal sections through leaf bases. Cell division has occurred in the abscission layer of walnut (*A, B*), whereas in cherry (*C*) the cells of this layer have begun to separate from each other. (*A,* ×13; *B,* ×98; *C,* ×17. After Esau, *Plant Anatomy,* John Wiley and Sons, 1953.)

being one of the important ones (Addicott and Lynch, 1955). Auxin, for example, may retard abscission by preventing loss of differential permeability and the related flooding of the intercellular spaces with liquid (Sacher, 1957). On the other hand, auxins are also known to accelerate abscission (Addicott and Lynch, 1955). Various chemical substances may retard or accelerate abscission, and the regulation of the separation of leaves, flowers, fruits, and even bark has become a common agricultural practice.

REFERENCES

Addicott, F. T., and R. S. Lynch. Physiology of abscission. *Annu. Rev. Plant Physiol.* 6:211–238. 1955.

Artiūshenko, Z. T., and S. IĀ. Sokolov. O roste plastinki lista u nekotorykh drevesnykh porod. [Growth of leaf blade of certain tree species.] *Bot. Zhur. SSSR* 37:610–628. 1952.

Aykin, S. Hygromorphic stomata in xeromorphic plants. *Istanbul Univ. Rev. Fac. Sci. Ser. B. Sci. Nat.* 18:75–90. 1952.

Eames, A. J. Neglected morphology of the palm leaf. *Phytomorphology* 3:172–189. 1953.

Esau, K. *Plant anatomy.* New York, John Wiley and Sons. 1953.

Facey, V. Abscission of leaves in *Picea glauca* (Moench.) Voss and *Abies balsamea* L. *North Dakota Acad. Sci. Proc.* 10:38–43. 1956.

Foster, A. S. Foliar venation in angiosperms from an ontogenetic standpoint. *Amer. Jour. Bot.* 39:752–766. 1952.

Girolami, G. Leaf histogenesis in *Linum usitatissimum. Amer. Jour. Bot.* 41:264–273. 1954.

Hara, N. On the types of the marginal growth in dicotyledonous foliage leaves. *Bot. Mag. Tokyo* 70:108–114. 1957.

Philpott, J. A blade tissue study of leaves of forty-seven species of *Ficus. Bot. Gaz.* 115:15–35. 1953.

Pray, T. R. Foliar venation of angiosperms. I. Mature venation of *Liriodendron. Amer. Jour. Bot.* 41:663–670. 1954. II. Histogenesis of the venation of *Liriodendron. Amer. Jour. Bot.* 42:18–27. 1955a. III. Pattern and histology of the venation of *Hosta. Amer. Jour. Bot.* 42:611–618. 1955b. IV. Histogenesis of the venation of *Hosta. Amer. Jour. Bot.* 42:698–706. 1955c.

Roth, I. Relation between the histogenesis of the leaf and its external shape. *Bot. Gaz.* 118:237–245. 1957.

Sacher, J. A. Relationship between auxin and membrane integrity in tissue senescence and abscission. *Science* 125:1199–1200. 1957.

Schneider, R. Histogenetische Untersuchungen über den Bau der Laubblätter, insbesondere ihres Mesophylls. *Österr. Bot. Ztschr.* 99:253–285. 1952.

Sussex, I. M. Morphogenesis in *Solanum tuberosum* L.: Experimental investigation of leaf dorsiventrality and orientation in the juvenile shoot. *Phytomorphology* 5:286–300. 1955.

Troll, W. Über den morphologischen Wert der sogenannten Vorläuferspitze von Monokotylenblättern. *Beitr. zur Biol. der Pflanz.* 31:525–558. 1955.

Venkatanarayana, G. On certain aspects of the development of the leaf of *Cocos nucifera* L. *Phytomorphology* 7:297–305. 1957.

Wylie, R. B. The bundle sheath extension in leaves of dicotyledons. *Amer. Jour. Bot.* 39:645–651. 1952.

19. THE LEAF:
VARIATIONS IN STRUCTURE

LEAF STRUCTURE AND ENVIRONMENT

Evolutionary adaptations of plants to different habitats, especially with regard to the availability of water, may be associated with distinct structural features. On the basis of their water relations plants are usually classified as xerophytes, mesophytes, and hydrophytes. The xerophytes are adapted to dry habitats; mesophytes require abundant available soil water and a relatively humid atmosphere; and hydrophytes (or hygrophytes) require a large supply of moisture or grow partly or completely submerged in water. The structural features typical of plants of the various habitats or plants having such features are referred to as xeromorphic, mesomorphic, or hydromorphic, respectively. The peculiarities distinguishing plants of the various habitats are most striking in leaves.

The features commonly interpreted as characteristic of mesophytes, and hydrophytes are pointed out in the descriptions of the various examples of leaves farther in the chapter. The xeromorphic characters, however, are given much attention in the literature and are, therefore, reviewed separately in some detail.

Xeromorphy. One of the most prevalent characteristics of xeromorphic leaves is the high ratio of volume to surface; that is, the leaves are small and compact (Shields, 1950; Stålfelt, 1956). This character is associated with distinct internal structural features such as thick mesophyll, with the palisade tissue more strongly developed than the spongy parenchyma, or present alone (fig. 19.1, *B, E, F*); small inter-cellular-space volume; compact network of veins, high stomatal frequency; and sometimes small cells. As would be expected (chapter 18), the dense venation is associated with a low frequency of bundle-sheath extensions (Philpott, 1956; Wylie, 1954).

A xerophytic flora may also have a high proportion of representatives with leaves having a hypodermis (Wylie, 1954), a tissue with few or no chloroplasts (fig. 19.1, *F*). Mechanical strengthening of leaves through abundant development of sclerenchyma, common among xerophytes, is thought to reduce the injurious effect of wilting (Stålfelt, 1956), and, actually, much sclerenchyma is found in plants of habitats with continuous or periodic dryness, such as the hot deserts (Vasilevskaîa, 1954). Trichomes are abundant in many xerophytes (fig. 19.1, *E, F*), and, if the same pubescent species has mesophytic and xerophytic forms, the latter usually have a denser covering of hairs. Studies on the effect of trichomes upon loss of water have given variable results, but it is likely that sometimes the trichomes play a role in insulating the mesophyll from excessive heat (e.g., Black, 1954). Thick cell walls, especially in the epidermis (fig. 19.1, *B*), and thick cuticles are often recorded in xerophytic plants, but generally cuticle thickness is variable. Stomata may occur in cavities, stomatal crypts (*Nerium oleander*), or in grooves (Ericales; Hagerup, 1953) lined with epidermal hairs. Some xerophytes are succulent plants with their own peculiar histologic features, especially the presence of a water-storage tissue (fig. 19.1, *D*) and paucity of vascular tissue.

Xeromorphic features (and other ecotypic features) show variable degrees of constancy, but they may be well fixed genetically in a given species. On the other hand, environmental factors may induce a degree of xeromorphy in normally mesomorphic leaves or intensify the xeromorphic characters in xerophytes (Shields, 1950; Vasilevskaîa, 1954). The deficiency of moisture is only one such factor. In fact, nutrient deficiencies and cold may induce stronger expression of xeromorphy than lack of moisture (Stålfelt, 1956). Succulence, for example, is increased when nitrogen is deficient. It may also develop in shore plants exposed to a spray of sea water (fig. 19.2, *A, B;* Boyce, 1954).

Another important formative factor is light. Leaves developing in light of high intensity show a higher degree of xeromorphy than those protected from the light. This developmental reaction is the basis for the differentiation between sun and shade leaves. It has frequently been observed that leaves developing in direct sunlight are smaller but thicker and have a more strongly differentiated palisade tissue than leaves developing in the shade (e.g., Wylie, 1949; fig. 19.2, *C–E*).

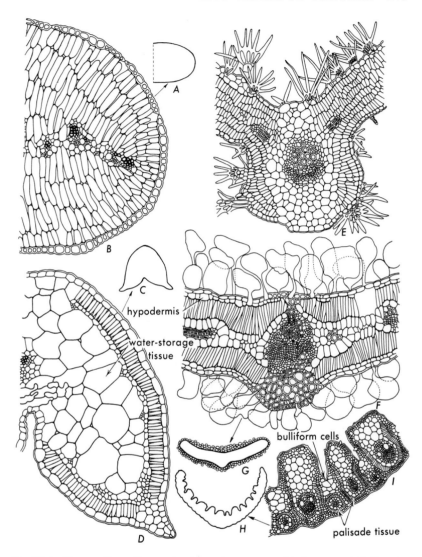

Fig. 19.1. Cross sections of leaves showing various xeromorphic features. *A, B, Greggia camporum,* low ratio of surface to volume, and the entire mesophyll is differentiated as palisade tissue. *C, D, Salsola kali,* succulent leaf with a large-celled water tissue inclosed by a single layer of palisade parenchyma. Some water cells to the left are shown in a wilted condition—response to depletion of water supply. *E, Sphaeralcea incana,* entire mesophyll differentiated as palisade tissue; also trichomes. *F, G, Atriplex canescens,* vesicular hairs and isobilateral mesophyll. *H, I, Sporobolus airoides,* partly involute leaf with grooves on the adaxial surface. Stomata (not shown) occur along the grooves. (After Shields; *A–G, Bot. Rev.,* 1950; *H, I, Phytomorphology,* 1951.)

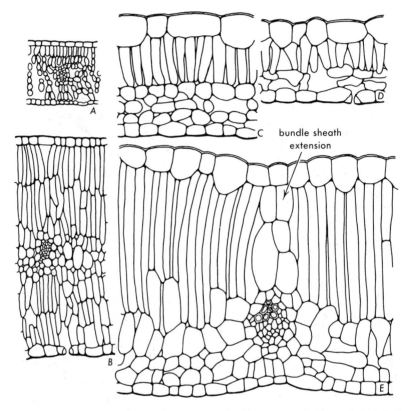

Fig. 19.2. Effect of environmental factors on leaf structure. *A, B, Baccharis halimifolia,* leaf blades in cross sections showing normal leaf (*A*) and a succulent leaf (*B*), both taken from the same plant, except that *B* was derived from the side exposed to ocean spray. (After Boyce, *Ecol. Monogr.,* 1954.) *C–E, Acer platanoides,* leaf blades in cross sections, all from one tree, showing effect of light upon structure of mesophyll: *C,* from interior of crown, moderately shaded; *D,* from deep interior, strongly shaded; *E,* from a sunny location. (After Wylie, *Proc. Iowa Acad. Sci.,* 1949.)

Leaf structure and position on plant. The foliage leaves developing at successive levels in a flowering plant show morphologic differences that may be interpreted as resulting from changes from young to adult forms of leaves. The differences concern the size as well as the shape of leaves. Shape differences depend on ratio of length to breadth, the amount of dissection in normally dissected leaves, and the number of leaflets in pinnate leaves.

The shape of a leaf is determined by various developmental phenomena. The principal of these are (*a*) the shape of the primordium at initiation; (*b*) the number, distribution, and orientation of subsequent cell divisions; (*c*) the amount and distribution of cell enlargement. The coordination between the three kinds of phenomena appears to vary in leaves inserted at different levels of the shoot. The mesophyll of the earliest leaves is less differentiated than that of the subsequent ones, a difference especially pronounced in the development of the palisade parenchyma. In later leaves the palisade tissue may undergo more anticlinal divisions, contain more cell layers, and have longer cells than in the earlier leaves (Schneider, 1952).

The structural modifications at the successively higher nodes may be interpreted as an increase in xeromorphy of leaves. The change is often regarded as resulting from a relative water shortage at the higher levels, either because the upper leaves are produced under drier microclimatic conditions than the lower leaves or because the lower leaves deprive the upper of adequate supply of water. Experimental studies exploring the interplay of developing parts in relation to microclimate indicate, however, that the structural gradients may be related to undetermined influences of immature leaves on still younger leaves and to some process of ageing in the apical meristem (Ashby, 1948; Ashby and Wangermann, 1950).

DICOTYLEDON LEAVES

Variations in mesophyll structure. Many herbaceous dicotyledons have leaves with a relatively undifferentiated mesophyll. The palisade tissue is absent or weakly developed; the intercellular volume is large; the leaf is often thin; the epidermis bears a thin cuticle; and the stomata are more or less raised. When strongly expressed such features characterize hydromorphic leaves, but they are also found, in varying degrees, in herbaceous plants growing in conditions of more moderate amounts of available moisture. Examples of leaves with relatively undifferentiated mesophyll are those of *Pisum sativum, Linum usitatissimum,* and *Lactuca sativa.* In the sugar beet the cell form in the mesophyll is associated with leaf thickness. In very thin leaves the mesophyll consists of short rounded cells; in thick leaves most of the cells are elongated.

A thin loosely organized mesophyll with a single row of palisade cells is found in *Ipomoea batatas, Pastinaca sativa* (fig. 19.3, *A*), *Raphanus sativus, Solanum tuberosum,* and *Lycopersicon esculentum.* The similarly constructed leaves of *Cannabis sativa* (fig. 18.6) and *Humulus lupulus* have cystolith-containing cells in the epidermis and numerous trichomes, glandular and nonglandular. In alfalfa (*Medicago sativa*) the palisade consists of two rows of rather short cells. The thin leaves of *Gossypium* (cotton) have long palisade cells that occupy approximately one-third to one-half of the blade thickness. The cotton leaf has lysigenous glands in the mesophyll and pit-like nectaries with club-shaped papillae on the abaxial surface of the main vein ribs.

Various shrubby and woody species furnish examples of leaves with well-differentiated palisade parenchyma on the adaxial side of the leaf, that is, typical dorsiventral mesomorphic leaves (fig. 19.3, *B; Vitis, Syringa, Ligustrum, Pyrus*), as well as leaves with various combinations of xeromorphic features. The leaves of *Citrus* are thick and leathery and have a thick cuticle with wax layers (Scott et al., 1948). The compact palisade contrasts strikingly with the loose thick spongy parenchyma (fig. 19.3, *C*). Lysigenous cavities occur in the mesophyll (fig. 13.4, *C*). *Ficus* leaves have a chlorophyll-free hypodermis derived from the epidermis (multiple epidermis). They also contain cystoliths in the epidermis (fig. 13.4, *D*) and laticifers in the mesophyll.

The isobilateral mesophyll, with palisade parenchyma on both sides of the leaf—a strongly xeromorphic character—is exemplified by *Artemisia, Atriplex* (fig. 19.1, *F*), *Chrysothamnus, Sarcobatus,* and many other genera in various families (cf. Metcalfe and Chalk, 1950, p. 1334). A modification of the isobilateral form is the centric (Metcalfe and Chalk, 1950, p. 1333) found in leaves that are very narrow or entirely cylindrical. In such leaves the abaxial and adaxial palisade cells form a continuous·layer (fig. 19.1, *B*). An isobilateral leaf may also result when the palisade and the spongy tissues are not clearly differentiated (fig. 19.3, *D*).

In *Salsola* the palisade tissue surrounds a parenchyma of large colorless cells interpreted as water storage tissue (fig. 19.1, *D*). Such tissue may occur also in median position in flat leaves (*Haplopappus spinulosus*, Shields, 1950) and also outside the mesophyll as hypodermis. In the fleshy leaves of *Peperomia* the hypodermis may be fifteen layers of cells in depth, exceeding the thickness of the mesophyll (Metcalfe and Chalk, 1950, p. 1122).

The leaves of aquatics vary in relation to their growth conditions. Some have leaves that are partially or completely submerged; others have floating leaves; and still others combinations of floating and submerged leaves. Common features of hydromorphic leaves are large air spaces (fig. 19.5, *A*), small amount of sclerenchyma, and a weak development of the vascular system (Hasman and Inanç, 1957).

Supporting tissue. In the dicotyledons the supporting tissue in leaves may be collenchyma or sclerenchyma, and the vascular bundles

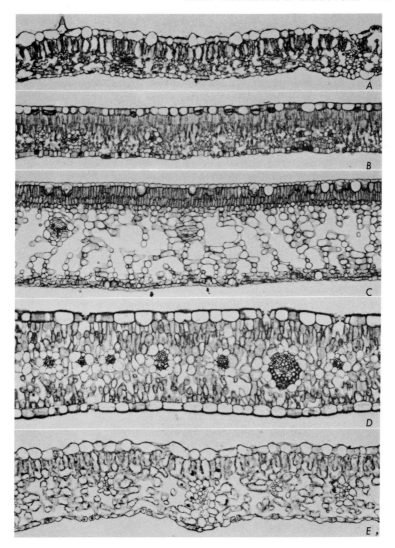

Fig. 19.3. Variations in structure of mesophyll as seen in cross sections of leaves. *A,* parsnip (*Pastinaca sativa*). *B,* peach (*Prunus Persica*). *C,* lemon (*Citrus Limon*). *D, Dianthus. E, Lilium.* (All, ×90.)

as such also contribute to the support of the blade. The collenchyma occurs along the larger veins, on one or both sides (fig. 18.6), and the nonconducting part of the xylem and phloem may be collenchymatously thickened. Sclerenchyma occurs in the form of bundle "caps," bundle sheaths, and bundle-sheath extensions composed of fibrous cells, and as sclereids in the mesophyll. Examples of vascular bundles accompanied by sclerenchyma are found in the Boraginaceae, Caryophyllaceae (fig. 19.3, *D*), Labiatae, Lauraceae, all tribes of Leguminosae, Monimiaceae, Proteaceae, some Rosaceae, and Sterculiaceae (Metcalfe and Chalk, 1950).

Petiole. The petiole of dicotyledon leaves contains the same tissues as the stem, often in similar arrangement. The epidermis has some stomata and the ground tissue contains chloroplasts. Collenchyma or sclerenchyma occur as supporting tissues. The vascular tissue shows a great variety of arrangements (fig. 18.4).

Certain plants have joint-like thickenings, the pulvini, at the base of the petiole and also at the bases of the petiolules in compound leaves. The pulvini are involved in movements of leaves which may be stimulated by environmental conditions or may be autonomous (gyrations, Datta, 1952–1953). The pulvini have a characteristic structure. The best known pulvini are those of the Leguminosae (Brauner and Brauner, 1947; Brown and Addicott, 1950; Datta, 1952–1953; Weintraub, 1952). The pulvinus is somewhat swollen, and its surface is wrinkled. The vascular system is concentrated in the center as one concentric strand even though above and below several bundles in cylindrical arrangement are present. The largest volume is occupied by a thin-walled parenchyma with intercellular spaces. Stomata are few or absent; trichomes may be present. The movement appears to be associated with changes in turgor and the concomitant changes in size and shape of the ground-parenchyma cells. Structural peculiarities in cell walls— apparent close association of the cytoplasm with the cell wall and little or no calcification of the middle lamella—are also mentioned as possible explanations of the partial inward collapse of walls during contraction of protoplasts (Weintraub, 1952).

MONOCOTYLEDON LEAVES

The leaves of the monocotyledons vary in form and structure, and some resemble those of the dicotyledons. Monocotyledon leaves may have petioles and blades (e.g., *Canna, Zantedeschia, Hosta*), but the majority are differentiated into blade and sheath, and the blade is relatively narrow. The venation is typically parallel.

The anatomic structure ranges from hydromorphic to extreme xeromorphic. Hydrophytes in the monocotyledons show the same basic features as those in the dicotyledons. An example of a dorsiventral leaf with the palisade parenchyma on the adaxial side is furnished by *Lilium* (fig. 19.3, *E*). The dorsiventral leaf of banana (*Musa sapientum*) is thick and has several layers of palisade and a wide region of spongy parenchyma with large lacunae (Skutch, 1927). The rigid leaves of *Carex* (fig. 19.4) have prominently developed sclerenchyma and, in the mesophyll, strands of large thin-walled cells.

The monocotyledons have many leaf forms that appear to be highly specialized and difficult to interpret morphologically. The *Iris* leaf, for example, has a blade flattened not parallel with the tangent of the axis but perpendicular to it. Its vascular bundles appear in one file in transverse section, but approximately half of these bundles are oriented with their xylem toward one side, the other half to the other, just as though the blade has been folded and the vascular bundles of one half have been wedged among those of the other half (fig. 19.5, *D, E*). Many species of *Allium* have tubular leaves (fig. 19.5, *B, C*). The palisade tissue appears beneath the epidermis around the entire circumference, and beneath it is the spongy parenchyma. The center of the leaf is occupied by a cavity surrounded by remnants of parenchyma cells that initially occupied the region of the cavity.

Many monocotyledon leaves develop large amounts of sclerenchyma that in some species serves as an important source of commercial hard leaf fibers (chapter 6). The fibers may be associated with the vascular bundles or may appear also as independent strands.

Grass leaves. The grass leaf is rather distinct in structure. Typically it consists of a more or less narrow blade and a sheath enclosing the stem. Auricles and a ligule commonly occur between the blade and the sheath. The vascular bundles of different size alternate rather regu-

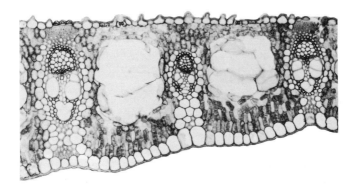

Fig. 19.4. Cross section of *Carex* leaf illustrating lacunae filled with large thin-walled cells. (×130. Slide courtesy of J. E. Sass.)

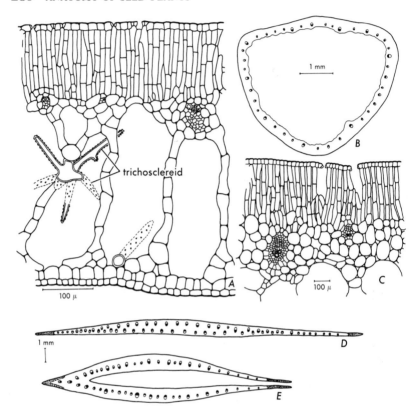

trichosclereid

Fig. 19.5. Variations in leaf structure as seen in cross sections. *A, Nymphaea,* a dicotyledon water plant. *B–E,* monocotyledons: *B, C,* hollow leaf of onion (*Allium Cepa*); *D, E, Iris* leaf cut through blade (*D*) and sheath (*E*).

larly with one another (fig. 19.6) and are interconnected by small transverse commissural strands (fig. 18.3, *A*). The median bundle may be the largest (fig. 19.6, *B*), or the median part of the blade is thickened on the adaxial side (fig. 19.6, *D*).

The mesophyll of grasses shows, as a rule, no distinct differentiation into palisade and spongy parenchyma, although sometimes the cell rows beneath both epidermal layers are more regularly arranged than the rest of the mesophyll. In the panicoid grasses the mesophyll cells often surround the vascular bundles in an orderly manner, each cell oriented with its longer diameter at right angles to the bundle so that in transverse sections the mesophyll cells appear to radiate from the bundles (fig. 19.6, *C*). Some festucoid grasses show similar structure of the mesophyll (fig. 19.1, *I*).

The epidermis of grasses contains a variety of cells.   The ground tissue consists of narrow elongated cells, often with strongly undulate anticlinal walls.   Enlarged bulliform cells (figs. 19.1, *I*, and 19.6, *B*, *C*) form bands of various widths and numbers.   These cells are often described as motor cells involved in involution and folding of leaves, but

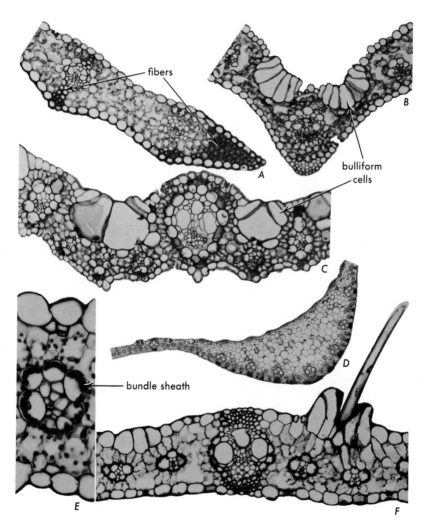

Fig. 19.6.   Leaves of grasses in cross sections.   *A*, margin of *Bromus* blade.   *B*, midvein of *Poa* blade.   *C*, blade of *Saccharum* (sugar cane).   *D*, midvein of *Zea* (corn) leaf. *E*, *F*, blade of *Zea*.   (*A–C*, ×150; *D*, ×14; *E*, ×320; *F*, ×98.   *A–C*, slides courtesy of J. E. Sass; *D*, *F*, from J. E. Sass, *Botanical Microtechnique*, 3rd ed.   The Iowa State College Press, 1958.)

experimental studies indicate that the shrinkage of other tissues is just as much concerned with these phenomena (Shields, 1951). Stomata are narrow cells associated with subsidiary cells, and silica, cork cells, and trichomes (fig. 19.6, *F*) may be present in certain parts of the leaf (chapter 7).

The bundle sheaths of grasses are characteristic structures used in grass systematics (Brown, 1958). Certain groups of grasses (festucoid) have double sheaths composed of an inner sheath with thickened walls and an outer thin-walled sheath (fig. 19.7). In contrast panicoid grasses have only one thin-walled sheath (fig. 19.6, *E*). In this respect the Bambusaceae are an exception among the panicoid grasses, for they have double sheaths, and certain other festucoid characters, combined with panicoid characters (Metcalfe, 1956). The bamboos, therefore, are thought to be rather primitive. The evolutionary advance appears to have involved a separation of festucoid and panicoid characters that are still combined in the bamboos.

Grass leaves have strongly developed sclerenchyma. Commonly fibers appear in longitudinal plates extending from the larger vascular bundles to the epidermis (fig. 19.7, *A*). The large vascular bundles may be enclosed in fibers as well and be associated with plates of fibers on both sides. Smaller bundles may be connected with only one fiber plate. In some species fibers occur in subepidermal strands or plates having no contact with the vascular bundles (fig. 19.7, *B*). The leaf margins have fibers (fig. 19.6, *A*), and so does the epidermis in some species.

GYMNOSPERM LEAVES

The gymnosperm leaves are less variable in structure than those of the angiosperms and are highly independent of environmental conditions. Most gymnosperms are evergreen. The well-known exceptions are *Ginkgo, Larix,* and *Taxodium.* The leaves of the conifers, which are represented by the largest number of species among the gymnosperms, have been studied most frequently, especially those of the pine. The pine leaf, therefore, is described first, and those of other conifers and other gymnosperms are treated comparatively.

Pine needles originate on dwarf branches (short shoots), singly or most commonly in groups of two to several. Depending on this number the transectional form varies (figs. 19.8, *A,* and 19.9, *A*) from approximately oval to triangular. The needle has a thick-walled epidermis with a heavy cuticle and deeply sunken stomata with overarch-

metaxylem   protoxylem   outer bundle sheath

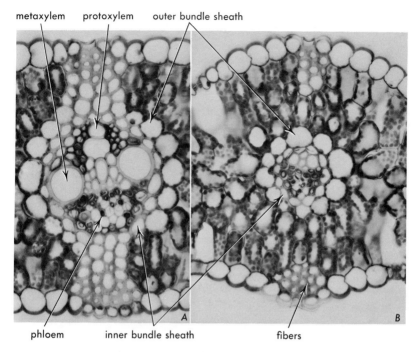

phloem   inner bundle sheath   fibers

Fig. 19.7. Cross sections of wheat (*Triticum*) leaf showing a larger (*A*) and a smaller (*B*) vascular bundle, each with two bundle sheaths, the inner and the outer. (Both, ×320.)

ing subsidiary cells (fig. 7.6). The stomata occur on all sides and are in vertical rows. A sclerified fibrous hypodermis occurs beneath the epidermis, except under the rows of stomata. The mesophyll consists of parenchyma cells with folds (plicate mesophyll) that protrude like vertical ridges into the lumen of the cell. The mesophyll is not differentiated into palisade and spongy parenchyma. Resin ducts occur in the mesophyll.

The vascular tissue usually forms one bundle or two bundles side by side and occurs in a central position in the needle. The xylem is on the adaxial, the phloem on the abaxial side. The xylem consists of protoxylem and metaxylem. The latter shows orderly radial seriation of cells, with rows of xylem parenchyma cells alternating with those of the tracheids. Possibly a small amount of secondary growth occurs after the termination of the primary extension growth of the needle, but the bulk of the xylem is metaxylem.

The vascular bundle is surrounded by a peculiar tissue known as transfusion tissue. This tissue is composed of nonliving tracheids and

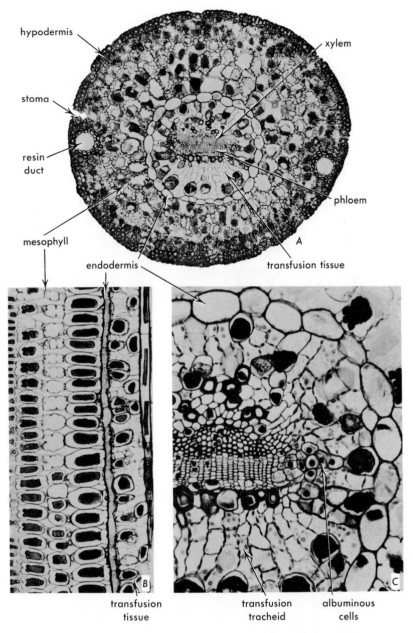

hypodermis

stoma

resin
duct

mesophyll

endodermis

xylem

phloem

A

transfusion tissue

transfusion
tissue

B

transfusion
tracheid

albuminous
cells

C

Fig. 19.8. Structure of pine (*Pinus monophylla*) leaf. *A,* cross section of entire needle (×50). *B,* longitudinal section through mesophyll and transfusion tissue (×90). *C,* cross section of part of vascular bundle, transfusion tissue, and endodermis (×150).

living parenchyma cells (fig. 19.8, *B, C*). The tracheids that occur next to the vascular bundle are elongated; those farther away have the same shape as the parenchyma cells. The walls of the tracheids, though provided with secondary thickenings, are relatively thin and lightly lignified and bear bordered pits. The tracheids usually appear somewhat deformed, probably because of the pressure of the associated living parenchyma cells. Some densely cytoplasmic cells interpreted as albuminous cells occur next to the phloem (fig. 19.8, *C*).

The vascular bundle and the associated transfusion tissue are surrounded by a thick-walled sheath of cells, the endodermis. No intercellular spaces are present in the endodermis and the tissues enclosed by it. The endodermis is often described as having Casparian strips in early stages of development and a suberin lamella later (e.g., Lederer, 1955) but there is no complete agreement on this point. In the mature state the endodermis has secondary lignified walls with, probably, the suberin confined to the anticlinal walls.

The structural features just described are found in many other conifers, usually with quantitative differences, but are absent in some (Lederer, 1955). The leaves of the conifers may be single veined and scale-like (Taxodiaceae, Cupressaceae, Podocarpaceae), or single veined and needle-like (*Abies, Larix, Picea, Pinus*), or broadly ovate, with many veins (Araucariaceae). Still other forms may be found.

The sclerified hypodermis is up to five layers in thickness in *Araucaria* but is entirely absent in some conifers (e.g., *Taxus*, fig. 19.9, *C; Torreya*). Most conifers do not have a plicate mesophyll. Palisade and spongy parenchyma are differentiated in such genera as *Abies* (fig. 19.9, *D*), *Cunninghamia, Dacrydium, Sequoia, Taxus* (fig. 19.9, *C*), *Torreya*, and in *Auracaria* and *Podocarpus* the palisade occurs on both sides of the leaf.

The boundary between the vascular region and the mesophyll is not equally clear in conifer leaves (Lederer, 1955). The Pinaceae (e.g., *Abies*, fig. 19.9, *D; Larix*, fig. 19.9, *B; Picea, Pinus*, fig. 19.9, *A*) have a distinctly differentiated endodermis. In species of *Taxus* (fig. 19.9, *C*), in *Sequoia sempervirens, Metasequoia glyptostroboides, Juniperus communis,* and *Araucaria excelsa* only a parenchymatous sheath is present.

The transfusion tissue is characteristic of the conifers, but it varies in amount and arrangement (Gathy, 1954; Lederer, 1955). It occurs right and left from the vascular bundle in such genera as *Cunninghamia, Cupressus, Juniperus, Thuja, Torreya, Sequoia,* and *Taxus* (fig. 19.9, *C*); as an arc about the xylem in *Araucaria, Dammara, Sciadopitys;* right and left from vascular bundle but most abundantly next to the phloem in *Larix* and in species of *Abies* and *Cedrus;* and completely surround-

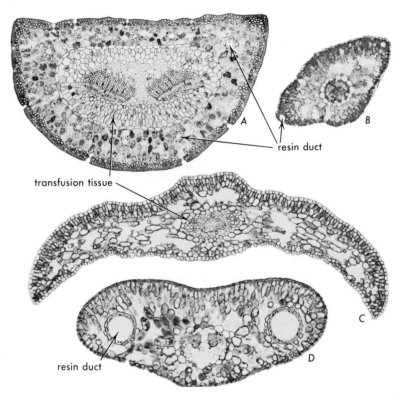

transfusion tissue

resin duct

resin duct

Fig. 19.9. Cross sections of various conifer leaves. *A*, pine (*Pinus nigra*). *B*, larch (*Larix*). *C*, ground-hemlock (*Taxus canadensis*). *D*, fir (*Abies*). (*A, C*, ×44; *B, D*, ×54.)

ing the vascular bundle in *Pinus* (fig. 19.9, *A*). The transfusion tracheids may have reticulate or pitted secondary walls. Evolution, ontogeny, and functions of the transfusion tissue are subjects of controversy in the literature. The tissue is often considered to be concerned with the translocation of water and food materials between the vascular bundle and the mesophyll.

In addition to the transfusion tissue associated with the vascular bundle a so-called accessory transfusion tissue has been described in *Podocarpus* (Griffith, 1957) and *Dacrydium* (Lee, 1952). This tissue consists of thick-walled elongated cells extending outward from the bundle sheath into the mesophyll. It is not in contact with the transfusion tissue next to the vascular bundle.

The arrangement and number of resin ducts are variable even in the same species. Among the single-veined genera only *Taxus* (fig. 19.9,

C) lacks resin ducts. As a basic pattern, the Cupressineae, the Taxineae (other than *Taxus*), *Sequoia, Podocarpus,* and most species of *Tsuga* have one resin duct located between the vein and the lower epidermis; the Abietineae, except *Tsuga,* have two resin ducts, right and left from the bundle (fig. 19.9, *D*); the many-veined genera (e.g., *Araucaria*) have one duct between each two bundles. In addition to the basic ducts, variable numbers of accessory ones may be present. In *Pinus* two lateral resin ducts occur almost invariably (figs. 19.8, *A,* and 19.9, *A*). The others vary in number and distribution.

To illustrate leaves of gymnosperms other than conifers those of *Cycas* and *Ginkgo* are described. The large leaves of the Cycadales are compound, and their broad single-veined pinnae are stiff and rigid. The leaf of *Cycas revoluta* (fig. 19.10, *A*) has a thick cuticle, a thick-walled epidermis, and sunken stomata on the abaxial side. The meso-phyll is composed of palisade and spongy parenchyma. A layer or two

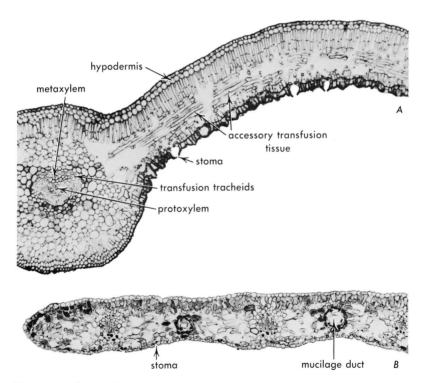

Fig. 19.10. Cross sections of gymnosperm nonconifer leaves. *A,* cycad (*Cycas revoluta*). *B,* maidenhair tree (*Ginkgo biloba*). (Both, ×44.)

of hypodermal sclerenchyma occur on the adaxial side. An endodermis with walls thickened next to the vascular tissue is present. The xylem shows an unusual—and primitive—arrangement in that the protoxylem is toward the abaxial, the metaxylem toward the adaxial side. A layer of parenchyma envelops the protoxylem, and some secondary xylem is present next to the phloem. A few transfusion tracheids occur on both flanks of the metaxylem. Accessory transfusion tissue composed of elongated parenchyma cells and nonliving tracheids with bordered pits is present (Lederer, 1955). Both kinds of cells appear to be in contact with intercellular spaces.

The *Ginkgo* (fig. 19.10, *B*) leaf is wide at the apex and narrow at the base. It has numerous dichotomously branching veins. The epidermis has relatively thin walls and hypodermal sclerenchyma is absent. The guard cells are slightly depressed and occur only on the abaxial side. There is a single palisade layer of rather short lobed cells and below it the spongy parenchyma. Each of the numerous vascular bundles has a uniseriate lignified endodermis. Tannins are often abundant in the endodermal sheath, especially in the small bundles. A few transfusion tracheids flank the xylem of each bundle. The vascular bundles alternate with mucilage ducts.

REFERENCES

Ashby, E. Studies in the morphogenesis of leaves. II. The area, cell size and cell number of leaves of *Ipomoea* in relation to their position on the shoot. *New Phytol.* 47:177–195. 1948.

Ashby, E., and E. Wangermann. Studies in the morphogenesis of leaves. IV. Further observations on area, cell size and cell number of leaves of *Ipomoea* in relation to their position on the shoot. *New Phytol.* 49:23–35. 1950.

Black, R. F. The leaf anatomy of Australian members of the genus *Atriplex*. I. *Atriplex vesicaria* Heward and *A. nummularia* Lindl. *Austral. Jour. Bot.* 2:269–286. 1954.

Boyce, S. G. The salt spray community. *Ecological Monographs* 24:29–67. 1954.

Brauner, L., and M. Brauner. Untersuchungen über den Mechanismus der phototropischen Reaktion der Blattfiedern von *Robinia Pseudoacacia*. *Istanbul Univ. Rev. Fac. Sci., Ser. B. Sci. Nat.* 12:35–79. 1947.

Brown, H. S., and F. T. Addicott. The anatomy of experimental leaflet abscission in *Phaseolus vulgaris*. *Amer. Jour. Bot.* 37:650–656. 1950.

Brown, W. V. Leaf anatomy in grass systematics. *Bot. Gaz.* 119:170–178. 1958.

Datta, M. Structure of the autonomic gyratory pulvini of *Desmodium gyrans* D. C. and *Oxalis repens* Thunb. as contrasted to *Marsilea quadrifolia* Linn. with irreversible movement. *Bose Res. Inst. Calcutta Trans.* 19:127–140. 1952–1953.

Gathy, P. Les feuilles de *Larix*. Étude anatomique. *Cellule* 56:331–353. 1954.

Griffith, M. M. Foliar ontogeny in *Podocarpus macrophyllus*, with special reference to the transfusion tissue. *Amer. Jour. Bot.* 44:705–715. 1957.

Hagerup, O. The morphology and systematics of the leaves in Ericales. *Phytomorphology* 3:459–464. 1953.

Hasman, M., and N. Inanç. Investigations on the anatomical structure of certain submerged floating and amphibious hydrophytes. *Istanbul Univ. Rev. Facul. Sci., Ser. B. Sci. Nat.* 22:137–153. 1957.

Lederer, B. Vergleichende Untersuchungen über das Transfusionsgewebe einiger rezenter Gymnospermen. *Bot. Studien* No. 4:1–42. 1955.

Lee, C. L. The anatomy and ontogeny of the leaf of *Dacrydium taxoides*. *Amer. Jour. Bot.* 39:393–398. 1952.

Metcalfe, C. R. Some thoughts on the structure of bamboo leaves. *Bot. Magazine (Tokyo)* 69:391–400. 1956.

Metcalfe, C. R., and L. Chalk. *Anatomy of the dicotyledons.* 2 vols. Oxford, Clarendon Press. 1950.

Philpott, J. Blade tissue organization of foliage leaves of some Carolina shrub-bog species as compared with their Appalachian mountain affinities. *Bot. Gaz.* 118:88–105. 1956.

Schneider, R. Histogenetische Untersuchungen über den Bau der Laubblätter, insbesondere ihres Mesophylls. *Österr. Bot. Ztschr.* 99:253–285. 1952.

Scott, F. M., M. R. Schroeder, and F. M. Turrell. Development, cell shape, suberization of internal surface, and abscission in the leaf of the Valencia orange, *Citrus sinensis. Bot. Gaz.* 109:381–411. 1948.

Shields, L. M. Leaf xeromorphy as related to physiological and structural influences. *Bot. Rev.* 16:399–447. 1950.

Shields, L. M. The involution mechanism in leaves of certain xeric grasses. *Phytomorphology* 1:225–241. 1951.

Skutch, A. F. Anatomy of leaf of banana, *Musa sapientum* L., var. hort. Gros Michel. *Bot. Gaz.* 84:337–391. 1927.

Stålfelt, M. G. Morphologie und Anatomie des Blattes als Transpirationsorgan. In: *Handbuch der Pflanzenphysiologie* 3:324–341. 1956.

Vasilevskaĭa, V. K. *Formirovanie lista zasukhoustoĭchivykh rasteniĭ.* [Formation of leaves of drought-resistant plants.] Akad. Nauk Turkmen SSR. 183 pp. 1954.

Weintraub, M. Leaf movements in *Mimosa pudica* L. *New Phytol.* 50:357–382. 1952.

Wylie, R. B. Differences in foliar organization among leaves from four locations in the crown of an isolated tree (*Acer platanoides*). *Iowa Acad. Sci. Proc.* 56:189–198. 1949.

Wylie, R. B. Leaf organization of some woody dicotyledons from New Zealand. *Amer. Jour. Bot.* 41:186–191. 1954.

20. THE FLOWER

THE FLOWER HAS BEEN THE OBJECT OF NUMEROUS STUDIES FROM MORphologic and anatomic aspects, but investigators are unable to come to an agreement regarding the nature of the flower and its phylogenetic relation to the other plant parts (cf. reviews in Leroy, 1955, and Tepfer, 1953). Some botanists, probably the large majority, regard the flower as a modified shoot and its parts as homologues of leaves; others reject the concept of homology between the flower and the vegetative shoot. In this book the flower is treated on the basis of the concept of homology between the flower and the shoot in their phylogeny and ontogeny.

STRUCTURE

Flower parts and their arrangement

Like the vegetative shoot, the flower consists of an axis (receptacle) and lateral appendages. These appendages are the floral parts or floral organs; they are commonly grouped into sterile and reproductive. The sepals and petals, composing the calyx and corolla, respectively, are the sterile parts; the stamens and carpels, the reproductive parts. All stamens together constitute the androecium; the carpels, free or united, compose the gynoecium. The stamens and carpels are concerned with sporogenesis. The reference to male and female in the terms androecium and gynoecium, respectively, has to do with the development of male gametophytes (pollen grains) from microspores originating in microsporangia (pollen sacs) on the stamens and of female gametophytes (embryo sacs) from megaspores originating in megasporangia (nucelli of the ovules) on the carpels.

The arrangement of the floral parts on the axis and the relation of parts to each other is highly variable and the variations are of particular concern in taxonomic and phylogenetic studies of flowers. If the flower is regarded as a modified shoot, the differences in floral structure may be interpreted as deviations of various degrees from the basic shoot form; and, in this sense, the greater the deviation, the more highly specialized is the flower.

The vegetative shoot is characterized by indeterminate growth. In contrast, the flower shows determinate growth, for its apical meristem ceases to be active after it has produced all the floral parts. The more highly specialized flowers have a shorter growth period and produce a smaller and more definite number of floral parts than the more primitive flowers. Further indications of increasing specialization are: whorled arrangement of parts instead of spiral, cohesion of parts within a whorl, adnation of parts of two or more different whorls, zygomorphy instead of actinomorphy, and epigyny (inferior ovary) instead of hypogyny (superior ovary).

Sepal and petal

The sepals and petals resemble leaves in their internal structure. They consist of ground parenchyma, a more or less branched vascular system, and an epidermis (fig. 20.1). Crystal-containing cells, laticifers, tannin cells, and other idioblasts may be present. Green sepals contain chloroplasts but rarely show a differentiation into palisade and spongy parenchyma. The color of petals results from pigments in chromoplasts (carotenoids) and in the cell sap (anthocyanins) and from various modifying conditions such as, for example, acidity of cell sap (Paech, 1955). The epidermal cells of petals often contain volatile oils that impart the characteristic fragrance to the flowers. In some plants the anticlinal walls of the epidermis of petals are wavy or bear internal ridges. The outer walls may be convex or papillate (fig. 20.1, *A*), especially on the adaxial side. The epidermis of both sepals and petals may have stomata and trichomes.

Stamen

A common type of stamen consists of a two-lobed four-loculed anther borne on a filament, a thin stalk with a single vascular bundle

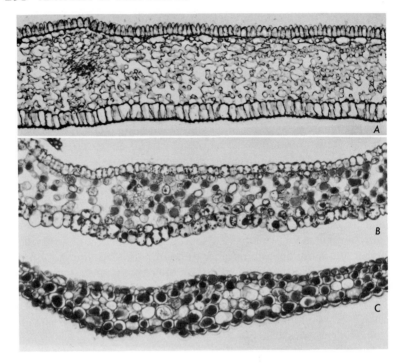

Fig. 20.1. Cross sections of petal of rose (*A*) and of petal (*B*) and sepal (*C*) of *Cassiope*. (*A*, ×90; *B*, *C*, ×150. Slides courtesy of, *A*, A. T. Guard; *B*, *C*, B. F. Palser.)

(fig. 20.2, *A*, *B*). Some of the more primitive families of dicotyledons have leaf-like stamens with three veins. The single-veined type is regarded as a derivative of the three-veined leaf-like type (e.g., Canright, 1952).

The filament is relatively simple in structure. Parenchyma surrounds the vascular bundle, which may be amphicribral. The cutinized epidermis may have trichomes, and on both filament and anther stomata may be present. The vascular bundle traverses the entire filament and ends blindly in the connective tissue located between the two anther halves.

The anther has special features in relation to the development of microsporangia and their wall layers. The outermost wall layer is the epidermis. The subepidermal layer, the endothecium, may have strips or ridges of secondary wall material mainly on walls not in contact with the epidermis (fig. 20.2, *D–F*). The innermost layer is the tapetum, a nutritive tissue frequently composed of multinucleate cells. The wall layers that occur between the endothecium and tapetum are often crushed and destroyed during the development of the pollen sacs. When the tapetum also disintegrates in connection with the maturation of pollen, the outer wall of the pollen sac may consist of only the epidermis and endothecium (fig. 20.2, *D, E*). Complex cytological phenomena are involved in the development of microspores (cf. Maheshwari, 1950) and in the maturation of the walls of pollen grains (cf. Erdtman, 1952).

The anthers commonly dehisce; that is, they open spontaneously. In many species the dehiscence is preceded by a breakdown of the partitions between locules of the same anther half (fig. 20.2, *B*). Later, the outer tissue in this region, which is sometimes reduced to a single epidermal layer (fig. 20.2, *E*), breaks also, and the pollen is released through the break. If a specialized endothecium is present in the locule walls, it is thought to be involved in dehiscence through a differential shrinkage of its unevenly thickened walls. The stomium, or opening resulting from the break, is often slit-like. In some species it is a pore formed either laterally or at the apex of the anther lobe.

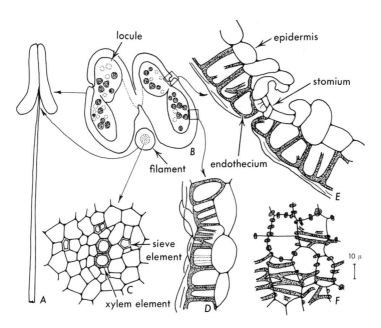

Fig. 20.2. Stamen of *Prunus* (*A*) and its parts: cross sections of anther (*B*), vascular bundle of filament (*C*), anther wall (*D, E*); and surface view of endothecium (*F*).

Gynoecium

The morphology of the gynoecium and the pertinent terminology are subjects of more controversy than those regarding any other part of the flower. The basic structural unit of the gynoecium is called the carpel. A flower may have one carpel or more than one. If two or more carpels are present, they may be united (syncarpous gynoecium) or free from one another (apocarpous gynoecium; fig. 21.1). A mono-carpellate gynoecium is also classified as apocarpous.

An old established term used with regard to the gynoecium is pistil. It refers to a single carpel in an apocarpous gynoecium (simple pistil) or to the entire gynoecium of a syncarpous gynoecium (compound pistil). Some botanists advocate the abandonment of the term pistil (e.g., Parkin, 1955).

The carpel is commonly interpreted as a foliar structure. In an apocarpous gynoecium it is described as being folded—conduplicately folded according to a modern view—so that the adaxial (ventral) sur-face is enclosed (fig. 20.3, *E*). The union of carpels in a syncarpous gynoecium follows two basic plans: the carpels are joined either in a folded condition, abaxial (dorsal) surface to abaxial surface (fig. 20.4, *B*), or in an unfolded or partially folded condition, margin to margin (fig. 20.4, *D*). The first type of junction results in a bilocular or multilocu-lar gynoecium; the second, in a gynoecium with one locule. Secondary modifications may bring about variations of the basic plans.

The carpel of an apocarpous gynoecium or the entire syncarpous gynoecium is commonly differentiated into the lower fertile part, the ovary, and the upper sterile part, the style (fig. 20.4, *A*, *C*). A more or less extensive peripheral part of the style is differentiated as a stigma, usually in the upper region of the style. If there is no extended structure that may be called style, the stigma is described as sessile, that is, sessile on the ovary. The differentiation into ovary, style, and stigma has resulted from phylogenetic specialization. In the less spe-cialized angiosperms the carpels appear as folded styleless structures with the stigmatic tissue covering the unsealed margins.

Within the ovary one distinguishes the ovary wall, the locule (cavity) or locules, and, in a multilocular ovary, the partitions. The ovules are borne on certain regions of the ovary wall located on its inner or adaxial side. An ovule-bearing region constitutes the placenta. A placenta may be a conspicuous outgrowth, sometimes almost occluding the lumen of the ovarian cavity (fig. 20.5, *A*, *B*). The position of the placentae is related to the method of union of carpels (cf. Puri, 1952*a*). In a given carpel the placenta occurs, as a rule, more or less close to

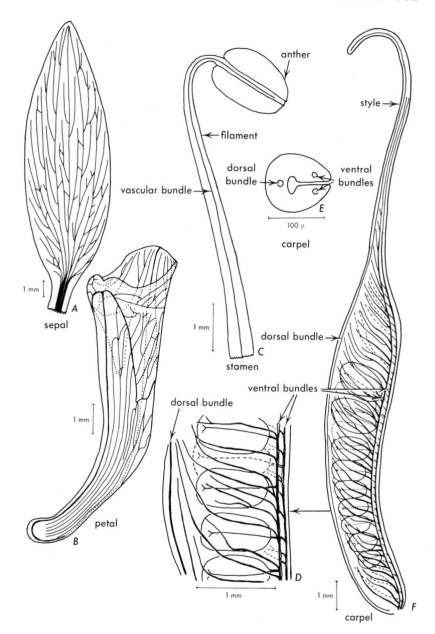

Fig. 20.3. Flower parts of *Aquilegia.* Longitudinal views of sepal (*A*), petal (*B*), stamen (*C*), and carpel (*D, F*), and cross section of carpel (*E*). (After Tepfer, *Calif. Univ., Pubs., Bot.,* 1953.)

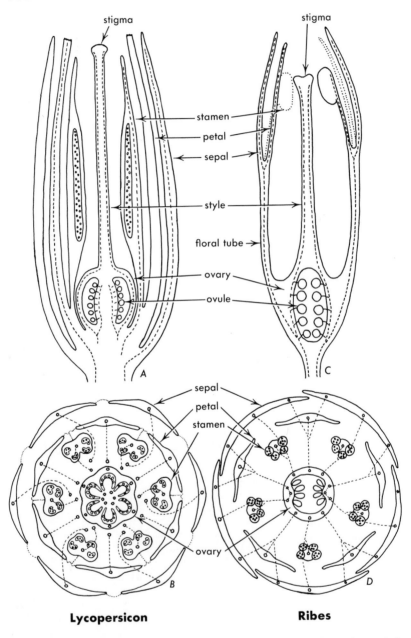

Lycopersicon

Ribes

Fig. 20.4. Diagram of flowers in longitudinal (*A, C*) and cross (*B, D*) sections. *A, B, Lycopersicon* (tomato); hypogynous, axile placentation. *C, D, Ribes;* epigynous, parietal placentation. Broken lines indicate course of vascular bundles and their interconnections.

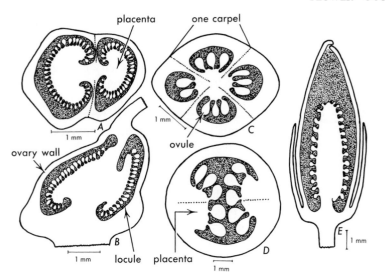

Fig. 20.5. Placentation. *A, B, Antirrhinum,* axile. *C, Fuchsia,* axile. *D, Ribes,* parietal. *E, Dodecatheon,* free central. (*A, C–E,* cross sections; *B,* longitudinal section; style not shown in *E.*)

the margin. Since there are two margins, the placenta is double in nature. The two halves may be separate, or they may be united, at least at the base. When the carpel margins are united at the base, the placenta has a U-shaped form (Leinfellner, 1951). In compound ovaries the number of double placentae usually corresponds to the number of carpels. If the carpels are joined in a folded condition, the ovary is multilocular, and the placentae occur in the center of the ovary where the carpellary margins meet (axile placentation, fig. 20.5, *A–C*). The partitions in this type of ovary may disappear, either in terms of phylogeny or ontogeny (Hartl, 1956*b*), so that a free central placentation results (fig. 20.5, *E*). If the carpels are joined margin to margin, the placentae are disposed on the ovary wall (parietal placentation, fig. 20.5, *D*). In such ovaries the two halves of each placenta are derived from two different carpels.

The understanding of placentation is materially aided by examination of the vascularization of the carpel. Most commonly the carpel has three veins, one dorsal or median and two ventral or lateral (fig. 20.3, *E, F*), and the vascular supply of the ovules is derived from the ventral bundles (fig. 20.3, *D*). In an ovary with axile placentation the ventral bundles appear in the center of the ovary (fig. 20.4, *B*), with the phloem turned inward, the xylem outward. The ventral bundles of two adjacent carpellary margins—of the same carpel or of two different carpels, depending on placentation—may be more or less completely fused (fig. 20.4, *B*).

The interpretation of the placentae as carpellary in origin may be readily accepted for many representatives of angiosperms. In some, however, the relation between carpels and ovules is obscure. It has been suggested, therefore, that the placentae may be of axial origin, at least in some groups of plants. Such placentae are supposed to occur on the receptacle that constitutes the bottom of the ovary (e.g., Gramineae; Barnard, 1957) or is prolonged upward in the center of the ovary (some representatives with central placentation). This view is contested by investigators who insist that despite mature appearances the primary relation of the placentae is with the carpels (Eckardt, 1957).

The delimitation of the carpels and the interpretation of origin of placentae is even more difficult in epigynous flowers (fig. 20.4, *C, D*). In such flowers the ovary is visualized as imbedded in extracarpellary tissue derived either from the receptacle (Leinfellner, 1954; Puri, 1952*b*) or from the floral tube, that is, fused bases of sepals, petals, and stamens (cf. Douglas, 1957). Some authors consider that the nature of the extracarpellary tissue varies in different groups of angiosperms. There is also the problem of recognizing whether the carpels actually line the extracarpellary tissue or whether they are limited to the upper part of the gynoecium, that is, the part that forms the top of the ovary and bears the styles. If the latter concept is accepted, the placentae would not be derived from the carpels. The nature of the cup enclosing the gynoecium but not joined to it in perigynous flowers is similarly subject to various interpretations.

The style is an upward prolongation of the carpel (fig. 20.3, *F*). In syncarpous gynoecia the style, if single, is derived from all the carpels composing the gynoecium (fig. 20.4, *A, C*). The carpels may be incompletely united, and as a result the style may be a single, but compound, structure at the base and a multiple structure at the top; or there may be as many stylar units (stylar branches or stylodes; Parkin, 1955) as there are carpels in the syncarpous ovary.

The stigma consists of a glandular tissue secreting materials that create a suitable medium for the germination of the pollen grains. The epidermal cells of the stigma are commonly elongated into papillae, short hairs, or long branched hairs. The stigmatic tissue is connected with the ovarian cavity by a rather similar tissue, the stigmatoid tissue, through which the pollen tube grows. In styles having canals the stigmatoid tissue lines the stylar canal. In solid styles the stigmatoid tissue forms one or more strands imbedded in the ground tissue or associated with the vascular bundles. The pollen tube passes through the stigmatoid tissue by intercellular (intrusive) growth.

Histologically the ovary and style are relatively simple at anthesis. They are composed of epidermis, ground tissue of parenchyma, and vascular bundles. The outer epidermis is cuticularized and may have stomata. When an ovary develops into a fruit, striking modifications occur in the structure of the ovary (chapter 21).

The ovule consists of a nucellus enclosing the sporogenous tissue, one or two integuments of epidermal origin, and a stalk, the funiculus. During anthesis the ovule consists largely of parenchyma and contains a more or less prominent vascular system, but during seed development it undergoes profound changes (chapter 22).

Vascular anatomy

The vascular system has been studied more often than any other histologic feature of the flower, and the results have been extensively used for purposes of establishing taxonomic relationships and for drawing conclusions regarding the morphologic nature of the parts of the flower (cf. Douglas, 1957; Leroy, 1955; Palser, 1954; Puri, 1952*a,b;* Tepfer, 1953).

In relatively unspecialized flowers the vascular organization is easily interpreted as comparable to that of the vegetative shoot. The vascular system of the floral axis can be described as a cylindrical complex of traces of the floral parts. At successive levels certain bundles assume an oblique course and become part of the vascular supply of the floral parts inserted at that level. The bundles may branch and combine with each other in the axis in a rather irregular manner (Sporne, 1958).

The number of traces of the different floral parts varies in the same flower. The usual pattern is as follows (fig. 20.4, broken lines). Each sepal has as many traces as the foliage leaf of the same plant. Each petal in a dicotyledon has one trace; each perianth member (tepal) in a monocotyledon, one to many. Within the sepal and petal the vascular bundles form more or less complex systems resembling those in foliage leaves (fig. 20.3, *A, B*). The specialized type of stamen has one trace, and it is continued as a single bundle in the filament and anther (fig. 20.3, *C*). The carpel has three traces, and their prolongations within the carpel may have branches (fig. 20.3, *D, F*). The ovules are supplied by branches from carpellary bundles, usually the two lateral ones (fig. 20.3, *F*). Carpellary vascular bundles are continuous through the style (fig. 20.4, *A, C*).

When floral parts are fused, the vascular bundles of these parts may also be fused. If carpels are united, the lateral bundles, either those of the same carpel or those of two adjacent carpels, may be fused in pairs. In some epigynous flowers certain bundles show an inverted orientation of xylem and phloem. Such arrangement is interpreted as an indication that the ovary is imbedded in receptacular (axial) tissue in which the inversion of bundles resulted from an invagination of the receptacle. The absence of inverted bundles, on the other hand, is regarded as evidence that the extracarpellary tissue is appendicular in nature (floral tube).

DEVELOPMENT

Floral meristem

When the apical meristem ceases to produce foliage leaves and initiates an inflorescence, or a flower, it undergoes more or less conspicuous morphologic changes (fig. 20.6). These changes are related, at least in part, to the cessation of indeterminate growth characteristic of the vegetative stage and to the altered mode of production of lateral appendages. The biochemical and physiological factors involved have not been fully explored, but much information is available on the effect of light on initiation of the flowering stage (Parker and Borthwick, 1950). In view of the morphologic and functional dissimilarities between the different floral parts, not one but a successive series of appropriate physiologic states are probably involved in the differentiation of the flower. This assumption is supported by some surgical experiments on flower primordia (Cusick, 1956). In these experiments results of bisection of primordia at different stages of development suggested that the ability of the meristem to produce the different floral parts is successively lost as the floral primordium grows older.

The often-observed morphologic change during the initiation of the reproductive stage is a sudden and rapid elongation of the axis accompanied or followed by an increase in width and a flattening of the apex (Popham and Chan, 1956; Rauh and Resnik, 1953; Tepfer, 1953). The sudden elongation of the axis in preparation for flowering is particularly striking in plants having a rosette-like habit during vegetative growth, as, for example, in grasses (e.g., Barnard, 1957). The elongated axis either bears a single flower or, probably, more frequently, an inflorescence. The terminal and axillary meristems may form flowers (fig. 20.6) in a sequence determined by the type of inflorescence.

During vegetative growth, the apical meristem grows upward before

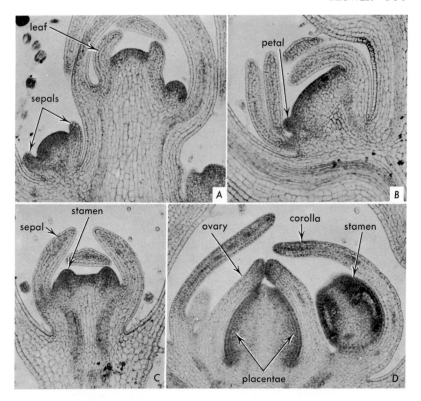

Fig. 20.6. Early stages of flower development in *Antirrhinum,* snapdragon. *A,* vegetative (above) and floral (right and left below vegetative) apices. *B,* flower with sepals and primordia of petals. (Petals are initiated as discrete units, later growth at their bases forms sympetalous corolla.) *C,* flower cut in plane exposing stamen primordia. *D,* flower with gynoecium; carpels inclose massive placenta but are not yet prolonged into style. (Part of young style in fig. 20.5, *B.*) (All, ×90.)

the beginning of a new foliar plastochron, as if it were restored after the emergence of each primordium. During the development of a flower, on the other hand, the area of the apical meristem gradually diminishes as the successive floral parts arise (fig. 20.6, *B, C*). In some flowers a small amount of apical meristem remains after the initiation of carpels, but it ceases to be active; in others, the carpels or the ovules appear to arise from the terminal part of the apical meristem (fig. 20.6, *D;* Barnard, 1957; Leroy, 1955).

The small depth and comparatively broad expanse of the meristematic tissue are common histologic features of the floral meristem. The broad apex is occupied by a mantle of meristematic cells, and beneath the mantle is a vacuolated core of ground tissue no longer concerned with growth. In other words, rib-meristem activity has ceased. These features may be encountered in meristems of single flowers (fig. 20.6, *B*) and those of inflorescences (fig. 20.7, *B*). The tunica and corpus organization may or may not be identifiable in the floral meristem. If it is present, the number of tunica layers may be the same as in the vegetative apex of the same species, or it may be changed, decreased or increased.

Origin and development of floral parts

The floral organs are initiated like the foliage leaves by periclinal divisions of cells located more or less deeply beneath the dermatogen (Tepfer, 1953), or also in the dermatogen itself (monocotyledons; Barnard, 1957). The depth of these divisions may be the same as in the origin of leaves of the same species or may be different; moreover, the depth may vary with regard to the different floral parts of the same flower.

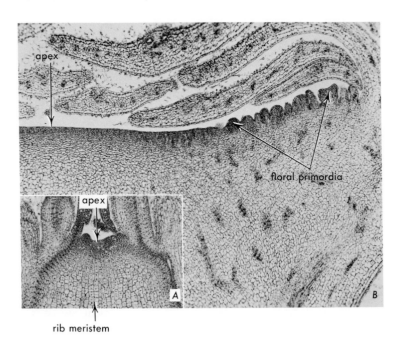

Fig. 20.7. *Helianthus annuus,* sunflower, vegetative shoot apex (*A*) and about half of floral apex (*B*). (Both, ×50. *A*, from Esau, *Amer. Jour. Bot.,* 1945. *B,* slide courtesy of A. T. Guard.)

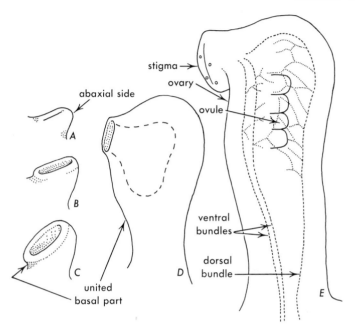

Fig. 20.8. Stages of development of carpel of *Drimys* in longitudinal views. (After Tucker, *Calif. Univ., Pubs., Bot.,* 1959.)

The initial periclinal divisions are followed by others, including anticlinal. As a result of this activity the primordium becomes a protuberance (e.g., fig. 20.6, *B*). Then follows growth in length and width, usually involving apical and marginal growth (except in staminal filaments) similar to those exhibited by foliage leaves. The growth is rather limited, however. Because of the relatively small size of floral organs and the close succession of their emergence, a buttress-like prominence is frequently not identifiable during their initiation.

The sepals resemble the foliage leaves most closely in their initiation and development. The growth pattern of the petals is more or less similar to that of the leaves. The stamens arise as stout short structures (fig. 20.6, *C, D*), the filament differentiating subsequently by intercalary growth (Trapp, 1956). If the perianth parts and stamens show cohesion and adnation, the union of parts may be congenital or ontogenetic or a combination of both. If congenital union is present, the united part develops by intercalary growth (e.g., Picklum, 1954).

The development of the gynoecium varies in detail in relation to the union of carpels with one another and with other floral parts. If carpels are not united, the individual carpel primordium arises as a horseshoe-shaped or circular mound which is somewhat higher on the abaxial side (fig. 20.8, *A–C*). Through upward growth a sac-like structure with free margins on the adaxial side is produced (fig. 20.8, *D*).

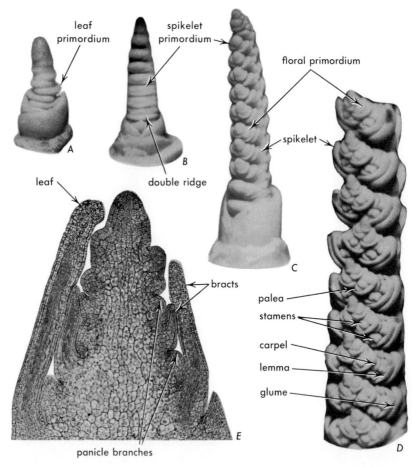

Fig. 20.9. Floral initiation in grasses. *A–D, Triticum* (wheat). *A,* vegetative apex with leaf primordia, somewhat elongated in advance of spike formation. *B,* young spike with double ridges, each composed of spikelet primordium and subtending leaf primordium. *C,* spike with spikelets developing first flower primordia. *D,* part of spike 3.5 millimeters long with spikelets having several florets each. *E,* longitudinal section of shoot apex of *Bromus* at transition to flowering stage: initiation of panicle branches. (*A,* ×35; *B,* ×24; *C,* ×26; *D,* ×30; *E,* ×75; *A–D,* from Barnard, *Austral. Jour. Bot.,* 1955. *E,* from Sass and Skogman, *Iowa State Coll. Jour. Sci.,* 1951.)

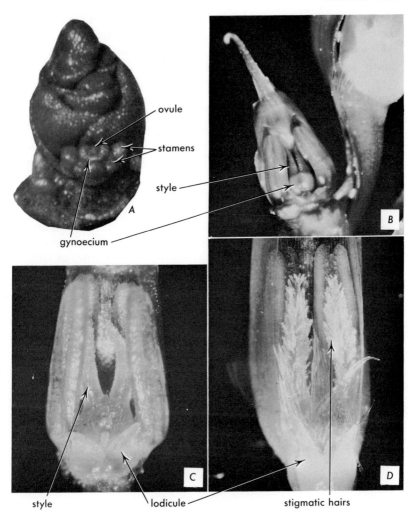

Fig. 20.10. Development of gynoecium. *A,* spikelet of *Scirpus* with spiral arrangement of glumes and flower primordia. Gynoecium still open, ovule exposed. *B,* floret of wheat (*Triticum*); young gynoecium with style. *C, D,* florets of oat (*Avena*) with younger (*C*) and older (*D*) gynoecia. (*A,* ×85; *B, D,* ×20; *C,* ×40. *A,* from Barnard, *Austral. Jour. Bot.,* 1957. *B–D,* from Bonnett, *Jour. Agr. Res., B,* 1936; *C, D,* 1937.)

The margins become more or less coherent during subsequent development. In many species a congenitally united basal part occurs beneath the open margins (fig. 20.8, *C, D*). This structural feature is interpreted as evidence of peltate nature of the carpel (Baum, 1952; Hartl,

1956*a*). The upper part of the carpel elongates into a style if one is present in the mature carpel or becomes a sessile stigma (fig. 20.8, *E*).

In syncarpous ovaries the carpels arise as individual primordia or jointly as a unit structure in which the delimitation of the individual carpels is obscured. Various degrees of congenital and ontogenetic unions are exhibited by the developing gynoecium in different species (e.g., Morf, 1950; Hartl, 1956*a*). The ontogenetic union may be so firm that the suture is not identifiable in the mature state. Such union involves enlargement and interpenetration of epidermal cells, sometimes accompanied by division of the epidermal cells (Hartl, 1956*a*). Depending on the degree of union of carpels the style of a syncarpous ovary grows out either as a unit structure or as styloidal prolongations of the individual carpels, free or partly united.

The development of inflorescences and flowers in Gramineae has been studied with especial care (e.g., Barnard, 1957). The rapidly elongating inflorescence of a spike type forms first a series of semicircular ridges—the leaf primordia (fig. 20.9, *A*). At a later stage double ridges appear (fig. 20.9, *B*). The doubleness results from the initiation of spikelet primordia in the axils of the leaf primordia. The leaf primordia develop less and less strongly toward the apex of the spike and are overtopped by the developing spikelets. In a panicle type of inflorescence branches develop in the axils of bracts (fig. 20.9, *E*). Within the spikelets the glumes, florets, and floral organs develop in a characteristic sequence (figs. 20.9, *C, D,* and 20.10, *B–D*). The gynoecium of the Gramineae and Cyperaceae is interpreted as a tricarpellate syncarpous structure. It is initiated as a single ring-like structure (fig. 20.10, *A*), and the single ovule arises at the center of the gynoecium. The margins of the grass gynoecium form two styles (fig. 20.10, *B, C*), which eventually develop stigmatic hairs (fig. 20.10, *D*).

REFERENCES

Barnard, C. Floral histogenesis in the monocotyledons. I. The Gramineae. *Austral. Jour. Bot.* 5:1–20. 1957.

Baum, H. Über die "primitivste" Karpellform. *Österr. Bot. Ztschr.* 99:632–634. 1952.

Canright, J. E. The comparative morphology and relationships of the Magnoliaceae. I. Trends of specialization in the stamens. *Amer. Jour. Bot.* 39:484–497. 1952.

Cusick, F. Studies of floral morphogenesis. I. Median bisections of flower primordia in *"Primula bulleyana"* Forrest. *Roy. Soc. Edinburgh Trans.* 63:153–166. 1956.

Douglas, G. E. The inferior ovary. II. *Bot. Rev.* 23:1–46. 1957.

Eckardt, T. Vergleichende Studie über die morphologischen Beziehungen zwischen Fruchtblatt, Samenanlage und Blütenachse bei einigen Angiospermen. *Neue Hefte zur Morphologie* No. 3:1–91. 1957.

Erdtman, G. *Pollen morphology and plant taxonomy; angiosperms.* I. Stockholm, Almqvist & Wiksell. 1952.

Hartl, D. Morphologische Studien am Pistill der Scrophulariaceen. *Österr. Bot. Ztschr.* 103:185–242. 1956a.

Hartl, D. Die Beziehungen zwischen den Plazenten der Lentibulariaceen und Scrophulariaceen nebst einem Excurs über Spezialisationsrichtungen der Plazentation. *Beitr. zur Biol. der Pflanzen.* 32:471–490. 1956b.

Leinfellner, W. Die U-förmige Plazenta als der Plazententypus der Angiospermen. *Österr. Bot. Ztschr.* 98:338–358. 1951.

Leinfellner, W. Die Kelchblätter auf unterständigen Fruchtknoten und Achsenbechern. *Österr. Bot. Ztschr.* 101:315–327. 1954.

Leroy, J. F. Étude sur les Juglandaceae. A la recherche d'une conception morphologique de la fleur femelle et du fruit. *Mém. du Muséum Nat. D'Hist. Naturelle., Ser. B., Bot.* 6:1–246. 1955.

Maheshwari, P. *An introduction to the embryology of angiosperms.* New York, McGraw-Hill Book Company. 1950.

Morf, E. Vergleichend-morphologische Untersuchungen am Gynoeceum der Saxifragaceen. *Schweiz. Bot. Gesell. Ber.* 60:516–590. 1950.

Paech, K. Colour development in flowers. *Annu. Rev. Plant Physiol.* 6:273–298. 1955.

Palser, B. F. Studies of the floral morphology in the Ericales—III. Organography and vascular anatomy in several species of the Arbuteae. *Phytomorphology* 4:335–354. 1954.

Parker, M. W., and H. A. Borthwick. Influence of light on plant growth. *Annu. Rev. Plant Physiol.* 1:43–58. 1950.

Parkin, J. A plea for a simpler gynoecium. *Phytomorphology* 5:46–57. 1955.

Picklum, W. E. Developmental morphology of the inflorescence and flower of *Trifolium pratense* L. *Iowa State Col. Jour. Sci.* 28:477–495. 1954.

Popham, R. A., and A. P. Chan. Origin and development of the receptacle of *Chrysanthemum morifolium*. *Amer. Jour. Bot.* 39:329–339. 1952.

Puri, V. Placentation in angiosperms. *Bot. Rev.* 18:603–651. 1952a.

Puri, V. Floral morphology and inferior ovary. *Phytomorphology* 2:122–129. 1952b.

Rauh, W., and H. Reznik. Histogenetische Untersuchungen an Blüten- und Infloreszenzachsen. II. Die Histogenese der Achsen köpfchenförmiger Infloreszenzen. *Beitr. zur Biol. der Pflanz.* 29:233–296. 1953.

Sporne, K. R. Some aspects of floral vascular systems. *Linn. Soc. London, Proc.* 169:75–84. 1958.

Tepfer, S. S. Floral anatomy and ontogeny in *Aquilegia formosa* var. *truncata* and *Ranunculus repens*. *Calif. Univ., Pubs., Bot.* 25:513–648. 1953.

Trapp, A. Zur Morphologie und Entwicklungsgeschichte der Staubblätter sympetaler Blüten. *Bot. Studien* 5. 1956.

21. THE FRUIT

AFTER THE EGG IS FERTILIZED THE OVARY COMMONLY DEVELOPS INTO a fruit, whereas the ovule becomes the seed. As was pointed out in the preceding chapter, in many plant groups the ovary is closely associated with extracarpellary tissues. These tissues may develop together with the ovary and form an intimate part of the final product. Such union of carpellary and extracarpellary parts raises terminological difficulties regarding the definition of the fruit (cf. Esau, 1953, p. 577). Strictly defined the fruit is the matured ovary. The modern tendency is to broaden the term fruit so as to include any of the extracarpellary parts that may be associated with the ovary at maturity. The term also refers to fruits without seeds, that is, parthenocarpic fruits.

Another source of terminological ambiguities is the occurrence of apocarpous and syncarpous gynoecia. If the definition of the fruit is based on a single ovary, an apocarpous bi- or multicarpellate flower would produce more than one fruit. In a syncarpous flower, on the other hand, the fruit would be derived from several carpels. The most satisfactory way of resolving this problem is to define the fruit as the product of the entire gynoecium and any floral parts that may be associated with the gynoecium in the fruiting stage. The product of an individual carpel in an apocarpous fruit would then constitute a fruitlet. Thus, in the strawberry, the fruit comprises the receptacle and the achenes imbedded in it, whereas each achene constitutes a fruitlet.

Four basic types of fruits may be distinguished according to Winkler's (cf. Esau, 1953, pp. 578–580) scheme: (1) aggregate free fruit derived from an apocarpous hypogynous flower (fig. 21.1, *A*); (2) united free fruit from a syncarpous hypogynous flower (fig. 21.1, *B*); (3) aggregate cup fruit from an apocarpous perigynous flower (fig. 21.1, *C*); and (4) united cup fruit from a syncarpous epigynous flower (fig. 21.1, *D*).

In a more detailed classification of fruits combinations of many characters are used, especially the arrangement and union of carpels and the nature of the fruit wall and its dehiscence (e.g., Baumann-Bodenheim, 1954).

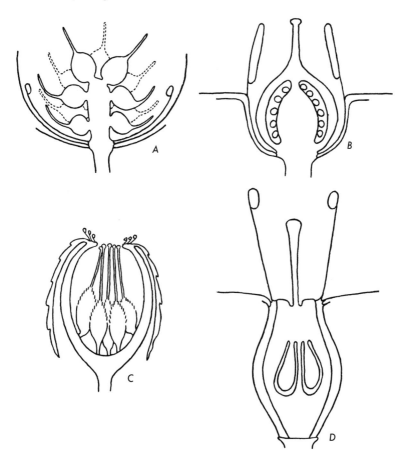

Fig. 21.1. Illustrations of flowers from which types of fruits named in parentheses are derived. *A, Ranunculus,* apocarpous, hypogynous (aggregate free fruit). *B, Solanum,* syncarpous, hypogynous (unit free fruit). *C, Rosa,* apocarpous, perigynous (aggregate cup fruit). *D, Cornus,* syncarpous, epigynous (unit cup fruit). (From Esau, *Plant Anatomy,* John Wiley and Sons, 1953.)

HISTOLOGY OF THE FRUIT WALL

The term fruit wall is used in this book to designate either the pericarp, that is, the wall of the matured ovary, or the pericarp together with any extracarpellary parts that may be united with the ovary in the fruit. In the flower the future fruit wall consists mainly of parenchyma as ground tissue interspersed with vascular tissue. During the development of the fruit profound histologic changes may occur, particularly in the ground tissue. These changes, coupled with the initial variations in the carpellary relations in the gynoecium, bring about the extensive structural variability of fruits in the flowering plants.

The fruit wall may be more or less highly differentiated, and frequently the pericarp proper shows two or three distinct layers. If such layers are recognizable, they are referred to—beginning with the outermost layer—as exocarp (or epicarp), mesocarp, and endocarp. These terms are commonly used for purposes of description without relation to the ontogenetic origin of the layers. The exocarp, for example, denotes sometimes the epidermis alone, sometimes the epidermis together with some subjacent tissue. A stricter terminology has been advocated (e.g., Sterling, 1953), but in this book the common descriptive approach is used.

Dry fruit wall

Dehiscent fruit wall. Dehiscent fruit walls commonly occur in fruits containing several seeds. A dehiscent fruit may develop from a single carpel (follicle, legume) or several carpels (e.g., capsule). The region and method of dehiscence are highly variable, and the phenomena involved have been studied in many species (e.g., Fahn and Zohary, 1955; Holden, 1956; Stopp, 1950). The break may occur where carpel edges of a given carpel are united, along the union of two carpels, longitudinally through the median (dorsal) parts of carpels, through a circular horizontal area involving all carpels in syncarpous gynoecia, or through formation of pores. Histologically, the zone of dehiscence may be visible earlier or later in the development of the fruit. Cell division may precede the dehiscence; the break then occurs through the band of thin-walled cells in the dehiscence region. Shape of cells and character of micellar structure of walls have also been related to dehiscence. A softening of the middle lamella and cell walls preceding the dehiscence has been reported (Holden, 1956).

A well-known example of a dry dehiscent fruit is the legume, the fruit of the Leguminosae. The exocarp may include only the epidermis

(*Pisum, Vicia*) or the epidermis and the subepidermal layer (*Phaseolus, Glycine*), both composed of thick-walled cells (fig. 21.2). The mesocarp is usually parenchymatous and thin walled, whereas the endocarp may consist of several rows of thick-walled cells. In the layers nearest the inner periphery of the pericarp the cells of the endocarp are oriented at an angle to the long axis of the fruit and opposite to that formed by the also diagonally oriented cells of the exocarp. The legume has two lines of dehiscence: one through the union of the carpel margin, the other along the median vascular bundle.

The capsule, another common dehiscent fruit, has parenchyma and sclerenchyma cells in various distributions. In *Linum usitatissimum,* for example, the pericarp may be divided into an exocarp of highly lignified cells and a mesocarp and endocarp of parenchyma cells. The capsule of *Nicotiana tabacum* has parenchymatous exocarp and mesocarp and a thick-walled endocarp, two or three cells in thickness.

Indehiscent fruit wall. An indehiscent fruit usually results from an ovary in which only one seed develops, though more than one ovule may be present. The pericarp of an indehiscent fruit often resembles a seed coat in structure. The actual seed coat in such fruits may become obliterated to a considerable degree (e.g., achene of the Compositae, fig. 21.3; cf. Esau, 1953, pp. 583–587) or fused with the pericarp (e.g., caryopsis of Gramineae).

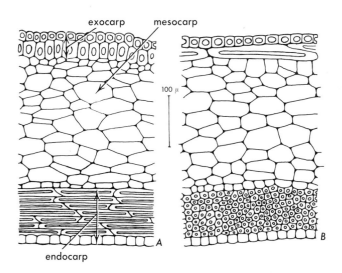

exocarp   mesocarp

100 μ

endocarp

Fig. 21.2. Pericarp of soybean (*Glycine*) pod. *A,* cross section; *B,* longitudinal section. Sclerified cells occur in exocarp and endocarp. (After Monsi, *Japanese Jour. Bot.,* 1948.)

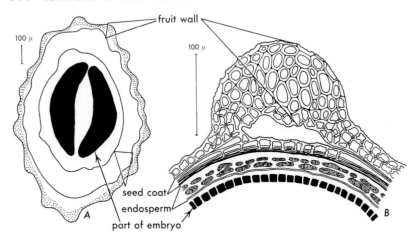

Fig. 21.3.   The fruit of lettuce (*Lactuca sativa*).   *A*, entire achene and, *B*, fruit coat, with subjacent layers, in cross sections.   (From Esau, *Plant Anatomy*, John Wiley and Sons, 1953.   After Borthwick and Robbins, 1928.)

The fruit of grasses, the caryopsis, has been extensively studied.   The covering layers of the wheat caryopsis are composed of the pericarp and the remains of the seed coat (fig. 21.4).   In the pericarp the layers from the outside in are (fig. 21.4, *B*): outer epidermis covered with a cuticle; one or more layers of parenchyma, partly compressed; partly resorbed parenchyma; cross cells, elongated transversely to the long axis of the grain and having thick lignified walls; and remains of the inner epidermis in the form of lignified cells elongated parallel with the long axis of the grain (tube cells).   In the development of the seed coat, the outer integument disintegrates, and the inner becomes altered and compressed.   The compressed inner integument contains pigment, gives a positive reaction for fatty materials, and is covered by a cuticle on both sides (Bradbury et al., 1956*b*).

In the grain of *Zea* (Kiesselbach and Walker, 1952), the outer pericarp consists of cells with thick pitted walls and is much compressed, especially at the distal end of the fruit.   The central pericarp disintegrates.   The inner pericarp remains thin walled and is variously stretched, torn, and compressed.   The integuments disintegrate completely.   A cuticle occurs between the thick-walled nucellar epidermis and the pericarp.   *Sorghum* (Sanders, 1955) shows less collapse in the pericarp than many other cereals.

In general, the degree of modification of the seed coat and pericarp during the development of the caryopsis shows a rather wide range of

variation (Narayanaswami, 1955). Of considerable interest physiologically is the development of a fatty membrane or membranes outside the nucellus. Such membranes are probably derived from the
integuments and the nucellar epidermis, although the inner pericarp
layer may also contribute fatty materials. All these tissue layers are
epidermal in origin. The membranes, therefore, may be interpreted
as cuticular layers. They are sometimes referred to as semipermeable
layers.

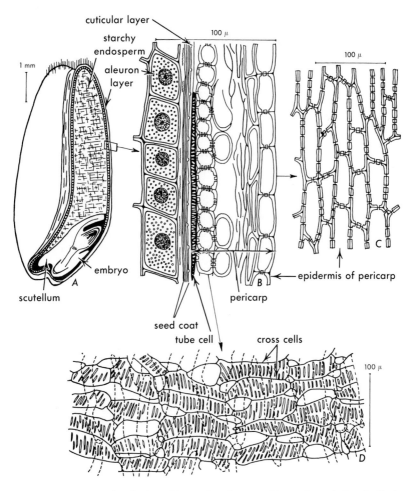

Fig. 21.4. *A,* caryopsis of wheat (*Triticum*) and parts of its pericarp in longitudinal
section (*B*) and surface views (*C, D*). (*A, B,* from Esau, *Plant Anatomy,* John Wiley and
Sons, 1953. *C, D,* drawn from photographs in Bradbury et al., *Cereal Chem.,* 1956.)

The caryopses of grasses have a large amount of endosperm. The outermost layer (or layers) of endosperm contains proteinaceous inclusions and is called the aleuron layer (fig. 21.4, *A, B*). The inner endosperm contains starch and various amounts of amorphous proteins, the glutens. The bran of wheat grains is composed of the pericarp and the outermost tissues of the seed, including the aleuron layer (Bradbury et al., 1956*a*).

Fleshy fruit wall

The fleshy fruits, like the dry fruits, may be derived from monocarpellate or multicarpellate gynoecia. Their walls may consist of the pericarp or of pericarp fused with extracarpellary tissues. The outer part of the fruit wall or the entire fruit wall may become fleshy by differentiating into soft, succulent parenchyma. Parts other than the wall, such as placentae and partitions in multilocular ovaries, may become fleshy.

Examples. Well-known examples of fleshy fruits are berries, in which the entire ground tissue is fleshy. In the grape the fleshy tissue originates from the pericarp; in the tomato, from the partitions and the placentae in addition. The citrus fruit, or hesperidium, with its leathery rind, is usually placed in a category close to the berry. It develops from an ovary with axile placentation. The pericarp of the hesperidium is differentiated into a compact collenchymatous exocarp, the flavedo, with oil glands (this is the yellow part of the rind); a spongy mesocarp, the albedo (this is the white part of the rind); and a compact endocarp which gives rise to the juice sacs. The juice sacs are comparable to multicellular hairs but originate subepidermally, with the epidermis forming a single layer (Hartl, 1957). The distal part of each hair is enlarged, and the interior broken down and filled with juice.

The pepo of Cucurbitaceae is another example of a berry-like fruit; it has a hard rind. Since the fruit develops from an inferior ovary the fruit wall consists of the pericarp and the extracarpellary tissue. No dividing line between the two is discernible. The fruit wall is massive and heterogeneous in structure (Matienko, 1957). Beneath the outer epidermis is a collenchymatous layer. The next region below is composed of parenchyma, part of which may contain chloroplasts. Some genera have a layer of sclereids, continuous or discontinuous, in this region. The third region consists of fleshy parenchyma. Then follows another region of parenchyma that contains the juice in the juicy species. Carotenoid pigments may be present in this tissue. Vascular bundles are located in the fleshy part of the fruit wall. The incurved carpel margins in the center of the fruit also contribute to the flesh of the fruit. The inner epidermis clings to the seed in many species as a thin transparent membrane.

A drupe (e.g., *Prunus*) is a fleshy fruit derived from a superior ovary (fig. 21.5, *A*) and characterized by a stony endocarp, a fleshy mesocarp, and a relatively thin exocarp composed of the epidermis and subepidermal collenchyma (fig. 21.5, *C*). Vascular bundles occur in the fleshy part and in the stony endocarp (fig. 21.5, *B*). In the fruit of prune the stony endocarp (pit wall) is derived from three regions of the ovary. The inner epidermis forms a multiseriate layer of vertically elongated sclereids; the next outer region is a multiseriate layer of tangentially elongated sclereids (fig. 21.5, *E, F*); and the two to four layers still farther outward differentiate into isodiametric sclereids.

The individual drupelets of the raspberry (*Rubus*) also have a stony endocarp composed of elongated curved sclereids oriented differently in the different layers (fig. 21.6; Reeve, 1954*b*). The parenchymatous mesocarp forms most of the succulent pulp. The exocarp bears epidermal hairs that hold the drupelets together at maturity (Reeve, 1954*a*).

The fruit of *Pyrus* (apple, pear) is derived from an inferior ovary and the extracarpellary part (floral tube according to many workers) composes the bulk of the fleshy part of the fruit wall (cf. Esau, 1953, p. 589). The boundary between the pericarp and the extracarpellary tissue may or may not be discernible. Most of the flesh of the apple is parenchyma. The vascular bundles of the sepals and petals and their anastomoses are imbedded in the extracarpellary region. The carpels are five in number, and each has three distinct vascular bundles. The pericarp is composed of the fleshy parenchymatous exocarp, more or less confluent with the extracarpellary parenchyma, and of the cartilaginous endocarp which consists of sclereids variously oriented in the different layers (chapter 6). The outer epidermis may contain anthocyanins and phlobaphenes which give the characteristic coloration to the skin of some apple varieties (Miličić, 1952).

Development. Fleshy fruits are often the object of quantitative developmental studies, the results of which have served for drawing conclusions about factors involved in growth and development of form in fruits (cf. Nitsch, 1953). In many species the ovary enlarges mainly through cell multiplication (e.g., cucurbits, tomatoes, cherries, apples). At the time of anthesis cell division gradually ceases, and the later part of fruit enlargement—this constitutes the longest period of growth —is determined by increase in cell size. In some species (cucurbits, some varieties of tomatoes) the shape of the fruit is determined at

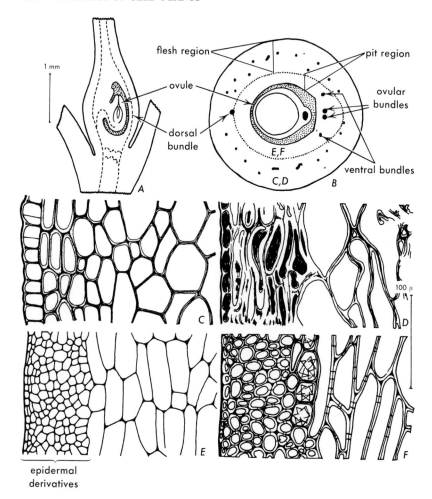

Fig. 21.5. *A, B,* fruit of prune (*Prunus*) in longitudinal (*A*) and cross (*B*) sections. *C, D,* outer part of pericarp from cross section of fruit in two stages of development. *C,* 6 weeks after full bloom; *D,* 2 weeks after fruit abscission. *E, F,* inner part of pericarp (pit) from cross section of fruit in two stages of development. *E,* 6 weeks after full bloom, at cessation of cell division in this region. *F,* 8 weeks after full bloom, at initiation of lignification. (Drawn from photographs in Sterling, *Torrey Bot. Club Bul.,* 1953.)

anthesis; in others (pepper, some varieties of tomato), during growth after anthesis (Kano et al., 1957). Polarity of cell division and of cell expansion are both related to the determination of fruit shape. The avocado fruit deviates from most of the investigated fruit in its method of development (Schroeder, 1953). Its initial period of growth involves cell division and cell enlargement; later, cell division becomes the major factor of enlargement and continues as long as the fruit remains on the tree.

Periderm and lenticels

Cork may replace the outer epidermis of the fruit wall (Meissner, 1952). In some fruits (e.g., *Aesculus*) a phellogen of subepidermal origin forms cork cells and lenticels. The phenomenon is obligate in some species; in others it depends on environmental conditions. In *Cucumis* the cork layer is discontinuous; it forms a corky net. *Cucumis* develops cracks during the later stages of fruit growth. These cracks may occur beneath the stomata and in stomata-free areas. A phellogen arises under the dead cells in the crack area and produces cork cells. The cork bulges out on the two sides of the crack but remains thin and continues to split in the bottom of the crack. The resulting structure resembles a lenticel. The russeting of apples results from periderm formation in parts of the fruit and suberization of cells under stomata and scars of trichomes. The formations developing under the stomata resemble lenticels but rarely have a phellogen (cf. Esau, 1953, p. 593).

Stresses resulting from growth of fruits may produce modifications

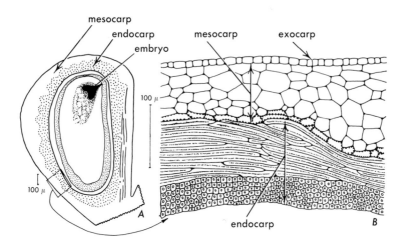

Fig. 21.6. *A*, drupelet of raspberry (*Rubus*) in longitudinal section and, *B*, a fragment of its pericarp. The sclereids of the two parts of endocarp are oriented crosswise to one another. (Drawn from photographs in Reeve, *Amer. Jour. Bot.*, 1954.)

of stomata (Miličić and Despot, 1957). The stomata may be torn, especially along the line of union of the guard cells. In flowering quince (*Chaenomeles*) the substomatal parenchyma assumes a looser structure after the tearing, and some cork cells are sometimes formed by tangential divisions beneath the stomata. In *Prunus* (cherry and nectarine) cork was not observed, but in this genus divisions may occur in cells next to and beneath stomata and thus prevent the tearing apart of the guard cells.

Abscission

Fruits may abscise in different stages of development; at maturity they may do so with the seeds enclosed or after the seeds are released. The abscission of fruits, as that of leaves, may be prepared by cell division, or the separation layer may differentiate without divisions. In fruit clusters often two to three separation layers develop; first the fruit abscises, then the stalks. In certain species of *Prunus* the fruit abscises above the pedicel, then the pedicel, finally the spur. The scar left by the spur is healed over by periderm. Some fruits (e.g., *Pyrus*) separate with their stalks.

In the separation layer of mature apple fruits (cf. Esau, 1953, p. 593) cells in several tiers increase in size; secondary walls in sclerenchyma cells lose their anisotropic qualities; and the middle lamellae, primary walls, and much of the secondary wall thickening dissolve. Vessels and fibers are ruptured.

The abscission of fruits, as that of leaves, is being studied from the physiological viewpoint, especially with regard to chemicals that retard or accelerate abscission (cf. chapter 18).

REFERENCES

Baumann-Bodenheim, M. G. Prinzipien eines Fruchtsystems der Angiospermen. *Schweiz. Bot. Gesell. Ber.* 64:94–112. 1954.

Bradbury, D., I. M. Cull, and M. M. MacMasters. Structure of the mature wheat kernel. I. Gross anatomy and relationships of parts. *Cereal Chem.* 33:329–342. 1956a.

Bradbury, D., M. M. MacMasters, and I. M. Cull. Structure of mature wheat kernel. II. Microscopic structure of pericarp, seed coat, and other coverings of the endosperm and germ of hard red winter wheat. *Cereal Chem.* 33:342–360. 1956b.

Esau, K. *Plant anatomy.* New York, John Wiley and Sons. 1953.

Fahn, A., and M. Zohary. On the pericarpical structure of the legumen, its evolution and relation to dehiscence. *Phytomorphology* 5:99–111. 1955.

Hartl, D. Struktur und Herkunft des Endocarps der Rutaceen. *Beitr. zur Biol. der Pflanz.* 34:35–49. 1957.

Holden, D. J. Factors in dehiscence of the flax fruit. *Bot. Gaz.* 117:294–309. 1956.

Kano, K., T. Fujimura, T. Hirose, and Y. Tsukamoto. Studies on the thickening growth of garden fruits. I. On the cushaw, egg-plant and pepper. *Kyoto Univ. Res. Inst. Food Sci. Mem.* 12:45–90. 1957.

Kiesselbach, T. A., and E. R. Walker. Structure of certain specialized tissues in the kernel of corn. *Amer. Jour. Bot.* 39:561–569. 1952.

Matienko, B. T. Ob anatomo-morfologicheskoĭ prirode tsvetka i ploda tykvennykh. [On the antomico-morphological nature of the flower and fruit of the cucurbit family.] *Akad. Nauk SSSR Bot. Inst. Trudy Ser.* 7. *Morf. i Anat. Rast.* 4:288–322. 1957.

Meissner, F. Die Korkbildung der Früchte von *Aesculus-* und *Cucumis-* Arten. *Österr. Bot. Ztschr.* 99:606–624. 1952.

Miličić, D. Zur Kenntnis der Phlobaphenkörper in Früchten einiger *Malus-*Arten. *Protoplasma* 41:327–335. 1952.

Miličić, D., and S. Despot. A pneumatodnim organima nekih plodova. [Über die Pneumatoden einiger Früchte.] *Jugoslav. Akad. Znan. i Umjetn. Od. Prirod. Nauk* 312:77–93. 1957.

Narayanaswami, S. The structure and development of the caryopsis in some Indian millets. V. *Eleusine coracana* Gaertn. *Papers Michigan Acad. Sci., Arts and Letters* 40:33–46. 1955.

Nitsch, J. P. The physiology of fruit growth. *Annu. Rev. Plant Physiol.* 4:199–236. 1953.

Reeve, R. M. Fruit histogenesis in *Rubus strigosus.* I. Outer epidermis, parenchyma, and receptacle. *Amer. Jour. Bot.* 41:152–160. 1954a.

Reeve, R. M. Fruit histogenesis in *Rubus strigosus.* II. Endocarp tissues. *Amer. Jour. Bot.* 41:173–181. 1954b.

Sanders, E. H. Developmental morphology of the kernel in grain sorghum. *Cereal Chem.* 32:12–25. 1955.

Schroeder, C. A. Growth and development of the Fuerte avocado fruit. *Amer. Soc. Hort. Sci.* 61:103–109. 1953.

Sterling, C. Developmental anatomy of the fruit of *Prunus domestica* L. *Torrey Bot. Club Bul.* 80:457–477. 1953.

Stopp, K. Karpologische Studien. I. Vergleichend-morphologische Untersuchungen über die Dehiszenzformen der Kapselfrüchte. *Abhandl. Akad. Wiss. Lit. Mainz. Math.-Nat. Kl.* 1950 (7):165–210. 1950.

22. THE SEED

THE SEED OF ANGIOSPERMS—THE SOLE TOPIC OF THIS CHAPTER—CON-sists of the embryo, variable amounts of endosperm (or none), and the seed coat, or testa. The various structural details of the ovule are more or less preserved during the transformation of the ovule into the seed. The embryo alone or the embryo and the endosperm come to occupy the largest volume of the seed, whereas the integuments in developing into the seed coat usually undergo a considerable reduction in thickness and a partial disorganization. The micropyle may remain visible as an occluded pore or may be obliterated. The funiculus, whole or in part, abscises from the ovule and leaves a scar, the hilum, which usually is looked upon as the part of the seed most permeable to water. In anatropous ovules the part of the funiculus that is adnate to the ovule remains recognizable as a longitudinal ridge, the raphe, on one side of the seed. The variability in structure of seeds throughout the angiosperms and their relative constancy in narrower groups permit the use of seed characteristics in classification of plants (e.g., McClure, 1957).

The embryo of monocotyledons and dicotyledons has been discussed in chapter 2. The present chapter deals with the other two basic structural components of the seed, the seed coat and the endosperm.

SEED COAT

Variations in the structure of the seed coat are dependent, on the one hand, on specific features of the ovule, especially the number and

326

thickness of integuments and the arrangement of the vascular tissues and, on the other, on developmental changes the integuments undergo as the seed matures. The most familiar seeds are those whose coats are dry. In seeds of certain nonangiospermic vascular plants the integument becomes a fleshy edible structure (Pijl, 1955). This type of development is considered to be primitive and to have given way to seeds with reduced edible parts (such as the aril, caruncle, and other restricted structures) and then to dry seeds enclosed in a fleshy pericarp.

The structure of the seed coat is best understood if it is studied developmentally. Such studies have been carried out, for example, on the seeds of *Asparagus, Beta,* and *Lycopersicon* (cf. Esau, 1953, pp. 604–609); *Ricinus* (Singh, 1954); *Phaseolus lunatus* (Sterling, 1954); representatives of Cruciferae (Černohorský, 1947) and Cucurbitaceae (Singh, 1953).

In *Ricinus,* the outer of the two integuments is represented in the seed by (*a*) the outer epidermis composed of tangentially elongated cells containing coloring matter, (*b*) the inner epidermis of cells columnar in shape, and (*c*) layers of crushed parenchyma cells between the two epidermal layers. The outer epidermis of the inner integument is palisade-like and sclerified. The other layers of this integument are crushed and become papery in consistency. A caruncle develops through divisions of cells of the outer integument near the micropyle.

The ovules of the Cruciferae have rather thick integuments. The outer has two to five cell layers, the inner up to ten. The epidermal cells of the outer integument become almost filled with mucilaginous material which appears in layers (fig. 22.1). The material swells when

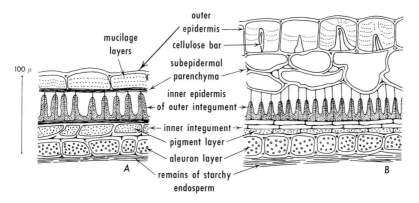

Fig. 22.1. Cross sections of seed coat and aleuron layer of *Brassica* (*A*), and *Sinapis* mustard (*B*). (After Černohorský, *Graines des Crucifères de Bohême,* Opera Botanica Cechica, 1947.)

it comes in contact with water and, in some species, bursts the outer wall of the cell. Heavy cellulosic deposit may form radially oriented bars, one in each cell, on the inner tangential wall (fig. 22.1, *B*). If subepidermal parenchyma is present in the outer integument, it may develop thickened walls (fig. 22.1, *B*) or may become crushed and absorbed (fig. 22.1, *A*). The inner epidermis of the outer integument is the strongest layer in most species, for its cells develop lignified thickenings on the radial and inner tangential walls (fig. 22.1). These cells are structurally most distinctive and most useful for systematic diagnosis. The inner integument dies and becomes compressed. In some species the inner epidermis of this integument becomes the pigment layer.

The seed coat of the Cucurbitaceae (fig. 22.2, *A*) is derived from two integuments, the outer of which contains two or three layers of cells, the inner a rather large variable number of layers. The inner integument disintegrates more or less early in the development of the seed, whereas the outer undergoes periclinal divisions and forms a complex system of structurally distinct layers. In the mature seed these layers and their structure are the following (fig. 22.3, *A*). The layer that remains the outermost after the periclinal divisions are completed—now the epidermis of the seed coat—consists of radially or tangentially elongated cells bearing rod-like lignified thickenings on the radial walls—at least in parts of the seed—and containing some coloring matter. The subepidermal tissue, variable in thickness, may be differentiated in two parts: the outer, containing pigments; and the inner, with thick walls. Beneath these subepidermal tissues is a distinct layer of sclerenchyma, one cell in thickness and often palisade-like in cell

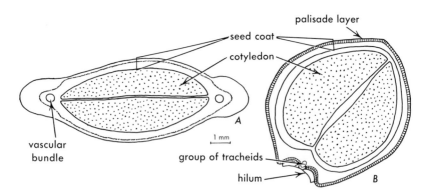

Fig. 22.2. Diagrams of cross sections of seeds of pumpkin, *Cucurbita* (*A*) and bean, *Phaseolus* (*B*).

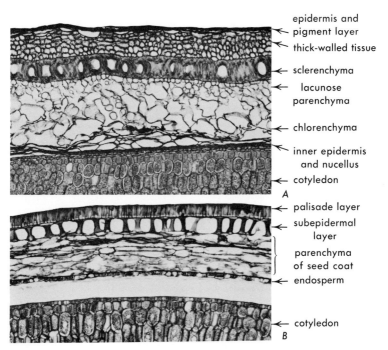

epidermis and
pigment layer

thick-walled tissue

sclerenchyma

lacunose
parenchyma

chlorenchyma

inner epidermis
and nucellus

cotyledon

A

palisade layer

subepidermal
layer

parenchyma
of seed coat

endosperm

cotyledon

B

Fig. 22.3. Cross sections of seed coats and parts of embryos of pumpkin, *Cucurbita* (*A*), and soybean, *Glycine* (*B*). (Both, ×90. *B*, slide courtesy of A. T. Guard.)

arrangement. Deeper inward is lacunose parenchyma composed of stellate cells, and then chlorenchyma, rather thin-walled compact tissue. The inner epidermis is composed of small cells and may be green. The nucellus is represented by two to four layers in the mature seed and its epidermis has a cuticle. Small amounts of endosperm may be present. In dry seeds the chlorenchyma usually separates from the seed coat as a thin green membrane and, together with the inner epidermis, closely envelops the embryo and the remains of nucellus and endosperm.

The seeds of Leguminosae (fig. 22.2, *B;* Papilonatae) have been frequently studied. The inner of the two integuments disappears during the ontogeny of the seed, whereas the outer integument differentiates into a variety of distinct layers. The outermost layer, the epidermis, remains uniseriate and develops into the palisade layer characteristic of leguminous seeds (figs. 22.3, *B*, and 22.4). This layer is composed of sclereids—macrosclereids (chapter 6), or Malpighian cells—with unevenly thickened walls. Two palisade layers occur in the hilum

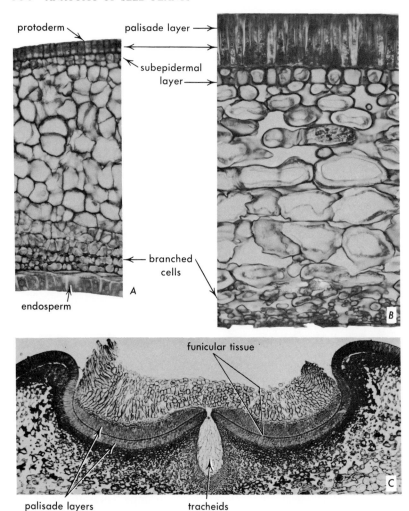

Fig. 22.4. Seed coat of bean (*Phaseolus*) in cross sections. Two stages of differentiation; *A*, young; *B*, almost mature; *C*, hilum region. (*A*, *B*, ×280; *C*, ×43.)

region. The outer of these is derived from the funiculus (fig. 22.4, *C*). The cells of the subepidermal layer differentiate into the so-called columnar cells, also termed pillar cells, hourglass cells, or osteosclereids, depending on the distribution of wall thickenings and shape of cells. Large intercellular spaces may occur between these cells (fig. 22.3, *B*). The deeper tissue is a lacunose parenchyma with large tangentially elongated cells in the outer part and smaller much branched cells in the inner (fig. 22.4, *B*). The subepidermal layer and the parenchyma beneath it have common origin (fig. 22.4, *A*). The vascular system of many legume seeds is well developed. From the funiculus the vascular bundle extends to the chalazal region where it branches. A compact group of tracheids of unknown role occurs in the hilum region (fig. 22.4, *C*).

The palisade layer has attracted much attention because its structure in certain hard legume seeds is assumed to be causally connected with their high degree of impermeability and thus their germinability. The so-called light line of the palisade cells is thought to be the particularly impermeable region. The light-line effect results from the high degree of refraction in a restricted region in epidermal walls. In sections of seeds this region is oriented tangentially, somewhat above the middle of each cell. The refractive regions of adjacent cells match in position and, therefore, form a continuous line through the epidermis as seen in section. In three-dimensional aspect the formation responsible for the line effect can be visualized as a sheath following the outline of the seed. The cell wall in the light-line region is described as being especially compact (Cavazza, 1950; Coe and Martin, 1920). Experiments with entry of dyes into uninjured seeds indicate that the light line presents a barrier to their passage. The report that the palisade layer may be covered with other highly impermeable cell layers (Cavazza, 1950) appears to be based on erroneous observations (cf. Steiner and Jancke, 1955).

The hard legume seeds achieve and maintain a very low per cent moisture which is not affected by fluctuations in moisture content of the surrounding air. The attainment of this high degree of desiccation is ascribed to a combination of intensely impermeable testa with valvular action of the hilum (Hyde, 1954). The hilum is said to act like a hygroscopic valve. A fissure occurs along the groove of the hilum. This fissure opens when the seed is surrounded by dry air and closes when the outside air is moist. Thus, the entry of moisture is prevented, but loss of moisture is permitted.

The occurrence of highly impermeable seed coats is one of the important factors in delayed germination of seeds not only in Leguminosae but also in many other angiosperms (cf. Toole et al., 1956). In this connection the occurrence of cuticular layers in seeds is of particular interest. These layers have their origin in the cuticles of the ovule (cf. Esau, 1953, p. 555). The young ovule bears a cuticle over the entire surface. After the development of the integument or integuments, two or three cuticular layers may be distinguished, those of the integuments and that of the nucellus. During the development of the seed coat the breakdown of integumentary tissue may result in juxtaposition, if not fusion, of the various cuticles, so that the embryo and endosperm may become enclosed in a rather prominent cuticular sheath interrupted only at the hilum. Moreover, the outer epidermis of the seed is also frequently reported to have a cuticle.

ENDOSPERM

As is well known, in the angiosperms the product of fusion of the two polar nuclei of the embryo sac with one sperm nucleus from the pollen tube gives rise to the endosperm initial. Two principal patterns of endosperm development from the initial are recognized (cf. Swamy and Ganapathy, 1957): (*a*) the nuclear, characterized by multiplication of nuclei without an immediately following cytokinesis; (*b*) the cellular, in which nuclear division is associated with cytokinesis. It has been reported that the nuclear endosperm is correlated with various primitive vegetative and floral characteristics (Sporne, 1954). On the other hand, statistical analysis of correlation between the type of endosperm development and the type of perforation plates in the xylem vessels indicates a positive correlation between the nuclear type and the simple (porous), that is, the most advanced, type of perforation plate (Swamy and Ganapathy, 1957).

The longevity of the endosperm and the amount and kind of storage material in this tissue vary greatly in different species. The most common storage material in the endosperm is starch (fig. 22.5, *A*). Other carbohydrates, such as the polysaccharides mannans (Meier, 1958) and hemicelluloses (Crocker and Barton, 1953), deposited as wall material, are also regarded as storage material (e.g., *Asparagus,* fig. 22.5, *B–D; Coffea, Diospyros, Iris, Phoenix*). The starch is formed in plastids, although nonplastid starch also has been described in cereals (Aleksandrov and Aleksandrova, 1954). The starchy endosperm appears to be nonliving at maturity, for the nuclei degenerate (Aleksandrov and Aleksandrova, 1954; Bradbury et al., 1956). According to microchemical tests (Müller, 1943), nonliving starchy endosperm and perisperm (nucellar storage tissue) occur in numerous families, for example, in the Juncaceae, Cyperaceae, Gramineae, Polygonaceae, Caryophyllaceae, Chenopodiaceae, and others. The starchy endosperm appears to be living in the Liliaceae, Amaryllidaceae, *Viola, Ricinus,* and *Daucus.* In cereal grains the proteinaceous aleurone layer located on the periphery of the endosperm is living.

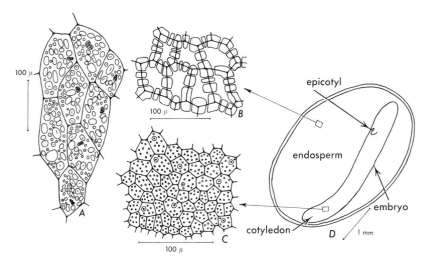

Fig. 22.5. Endosperm and embryo. *A,* starchy endosperm of rye kernel. *B–D, Asparagus officinalis;* thick-walled endosperm (*B*), parenchyma of cotyledon (*C*), and longitudinal section of seed (*D*). (*B–D,* after Robbins and Borthwick, *Bot. Gaz.,* 1925.)

A rather unfamiliar endosperm structure has been found in certain grasses (Matlakówna, 1912; Dore, 1956). It is a starchy nonliving endosperm, rich in oil, devoid of cell walls, and having a mushy, oily consistency. In species having such endosperm the scutellum bears an apical prolongation that is deeply imbedded in the endosperm, and the scutellar epithelial cells (epidermis in contact with the endosperm) grow out into long hairs.

Seeds having no endosperm, or only small remains of one, are called *exalbuminous* seeds; those with endosperm, *albuminous* seeds. In exalbuminous seeds the embryo is large in relation to the seed as a whole and stores food reserves in its body parts, particularly the cotyledons (e.g., Leguminosae, Cucurbitaceae, Compositae). As in the endosperm, various substances may be stored in the cotyledons (fig. 22.5, *C*). Starch is a common reserve product, but carbohydrates may be stored also as hemicelluloses in the cell walls of cotyledons (e.g., *Tropaeolum, Primula, Impatiens, Lupinus;* Crocker and Barton, 1953). In albuminous seeds the size relations between the embryo and endosperm vary greatly.

A nutritional relation is assumed to exist between the embryo and endosperm. Commonly the development of the embryo depends on the presence of endosperm even, apparently, in pseudogamous species in which the egg cell develops into the embryo without fertilization

(Rutishauser, 1954). The digestion of the endosperm by the developing embryo is part of a complex series of digestive phenomena that bring about a transfer of food materials from the old sporophyte to the new. The ovular tissues have been observed to accumulate starch after pollination and then undergo destruction in relation to embryo development. In the early stages of embryo development the endosperm appears to be a tissue that transmits the food materials from the ovular tissues to the embryo. Later the endosperm itself is destroyed, in part or completely, as the embryo advances in the embryo sac during its growth. The endosperm which is retained in albuminous seeds is utilized during germination.

Studies on living material have shown that chloroplasts may be present in developing embryos and also in the endosperm (cf. Ioffe, 1957). These observations are still to be correlated with what is known of the nutrition of the developing embryo, and they sharply underline the inadequacy of our understanding of the biochemical aspects of the development of the young sporophyte.

REFERENCES

Aleksandrov, V. G., and O. G. Aleksandrova. Ob otmiranii i razrushenii ĩader v kletkakh endosperma zlakov kak odnom iz vazhneĩshikh faktorov, obuslavlivaĩushchikh naliv zernovki. [Death and disintegration of nuclei in cells of endosperm of cereals as one of the most important factors determining the filling of the grain.] *Izvest. Akad. Nauk SSSR Ser. Biol.* 1954:88–103. 1954.

Bradbury, D., M. M. MacMasters, and I. M. Cull. Structure of the mature wheat kernel. III. Microscopic structure of the endosperm of hard red winter wheat. *Cereal Chem.* 33:361–373. 1956.

Cavazza, L. Recherches sur l'imperméabilité des graines dures chez les Légumineuses. *Schweiz. Bot. Gesell. Ber.* 60:596–610. 1950.

Černohorský, Z. Grains des Crucifères de Bohême. Étude anatomique et morphologique. *Opera Bot. Čech.* Vol. 5. 92 pp. 1947.

Coe, H. S., and J. N. Martin. Sweet-clover seed. *U. S. Dept. Agric. Bul.* 844. 1920.

Crocker, W., and L. V. Barton, Editors. *Physiology of seeds. An introduction to the experimental study of seed and germination problems.* Vol. 29. Waltham, Mass., Chronica Botanica Company. 1953.

Dore, W. G. Some grass genera with liquid endosperm. *Torrey Bot. Club Bul.* 93:335–337. 1956.

Esau, K. *Plant anatomy.* New York, John Wiley and Sons. 1953.

Hyde, E. O. C. The function of the hilum in some Papilionaceae in relation to ripening of the seed and the permeability of the testa. *Ann. Bot.* 18:241–256. 1954.

Ioffe, M. D. Razvitie zarodysha i endosperma u pshenitsy, konskikh bobov i redisa. [Development of embryo and endosperm in wheat, horse beans, and radishes.] *Akad. Nauk SSSR Bot. Inst. Trudy Ser.* 7. *Morf. i Anat. Rast.* 4:211–269. 1957.

Matlakówna, M. Ueber Gramineenfrüchte mit weichem Fettendosperm. *Acad. Sci. Cracovie Bul. Ser. B* 1912:405–416. 1912.

McClure, D. S. Seed characters of selected plant families. *Iowa State Coll. Jour. Sci.* 31:649–682. 1957.

Meier, H. On the structure of cell walls and cell wall mannans from ivory nuts and from dates. *Biochem. et Biophys. Acta* 28:229–240. 1958.

Müller, D. Tote Speichergewebe in lebenden Samen. *Planta* 33:721–727. 1943.

Pijl, L. van der. Sarcotesta, aril, pulpa and the evolution of the angiosperm fruit. I and II. *Nederland. Akad. Wetenschap. Proc. Ser. C Biol. and Med. Sci.* 58:154–161; 307–312. 1955.

Rutishauser, A. Entwicklungserregung der Eizelle bei pseudogamen Arten der Gattung *Ranunculus. Schweiz. Akad. Wiss. Bul.* 10:491–512. 1954.

Singh, B. Studies on the structure and development of seeds of Cucurbitaceae. *Phytomorphology* 3:224–239. 1953.

Singh, R. P. Structure and development of seeds in Euphorbiaceae: *Ricinus communis* L. *Phytomorphology* 4:118–123. 1954.

Sporne, K. R. A note on nuclear endosperm as a primitive character among dicotyledons. *Phytomorphology* 4:275–278. 1954.

Steiner, M., and I. Jancke. Sind die Malpigischen Zellen die Epidermis der Leguminosentesta? *Österr. Bot. Ztschr.* 102:542–550. 1955.

Sterling, C. Development of the seed coat of lima bean (*Phaseolus lunatus* L.). *Torrey Bot. Club Bul.* 81:271–287. 1954.

Swamy, B. G. L., and P. M. Ganapathy. On endosperm in dicotyledons. *Bot. Gaz.* 119:47–50. 1957.

Toole, E. H., S. B. Hendricks, H. A. Borthwick, and V. K. Toole. Physiology of seed germination. *Annu. Rev. Plant Physiol.* 7:299–324. 1956.

GLOSSARY

 T HIS GLOSSARY DEFINES SELECTED TERMS USED IN THE LANGUAGE OF plant anatomy. A few cytological and morphological terms are explained also, mostly those that are apt to be forgotten after an introductory course in botany, or those the definitions of which are varied. Although a uniformity in the use of terms is desirable, some terms are bound to change in meaning with the increase in our understanding of the phenomena they indicate. Definitions in this glossary that may deviate from those commonly given for the terms in question will inform the reader how these terms are used in the present book.

Abaxial. Directed away from the axis.

Abscission layer. In abscission zone. Layer of cells the disjunction or breakdown of which separates a plant part, such as leaf, fruit, flower, etc., from the plant. **Syn.** separation layer.

Abscission zone. Zone at base of leaf, or fruit, or flower, or other plant part, that contains the *abscission layer* and the *protective layer,* both of which play a role in the separation of the plant part from the plant.

Accessory cell. Of stoma. See *subsidiary cell.*

Accessory parts (in fruit). Parts not derived from the ovary but associated with it in fruit.

Accessory transfusion tissue. In leaves of certain gymnosperms. Transfusion tissue located within the mesophyll rather than associated with the vascular bundle.

Acicular crystal. Needle-shaped crystal.

Acropetal (development). Produced in a succession toward the apex, as applied to organs, tissues, or cell series. The opposite of *basipetal,* but similar to *basifugal.*

Adaxial. Directed toward the axis.

Adaxial meristem. Meristematic tissue on the adaxial side of the leaf that contributes to the increase in thickness of the petiole and midrib.

Adult (plant). A term of convenience for contrasting a plant in early stages of growth, such as embryo or seedling, with one in later stages of growth.

Adventitious. Refers to a structure arising not in its usual place: to roots arising from stems or leaves or old parts of a root; to buds arising other than as terminal or axillary structures.

Aerenchyma. Parenchyma tissue characterized by particularly large intercellular spaces of *schizogenous, lysigenous,* or *rhexigenous* origin.

Aggregate cup fruit. A fruit derived from an apocarpous (free carpels) perigynous (ovary superior but with a cup-shaped structure bearing perianth and stamens) flower. Fruit composed of fruitlets, each with its own pericarp, but whole fruit may have a fruit wall of accessory parts.

Aggregate free fruit. A fruit derived from an apocarpous (free carpels) hypogynous (superior ovary) flower. Fruit composed of fruitlets, each with its own pericarp.

Aggregate ray. In secondary vascular tissues. A group of small rays arranged so as to appear to be one large ray.

Albedo. In citrus fruit. White tissue of the rind.

Albuminous cells. In gymnosperm phloem. Certain ray and phloem-parenchyma cells that appear closely associated with the sieve elements, morphologically and physiologically. In contrast to the companion cells of angiosperms, are commonly not derived from the same cells as the sieve elements.

Albuminous seed. A seed that contains endosperm in the mature state.

Aleuron (or aleurone). Granules of protein present in the seeds of numerous plants. Usually restricted to the external part, the *aleuron layer,* of endosperm, as in wheat or other cereals.

Aliform paratracheal parenchyma. In wood. Vasicentric groups of parenchyma cells have wing-like extensions, as seen in cross sections. See also *paratracheal parenchyma* and *vasicentric paratracheal parenchyma.*

Alternate pitting. In tracheary elements. Pits in diagonal rows.

Amphicribral vascular bundle. Concentric vascular bundle in which the phloem surrounds the xylem.

Amphiphloic siphonostele. A stele having a pith and two phloem regions, one external and the other internal to the xylem.

Amphivasal vascular bundle. Concentric vascular bundle in which the xylem surrounds the phloem.

Amyloplast. A colorless plastid (leucoplast) forming storage starch.

Analogy. Pertains to having the same function but a different phylogenetic origin; usually also a different structure.

Anastomosis. Interconnection of elongated structures, such as veins or other cell strands, the whole forming a net, or reticulum.

Anatomy. The area of morphology that treats of the internal structure of organisms.

Ångström. Tenth millimicron. Symbol Å. or Å.U. (Ångström unit).

Angular collenchyma. Collenchyma tissue in which the cell-wall thickening is deposited mainly in the angles where several cells are joined together.

Anisocytic stoma. A type in which three subsidiary cells, one distinctly smaller than the other two, surround the stoma. Old term, *cruciferous.*

Anisotropic. Having different properties along different axes. Optically anisotropic, affecting the light passing along certain axes; causing polarization of light and double refraction.

Annual ring. In wood. See *growth layer.* This term is deprecated because more than one growth layer may be formed during one year.

Annular cell-wall thickening. In tracheary elements. Secondary wall material deposited on the primary wall as rings.

Anomalous secondary growth. Commonly used with reference to unusual types of secondary growth.

Anomocytic stoma. A type in which no subsidiary cells are associated with the guard cells. Old term, *ranunculaceous.*

Anthesis. The time of full expansion of the flower, or the time when fertilization occurs.

Anticlinal. Perpendicular to the surface.

Apex (pl. apices or apexes), or summit. Tip, topmost part, or pointed end of anything. In shoot or root the part containing the apical meristem.

Apical cell. The single initial cell in an apical meristem of root or shoot. Characteristic of many lower vascular plants.

Apical meristem. A group of meristematic cells at the tip of root or shoot that by division produce the precursors of the primary tissues of root or shoot. May be vegetative (that is, initiating vegetative tissues and organs) or reproductive (that is, in angiosperms, floral meristem initiating floral tissues and organs, including the reproductive cells).

Apotracheal parenchyma. In wood. Axial parenchyma typically independent of the pores, or vessels. Includes *boundary* (*initial* or *terminal*), *banded,* and *diffuse* apotracheal.

Apposition (of cell wall). Growth by successive deposition of wall material, layer upon layer.

Areole. As used with regard to leaves, a small area of mesophyll delimited by intersecting veins.

Aril. A fleshy outgrowth arising at the base of the ovule and enveloping the seed.

Articulated laticifer. Compound laticifer. A laticifer composed of more than one cell. The walls between contiguous cells may or may not be partly or entirely removed during ontogeny. May be anastomosing or nonanastomosing.

Aspirated pit. In wood. Bordered pit in which the pit membrane is laterally displaced and the torus blocks the aperture.

Astrosclereid. A branched (ramified) type of sclereid.

Atactostele. A stele with the vascular bundles as though scattered within the ground tissue.

Auricle. A small lobe or appendage of a leaf.

Axial parenchyma. In secondary vascular tissues. Parenchyma cells in the axial system; as contrasted with ray-parenchyma cells.

Axial system. In secondary vascular tissues. All cells derived from the fusiform cambial initials and oriented with their longest diameter parallel with the main axis of stem or root. Other terms: *vertical system* and *longitudinal system.*

Axial tracheid. Tracheid in the axial system of the wood; as contrasted with ray tracheid.

Axillary meristem. Meristem located in the axil of a leaf and giving rise to an axillary bud.

Banded apotracheal parenchyma. In wood. Concentric bands of axial parenchyma, as seen in cross section, typically independent of the pores, or vessels. See also *apotracheal parenchyma.*

Bark. A nontechnical term applied to all tissues outside the vascular cambium or the xylem. In older trees may be divided into dead outer bark and living inner bark (usually consisting of phloem). See also *rhytidome.*

Bars of Sanio. See *crassulae*.

Basifugal. See *acropetal*.

Basipetal (development). Produced in a succession toward the base, as applied to organs, or tissues, or cell series. The opposite of *acropetal* and *basifugal*.

Bast fiber. Originally phloem fiber, now any extraxylary fiber.

Bicollateral vascular bundle. A bundle having the phloem on two sides of the xylem.

Bifacial leaf. A leaf with palisade parenchyma on one side of the blade and spongy parenchyma on the other side. A dorsiventral leaf.

Biseriate ray. In secondary vascular tissues. A ray two cells wide.

Blind pit. A pit without a complementary pit in an adjacent wall, which may face the lumen of a cell or an intercellular space.

Bordered pit. A pit in which the pit membrane is overarched by the secondary wall.

Bordered pit-pair. An intercellular pairing of two bordered pits.

Boundary apotracheal parenchyma. In wood. Axial parenchyma cells either occur singly or form a more or less continuous layer at the beginning of a season's growth (*initial*) or at the close of one (*terminal*). See also *apotracheal parenchyma*.

Brachysclereid. A short, roughly isodiametric sclereid, resembling a parenchyma cell in shape. Stone cell.

Branched pit. See *ramiform pit*.

Branch gap. In the nodal region of a stem. An area of parenchyma in the vascular cylinder of the main stem occurring where the branch traces are bent away from the vascular region of the main stem toward the branch. Usually confluent with the leaf gap of the leaf subtending the branch.

Branch root. A root arising from another, older root; also called secondary root if the older root is the taproot, or primary root.

Branch trace. A vascular bundle in the main stem extending between its connection with the vascular tissue of the branch and that with another vascular unit in the main stem. Actually a leaf trace of one of the first leaves (prophylls) on the branch.

Bulliform cell. An enlarged epidermal cell occurring in longitudinal rows of similar cells in leaves of grasses. Also called *motor cell* because it is often thought to play a role in the rolling and unrolling of leaves.

Bundle cap. Sclerenchyma or thick-walled parenchyma associated with a vascular bundle and appearing like a cap on the phloem or xylem side, as seen in cross section.

Bundle sheath. Layer or layers of cells enclosing a vascular bundle. May consist of parenchyma or sclerenchyma.

Bundle-sheath extension. A plate of ground tissue extending from a bundle sheath of a smaller vein, located in the mesophyll, to the epidermis. May be present on both sides of the bundle or only one. May consist of parenchyma or sclerenchyma.

Callose. An apparently amorphous polysaccharide giving glucose on hydrolysis. In seed plants, common wall constituent in the sieve areas of sieve elements, but observed also in parenchymatous cells after injury.

Callus. A tissue composed of large thin-walled cells usually developing as a result of injury. (Sometimes used also for accumulations of callose, a usage that should be abandoned.)

Callus tissue. See *callus*.

Calyptrogen. In apical meristem of root. Meristem giving rise to the rootcap independently of the other initials in the apical meristem.

Cambial initials. In vascular cambium and phellogen. Cells forming derivatives by periclinal divisions in two directions. In vascular cambium classified into *fusiform* *initials* (source of the axial cells of xylem and phloem) and *ray initials* (source of the ray cells).

Cambium. A meristem with products of divisions arranged orderly in parallel files. Preferably applied only to the two lateral meristems, the *vascular cambium* and the *cork cambium*, or *phellogen*. Consists of one layer of initials and their undifferentiated products, or derivatives.

Caruncle. A fleshy protuberance near the hilum of a seed.

Casparian dot. See *Casparian strip.*

Casparian strip, or band. A band-like wall formation within primary walls and containing lignin and suberin. Typical of endodermal cells in roots, in which it occurs in anticlinal walls, radial and transverse. Appears as dots in sectional views of walls.

Cataphylls. Leaves inserted at low levels of plant or shoot, as bud scales, rhizome scales, etc.

Cartilaginous. Like cartilage. A firm and elastic tissue, transluscent in color.

Cauline. Belonging to the stem or arising from it.

Caulis. Stem.

Cell. Structural and physiological unit of a living organism. The plant cell consists of cell wall and protoplast; in nonliving state, of cell wall only or cell wall and some nonliving inclusions.

Cell membrane. Translation of the German *Zellmembran* which refers in that language to the cell wall.

Cell plate. Structure appearing at telophase between the two new nuclei formed during mitosis and indicating the beginning of the division of a cell (cytokinesis) by means of a new cell wall. Formed in the *phragmoplast* and possibly consists of middle lamella substances.

Cellulose. A carbohydrate, more exactly a polysaccharide hexosan, which is the chief component of cell walls of plants. Consists of long chain-like molecules the basic units of which are anhydrous glucose residues of the formula $C_6H_{10}O_5$.

Cell wall. A more or less rigid membrane enclosing the protoplast of a cell. In higher plants composed of polysaccharides, chiefly cellulose, and other organic and inorganic substances. Term has triple usage: (1) cell wall of an individual cell, (2) partition between two cells composed of intercellular substance and two walls belonging to the two adjacent cells, (3) primary or secondary wall layer.

Cell-wall check. A fissure in the secondary cell wall, as in the tracheids of compression wood.

Central cylinder. See *vascular cylinder.*

Central mother cells. In apical meristem of shoot. Somewhat large and vacuolated cells in subsurface position. Commonly used with regard to gymnosperms in which these cells are derived from surface initials.

Centric mesophyll. A modification of isobilateral mesophyll in which the abaxial and adaxial palisade layers form a continuous layer. Found in narrow or cylindrical leaves.

Centrifugal (development). Produced or developing successively outward from the center. (Means sometimes also downward from the apex.)

Centripetal (development). Produced or developing successively inward toward the center. (Means sometimes also upward toward the apex.)

Chimera. A combination of tissues of different genetic constitution in the same part of the plant.

Chlorenchyma. Parenchyma tissue containing chloroplasts. Leaf mesophyll and other green parenchyma.

Chloroplast. Chlorophyll-containing specialized protoplasmic body in which sugar or starch is synthesized.

Chondriosomes. Most commonly used as another term for *mitochondria.*

Chromoplast. A specialized protoplasmic body containing pigments other than chlorophyll, usually yellow or orange carotenoid pigments.

Cicatrice. Scar left by the separation of one part from another, as of leaf from stem. Characterized by presence of substances protecting the new surface.

Circular bordered pit. A bordered pit with a circular aperture.

Closed venation. In leaf blade. Venation characterized by anastomosing veins.

Closing layer. In lenticel. One of the compact layers of cells formed periodically in alternation with the loose filling tissue.

Coenocyte. An aggregation of protoplasmic units; a multinucleate structure. In seed plants, sometimes applied to multinucleate cells.

Coleoptile. In Gramineae. Sheath enclosing the apical meristem with leaf primordia in the embryo. Often interpreted as the first leaf.

Coleorhiza. In Gramineae. Sheath enclosing the radicle in the embryo.

Collateral vascular bundle. A bundle having the phloem on one side of the xylem only.

Collenchyma. Supporting tissue composed of more or less elongated living cells with unevenly thickened walls, which are usually interpreted as primary walls.

Colleter. A multicellular glandular appendage (often a trichome) which secretes a sticky substance. Found on buds of many woody species.

Commissural vascular bundle. Used with reference to a small bundle interjoining larger bundles, as in leaves of grasses.

Companion cell. A specialized parenchyma cell in the phloem of an angiosperm associated with a sieve-tube member and arising from the same mother cell as the sieve-tube member.

Complementary tissue. In lenticel. See *filling tissue.*

Compound laticifer. A laticifer composed of more than one cell. *Articulated laticifer.*

Compound middle lamella. Collective term applied to two primary walls and the middle lamella; sometimes includes also the first secondary wall layers deposited on the primary walls.

Compound sieve plate. A sieve plate composed of several sieve areas in either scalariform or reticulate arrangement.

Compression wood. Reaction wood in conifers. Formed on the lower sides of branches and leaning or crooked stems, and characterized by dense structure, strong lignification, and certain other features.

Concentric vascular bundle. A vascular bundle with either the phloem surrounding the xylem (*amphicribral*) or the xylem surrounding the phloem (*amphivasal*).

Conducting tissue. See *vascular tissue.*

Confluent paratracheal parenchyma. In wood. Aliform groups of parenchyma cells are coalesced and thus form irregular tangential or diagonal bands, as seen in cross section. See also *paratracheal parenchyma* and *aliform paratracheal parenchyma.*

Connecting strands. In sieve elements. Protoplasmic strands which occur in the callose-lined pores of sieve areas and interconnect the protoplasts of contiguous sieve elements.

Contractile root. A root that undergoes contraction some time during its development and thereby brings about a change in position of the shoot parts with regard to the soil.

Coordinated growth. Growth of cells in a manner that involves no separation of walls, as opposed to *intrusive growth.* Sometimes called *symplastic growth.*

Copal. A resinous substance exuding from various tropical trees and hardening in air into roundish or irregular pieces, colorless, yellowish, reddish, or brown.

Cork. See *phellem.*

Cork cambium. See *phellogen.*

Cork cell. Nonliving cell with suberized walls derived from phellogen. Protective in function because its walls are highly impervious to water.

Corpus. In apical meristem of shoot. Group of cells located beneath anticlinally dividing peripheral layers (tunica), and dividing in various planes. Thus undergoing volume growth.

Cortex. The ground-tissue region between the vascular system and the epidermis. A primary tissue region.

Cotyledonary trace. The leaf trace of a cotyledon, located within the hypocotyl.

Crassulae (sing. crassula). In tracheids of gymnosperms. Thickenings of intercellular material and primary wall along the upper and lower margins of a pit-pair. Also called *bars of Sanio.*

Cross-field. Mainly in wood of conifers. A term of convenience for the rectangle formed by the walls of a ray cell and an axial tracheid, as seen in radial section.

Crystalloid. Protein crystal that is less angular than a mineral crystal, and swells in water.

Crystal sand. A mass of very fine free crystals.

Cuticle. A layer of waxy material, cutin, more or less impervious to water, on the outer wall of epidermal cells.

Cuticularization. The process of formation of cuticle.

Cutin. A wax-like highly complex fatty substance present in plants as an impregnation of epidermal walls and as a separate layer, the cuticle, on the outer surface of the epidermis. Makes the walls more or less impervious to water.

Cutinization. The process of impregnation with cutin.

Cystolith. A concretion of calcium carbonate on an outgrowth of the cellulose wall.

Cytochimera. A combination of tissues, the cells of which have different numbers of chromosomes in the same plant part.

Cytohistological zonation. In apical meristem. Presence of areas with distinctive cytological characteristics.

Cytokinesis. The process of division of a cell as distinguished from the division of the nucleus, or *karyokinesis.*

Cytology. The science dealing with the cell.

Cytoplasm. The visibly least differentiated part of the protoplasm that constitutes the groundmass enclosing all the other components of the protoplast.

Decussate. Leaf arrangement. In pairs alternately at right angles.

Dedifferentiation. A reversal in differentiation of a cell or tissue. Assumed to occur when more or less mature cells resume meristematic activity.

Dendroid venation. A type of venation in which the minor veins do not form closed meshes about small areas of mesophyll.

Derivative. In meristems. A cell produced by a meristematic cell by division.

Dermal tissue. Outer covering tissue of plant, that is, epidermis or periderm. Also *dermal tissue system.*

Dermal tissue system. The epidermis or the periderm.

Dermatogen. The meristem forming the epidermis. Arises from independent initials. One of the three histogens according to Hanstein.

Detached meristem. Meristem giving rise to an axillary bud and appearing detached from the apical meristem because of intervening vacuolated cells. Axillary meristem.

Determinate growth. Of apical meristem. Forms a restricted number of lateral organs and terminates growth. Characteristic of floral meristem.

Development. The process of change of a plant or plant part from its beginning to maturity.

Diacytic stoma. A type in which one pair of subsidiary cells, with their common walls at right angles to the long axis of the guard cells, surrounds the stoma. Old term, *caryophyllaceous.*

Diaphragmed pith. A pith in which transverse layers (diaphragms) of firm-walled cells alternate with regions of soft cells. The latter may collapse with age. Then the diaphragms divide the pith into small compartments.

Diarch. Primary xylem of root. Having two protoxylem strands, or protoxylem poles.

Dictyostele. A stele in which large overlapping leaf gaps dissect the vascular system into strands, each with the phloem surrounding the xylem.

Differentiation. Physiological and morphological change occurring in a cell, a tissue, an organ, or a plant during development from a meristematic or juvenile to a mature or adult state. Usually associated with an increase in specialization.

Diffuse apotracheal parenchyma. In wood. Axial parenchyma as single cells or as strands distributed irregularly among fibers, as seen in cross sections. See also *apotracheal parenchyma.*

Diffuse porous wood. Wood in which pores (vessels) are distributed fairly uniformly throughout a growth layer or change only gradually in size from early to late wood.

Dilatation. Growth of parenchyma by cell division in pith, rays, or axial system, causing a spread of tissue after it reached maturity.

Distal. Farthest from the place of origin or attachment.

Distichous. Leaf arrangement. Disposed in two vertical rows. Two ranked.

Dorsiventral. Possessing distinct upper and lower surfaces, as a leaf. Used also for *bifacial leaf.*

Druse. A globular compound crystal with many component crystals projecting from its surface.

Duct. An elongated space formed by separation of cells from one another (schizogenous origin), by dissolution of cells (lysigenous origin), or by a combination of the two processes (schizolysigenous origin).

Early wood. The wood formed in the first part of a growth layer. Less dense and has larger cells than the late wood. Replaces *spring wood.*

Ecotypic. From ecotype, or ecological type.

Ectophloic siphonostele. A stele having a pith and one phloem region, externally to the xylem.

Ectoplast. Outer boundary of the protoplast, next to the cell wall. Also called *plasma membrane.*

Elaioplast. A leucoplast type of plastid forming and storing oil.

Endarch xylem. Xylem strand in which the maturation of the cells progresses centrifugally; that is, the oldest elements (the protoxylem) are closest to the center of the axis. Typical of stems in seed plants; also of leaves, in which the oldest xylem is on the adaxial side.

Endocarp. The inner layer of the pericarp.

Endodermis. The layer of ground tissue forming a sheath around the vascular region and having specific wall characteristics—Casparian strips or secondary thickenings. In stem and root of a seed plant it is the innermost layer of the cortex.

Endodermoid. Resembling endodermis.

Endogenous. Arising from deep-seated tissues, as a lateral root.

Endosperm. The nutritive tissue formed within the embryo sac of seed plants.

Endothecium. In anther. A wall layer, usually with secondary wall thickenings.

Enucleate. Lacking a nucleus.

Epiblast. A small structure present opposite the scutellum in the embryo of some grasses. Sometimes considered to be a rudimentary cotyledon.

Epicarp. See *exocarp*.

Epidermis. The outer layer of cells primary in origin. If it is multiseriate (multiple epidermis), only the outermost layer differentiates epidermal characteristics.

Epithelial cell. A cell in the compact, apparently physiologically specialized tissue covering a free surface or lining a cavity. May be secretory.

Ergastic substances. Passive products of protoplasm, such as starch grains, fat droplets, crystals, and fluids. Occur in cytoplasm, vacuoles, and cell walls.

Eumeristem. Meristem composed of relatively small cells, approximately isodiametric in shape, compactly arranged, and having thin walls, dense cytoplasm, and large nuclei. Word means "true meristem."

Eustele. A stele typical of dicotyledons and gymnosperms, characterized by a cylindrical vascular system composed of anastomosing vascular bundles, collateral or bicollateral.

Exalbuminous seed. A seed without endosperm in the mature state.

Exarch xylem. Xylem strand in which the maturation of the cells progresses centripetally; that is, the oldest elements (the protoxylem) are farthest from the center of the axis. Typical of roots in seed plants.

Exocarp. The outer layer of the pericarp. Also called *epicarp*.

Exodermis. The outermost layer, one or more cells in depth, of the cortex in some roots. Structurally similar to the endodermis in having suberized, more or less thickened walls. A type of *hypodermis*.

Exogenous. Arising in superficial tissue, as an axillary bud.

External phloem. Primary phloem located externally to the primary xylem.

Extrafloral nectary. See *nectary*.

Extraxylary fibers. Fibers of the various tissues other than the xylem; outside the xylem.

False annual ring. In wood. One of more than one growth layers formed during one growth season, as seen in cross section.

Fascicle. A bundle.

Fascicular cambium. Vascular cambium originating from procambium within vascular bundles, or fascicles.

Fascicular tissue system. The vascular tissue system.

Festucoid. Pertaining to the subfamily of grasses Festucoideae.

Fiber. An elongated tapering sclerenchyma cell with a more or less thick secondary wall, with or without lignin. May or may not have a living protoplast at maturity.

Fiber-sclereid. A cell with characteristics intermediate between those of a fiber and a sclereid.

Fiber-tracheid. In wood. A fiber-like tracheid, commonly thick walled, with pointed ends and bordered pits having lenticular to slit-like apertures.

Fibril. See *macrofibril*.

Fibrous root system. A root system composed of many roots approximately equal in length and thickness. As in grasses and many other monocotyledons.

File meristem. See *rib meristem*.

Filiform. Thread-shaped.

Filiform sclereid. A much elongated slender sclereid resembling a fiber.

Filling tissue. In lenticel. The loose tissue formed by the lenticel phellogen toward the outside. May or may not be suberized. Also called *complementary tissue*.

Flavedo. In citrus. Yellow-colored tissue of the rind.

Floral apical meristem. See *apical meristem.*

Floral meristem. Floral apical meristem. See *apical meristem.*

Floral nectary. See *nectary.*

Follicle. A dry, dehiscent, many-seeded fruit derived from one carpel and splitting along one suture.

Fruit wall. The outer part of the fruit derived either from the ovary wall (*pericarp*) or from the ovary wall plus accessory parts associated with the ovary in fruit.

Fundamental tissue. See *ground tissue.*

Fundamental-tissue system. See *ground-tissue system.*

Funiculus. A stalk of the ovule.

Fusiform cell. Elongated and tapering at the ends.

Fusiform initial. In vascular cambium. An elongated cell with wedge-shaped ends that gives rise to the elements of the axial system in the secondary xylem and secondary phloem. Means "spindle shaped."

Gelatinous fiber. A fiber with little or no lignification and having a gelatinous appearance.

Gland. A multicellular secretory structure.

Glandular hair. A trichome having a unicellular or multicellular head composed of secretory cells. Usually borne on a stalk of nonglandular cells.

Gluten. Amorphous protein occurring in the starchy endosperm of cereals.

Grana (sing. granum). In chloroplasts. Wafer-like bodies composed of stacked membranes and containing the chlorophylls and carotenoids associated with photosynthesis.

Ground meristem. A primary meristem, or meristematic tissue, derived from the apical meristem and giving rise to the ground tissues.

Ground tissue. One of the tissues composing the *ground-tissue system.*

Ground-tissue system. The entire complex of ground tissues, that is, tissues other than the epidermis (or periderm) and the vascular. Fundamental-tissue system.

Growth. Increase in size by cell division and/or cell expansion.

Growth layer. A layer of secondary xylem or secondary phloem produced apparently during one growth season. (See also *false annual ring.*) The xylem is frequently divisible in early and late wood. Distinction between early and late phloem may also be present.

Growth ring. In secondary xylem and phloem. A growth layer as seen in cross section.

Guard cells. In stoma. Two cells which by changes in turgescence open or close the stomatal opening.

Gum. A nontechnical term applied to material resulting from breakdown of plant cells, mainly their carbohydrates.

Gum duct. A duct which contains gum.

Gummosis. A symptom of a disease characterized by the formation of gum, which may accumulate in interior cavities or ducts, or appear on the surface of the plant.

Guttation. Exudation of water from plants in liquid form.

Hadrom. In the seed plants, refers to the conducting and the living cells of the xylem, that is, to the tracheary elements and the parenchyma cells, but excludes the specifically supporting cells.

Half-bordered pit-pair. An intercellular pairing of a bordered and a simple pit.

Haplocheilic. Stomatal type in gymnosperms. Subsidiary cells are not related to the guard cells ontogenetically.

Hardwood. Wood produced by woody dicotyledons.

Heartwood. The inner layers of wood which, in the growing tree, have ceased to conduct and contain no living cells and in which reserve materials have been removed or converted into heartwood substances; generally darker in color than sapwood.

Helical (spiral) cell-wall thickening. In tracheary elements. Secondary wall material deposited on the primary or secondary wall as a continuous helix (spiral).

Heterocellular ray. In secondary vascular tissues. A ray composed of cells of more than one form; in dicotyledons, of procumbent and square or upright cells (square and upright are classified as one type); in conifers, of parenchyma cells and ray tracheids.

Heterogeneous ray tissue system. In secondary vascular tissues. Rays all heterocellular, or homocellular and heterocellular rays are combined. (Not used for conifers.)

Hilum. (1) The central part of a starch grain around which the layers of starch are arranged more or less concentrically. (2) Scar left by the funiculus on a seed.

Histogen. Hanstein's term for a meristem in shoot or root tip that forms a definite tissue system in the plant body. Three histogens were recognized: *dermatogen, periblem,* and *plerome.* (See their definitions.)

Histogen concept. Hanstein's concept stating that the three tissue systems in the plant body, the epidermis, the cortex, and the vascular system with associated ground tissue, originate from distinct meristems, the histogens, in the apical meristems.

Histogenesis. The formation of tissues; hence, *histogenetic,* having to do with origin, or formation, of tissues.

Homocellular ray. In secondary vascular tissues. A ray composed of cells of one form only; in dicotyledons, of procumbent or square or upright cells (square and upright are classified as one type); in conifers, of parenchyma cells only.

Homogeneous ray tissue system. In secondary vascular tissues. Rays all homocellular, procumbent cells only. (Not used for conifers.)

Homology. Pertains to having the same phylogenetic origin but not necessarily the same structure and/or function.

Horizontal parenchyma. See *ray parenchyma.*

Hourglass cell. A cell in the seed coat of some Leguminosae having uneven deposition of secondary wall so that the lumen has the shape of an hourglass.

Hydathode. Water pore or water gland, a structure that releases water. Usually on leaves. Varies in degree of differentiation.

Hydromorphic. Refers to structural features typical of plants (hydrophytes) that require a large supply of water and may grow partly or entirely submerged in water. **Syn.** *hygromorphic.*

Hyperplasia. Refers most commonly to an excessive multiplication of cells.

Hypertrophy. Refers most commonly to an abnormal enlargement. Hypertrophy of a cell or its parts involves no division. Hypertrophy of an organ may involve both enlargement of cells and their abnormal multiplication (hyperplasia).

Hypocotyl-root axis. In embryo. The axis below the cotyledon or cotyledons comprising the hypocotyl and the root meristem or also the radicle.

Hypodermis. General term for layer or layers of cells beneath the epidermis. So called if morphologically distinct from underlying cortical layers. *Exodermis* is a specialized hypodermis in a root.

Hypophysis. The uppermost cell of suspensor from which part of the root and the root-cap in the embryo of angiosperms are derived.

Hypsophylls. Leaves inserted at high levels of a plant as floral bracts. As contrasted with *cataphylls.*

Idioblast. A special cell in a tissue which markedly differs in form, size, or contents from other cells in the same tissue.

Included phloem. Secondary phloem included in the secondary xylem of certain dicotyledons. Replaces *interxylary phloem.*

Increment. In growth. An addition to the plant body by the activity of a meristem.

Indeterminate growth. Of apical meristem. Produces an unrestricted number of lateral organs indefinitely. Characteristic of vegetative apical meristem.

Initial apotracheal parenchyma. See *boundary apotracheal parenchyma.*

Initials. In meristems. Cells which by division perpetuate themselves and at the same time form new body cells.

Integumentary tapetum. In ovule. The epidermis of the integument (or the inner integument) located next to the embryo sac, and which consists of deeply staining cells. Appears to play a role in the nutrition of the embryo.

Intercalary meristem. Meristematic tissue derived from the apical meristem but continuing meristematic activity some distance from that meristem. May be intercalated between tissues that are more or less mature.

Intercalary growth. Growth by cell division that occurs some distance from the meristem in which the cells originated.

Intercellular space. A space among cells within a tissue. Varied in origin. (See *lysigenous, rhexigenous,* and *schizogenous.*)

Intercellular substance. See *middle lamella.*

Interfascicular cambium. Vascular cambium arising between vascular bundles, or fascicles, in the interfascicular parenchyma.

Interfascicular region. Tissue region located between vascular bundles in stem. Also called medullary, or pith, ray.

Internal phloem. Primary phloem located internally to the primary xylem. Replaces *intraxylary phloem.*

Interpositional growth. See *intrusive growth.*

Intervascular pitting. In tracheary elements. Pitting between tracheary elements.

Interxylary cork. Cork that develops among the elements of the xylem tissue.

Interxylary phloem. See *included phloem.*

Intraxylary phloem. See *internal phloem.*

Intrusive growth. A type of growth in which a growing cell intrudes among other cells, which separate from each other along the middle lamella before the growing cell. Also called *interpositional growth.*

Intussusception (of cell wall). Growth by interpolation of new wall material within previously formed wall.

Isobilateral leaf. A leaf in which the palisade parenchyma occurs on both sides of the blade. *Isolateral leaf.*

Isobilateral mesophyll. See *isobilateral leaf.*

Isodiametric. Regular in form, with all diameters equally long.

Isolateral leaf. See *isobilateral leaf.*

Isotropic. Having the same properties in all directions. Optically isotropic, not affecting the light.

Karyokinesis. The process of division of a nucleus as distinguished from the division of the cell, or *cytokinesis.*

Lacuna (pl. lacunae). Space. Usually air space. Varies in origin. (See *lysigenous, rhexigenous,* and *schizogenous.*)

Lacunar collenchyma. Collenchyma characterized by the presence of intercellular spaces and thickenings of cell wall facing the spaces.

Lamella. A thin plate or layer.

Lamellar collenchyma. Collenchyma in which thickenings of the cell wall are deposited mainly on the tangential walls.

Lamina. Of leaf. Blade or expanded part of a leaf.

Laminated cuticle. Composed of thin plates.

Late wood. The wood formed in the later part of a growth layer. Denser and has smaller cells than the early wood. Replaces *summer wood.*

Latex (pl. latices). A fluid, often milky, contained in laticifers. Consists of a variety of organic and inorganic substances, often including rubber.

Laticifer. A cell or a cell series with a characteristic fluid content called latex.

Laticiferous cell. A nonarticulated, or simple, laticifer.

Laticiferous vessel. An articulated, or compound, laticifer in which the cell walls between contiguous cells are partly or completely removed.

Leaf buttress. Lateral protrusion below the apical meristem constituting the initial stage in the development of a leaf primordium.

Leaf fiber. Economic designation of fibers derived from monocotyledons, chiefly from their leaves.

Leaf gap. In the nodal region of a stem. A region of parenchyma in the vascular cylinder occurring where a leaf trace is bent away from the vascular system of the stem toward the leaf.

Leaf sheath. The lower part of the leaf which invests the stem more or less completely.

Leaf trace. A vascular bundle in the stem extending between its connection with a leaf and that with another vascular unit in the stem. A leaf may have one or more traces. Sometimes the whole complex of traces of one leaf is called a leaf trace.

Lenticel. Special formation in the periderm distinguished from the phellem in having intercellular spaces. Lenticel tissue may or may not be suberized.

Leptom. In the seed plants, refers to the conducting and parenchyma cells of the phloem, that is, sieve elements, companion cells, and other parenchyma cells, but excludes the supporting cells.

Leucoplast. A colorless plastid.

Libriform fiber. A wood fiber commonly with thick walls and simple pits. Usually the longest cell in the wood.

Light line. A continuous line through the epidermis seen in sections of certain legume seeds. The result of matching of regions of high degree of refraction in the walls of adjacent epidermal cells. These regions are supposedly highly impermeable.

Lignification. Impregnation with lignin.

Lignin. An organic substance or mixture of substances of high carbon content, but distinct from carbohydrates; associated with cellulose in the walls of many cells.

Lithocyst. A cell containing a cystolith.

Longitudinal parenchyma. See *axial parenchyma.*

Lumen (in a plant cell). The space bounded by the cell wall.

Lysigenous. As applied to an intercellular space. Originating by dissolution of cells.

Macrofibril. An aggregation of *microbfibrils;* visible under light microscope.

Macrosclereid. A somewhat-elongated sclereid with unevenly distributed secondary walls. Common in seeds of Leguminosae, in which macrosclereids form the epidermis. Also called malpighian cell.

Malpighian cell. See *macrosclereid.*

Marginal growth. In leaf. See *marginal meristem.*

Marginal initials. In leaf. Cells along the margins of a growing leaf lamina that contribute cells to the protoderm. Components of *marginal meristem* which is concerned with *marginal growth.*

Marginal meristem. In leaf. Meristem located along the margin of a leaf primordium and forming the blade. May have distinct *marginal* and *submarginal initials.* Concerned with *marginal growth.*

Mass meristem. A meristematic tissue in which the cells divide in various planes so that the tissue increases in volume.

Matrix. Generally refers to a medium in which something is imbedded.

Mature. A term of convenience applied to cells or tissues that have completed their differentiation and thus have assumed the function or state characteristic of their kind in a fully developed part of the plant body.

Mechanical tissue. See *supporting tissue.*

Medulla. See *pith.*

Medullary bundle. Vascular bundle located more or less close to the center of the stem in the pith region.

Medullary ray. See *interfascicular region.*

Medullary sheath. See *perimedullary region.*

Meristem. Tissue primarily concerned with protoplasmic synthesis and formation of new cells by division.

Meristematic cell. A cell synthesizing protoplasm and dividing; thus, a source of other cells. Varies in form, size, wall thickness, and degree of vacuolation. According to one definition of wall layers, secondary walls do not occur in meristematic cells.

Mesarch xylem. Xylem strand in which the maturation of the cells begins in the center and then progresses both centripetally and centrifugally; that is, the oldest elements (the protoxylem) are in the center of the strand.

Mesocarp. The middle layer of the pericarp.

Mesocotyl. The internode between the scutellar node and the coleoptile in the embryo and seedling of a grass.

Mesomorphic. Refers to structural features typical of plants (mesophytes) requiring abundant available soil water and a relatively humid atmosphere.

Mesophyll. Photosynthetic parenchyma tissue of a leaf located between epidermal layers.

Mestom (or mestome) sheath. A vascular bundle sheath with thickened walls; the inner of two sheaths in leaves of grasses, mainly those of the festucoid subfamily. An endodermoid sheath.

Metaphloem. Part of the primary phloem that differentiates after the protophloem; also before the secondary phloem if the latter is present.

Metaxylem. Part of the primary xylem that differentiates after the protoxylem; also before the secondary xylem if the latter is present.

Micelle. In present usage refers to parts of cellulose microfibrils in which the cellulose molecules are arranged parallel to each other so that the atoms form a crystalline lattice structure.

Microfibril. A thread-like component of cell wall composed of cellulose molecules and visible only with the electron microscope.

Micromicron. Millionth micron. Hundredth Ångström. Symbol $\mu\mu$.

Micron. Thousandth millimeter. 10,000 Ångströms. Symbol μ.

Microsomes. Definitions vary. Widely accepted definition: small cytoplasmic particles (about 250 Ångström in diameter) associated with protein synthesis.

Middle lamella. Between cell walls. Layer of intercellular material, chiefly pectic substances, cementing together the primary walls of contiguous cells.

Millimicron. Thousandth micron or millionth millimeter. 10 Ångströms. Symbol $m\mu$.

Mitochondria (sing. mitochondrion). Small protoplasmic bodies in the cytoplasm ranging in size from a fraction of a micron in diameter to elongated forms that are several microns long. Considered to be the site of enzymes involved in respiration. Also frequently called *chondriosomes*.

Morphogenesis (of plants). Origin of form. Sum of phenomena of differentiation and development of tissues and organs.

Morphology (of plants). The area of science concerned with form, structure, and development of plants.

Mother cell. See *precursory cell*.

Motor cell. See *bulliform cell*.

Mucilage cell. Cell containing mucilages or gums or similar carbohydrate material. Mucilages have the property of swelling in water.

Mucilage duct. A duct containing mucilage or gum or similar carbohydrate material.

Multilacunar node. In stem. A node with numerous leaf gaps related to one leaf.

Multiperforate perforation plate. In vessel member. A perforation plate having more than one perforation. See also *perforation plate*.

Multiple epidermis. A tissue several cell layers deep derived from the protoderm; only the outer layer differentiates as a typical epidermis.

Multiseriate. Consisting of many layers of cells.

Multiseriate ray. In secondary vascular tissues. A ray a few to many cells wide.

Mycorrhiza (pl. mycorrhizae). The symbiotic union of fungi and roots of plants. May be ectotrophic (web of hyphae invests the root of the host) or endotrophic (the hyphae are entirely within cells).

Myrosin cell. Cell producing myrosin, a glucoside. Occur in vegetative parts and seeds of certain Cruciferae.

Nacreous wall. A wall thickening that occurs in sieve elements of certain plants. Not yet analyzed in terms of classification into primary and secondary walls.

Nacré wall. See *nacreous wall*.

Nectary. A multicellular glandular structure secreting a sugary liquid. Occurs in flowers (*floral nectary*) or on vegetative plant parts (*extrafloral nectary*).

Netted venation. See *reticulate venation*.

Nodal diaphragm. In hollow stems. A septum of tissue extending across the hollow of the stem at a node.

Node. That part of the stem at which one or more leaves are attached. Not clearly delimited anatomically.

Nodular end wall. In xylem parenchyma cells. The cell wall at right angles to the longitudinal axis of the cell appears beaded because of deeply depressed pits.

Nonarticulated laticifer. Simple laticifer. A laticifer that is a single cell, commonly multinucleate. May be branched or unbranched.

Nonporous wood. Secondary xylem having no vessels.

Nonstoried cambium. Vascular cambium in which the fusiform initials and rays are not arranged in horizontal series on tangential surfaces. *Nonstratified cambium*.

Nonstoried wood. Wood in which the axial cells and rays are not arranged in horizontal series on tangential surfaces. *Nonstratified wood*.

Nonstratified cambium. See *nonstoried cambium*.

Nonstratified wood. See *nonstoried wood*.

Ontogeny. The life history, or development, of an individual organism (or part of it).

Open venation. In leaf blade. Venation in which large veins end freely in the mesophyll, that is, without being connected with other veins by anastomoses.

Opposite pitting. In tracheary elements. Pits in horizontal pairs or in short horizontal rows.

Organ (of plant). A distinct and visibly differentiated part of a plant, such as, root, stem, leaf, and parts of a flower.

Orthostichy. In phyllotaxy. A vertical line along which is attached a row of leaves or scales on an axis. Usually a steep helix, rather than a straight line.

Osteosclereid. Sclereid having the shape of a bone, with a columnar middle part and enlargements at both ends.

Palisade parenchyma. Leaf-mesophyll parenchyma characterized by elongated form of cells and their arrangement perpendicular to the surface of the leaf.

Panicoid. Pertaining to the subfamily of grasses Panicoideae.

Papilla (pl. papillae). A type of trichome. A soft protuberance.

Paracytic stoma. A type in which one or more subsidiary cells flank the stoma parallel with the long axis of the guard cells. Old term, *rubiaceous.*

Paradermal. Refers to section made parallel with the surface of a flat organ, as leaf. Also referred to as *tangential.*

Parallel evolution (in plants). Evolution in a similar direction in different groups.

Parallel venation. In leaf blade. Main veins are arranged approximately parallel. (Converging at apex and base of leaf, however.)

Parastichy. In phyllotaxy. A curved line (helix) along which is attached a row of leaves or scales on an axis. See also *orthostichy.*

Paratracheal parenchyma. In wood. Axial parenchyma associated with vessels and other tracheary elements. Includes *vasicentric, aliform,* and *confluent.*

Parenchyma. Tissue composed of parenchyma cells.

Parenchyma cell. Usually refers to a cell with live nucleate protoplast concerned with one or more of the various physiologic activities in plants. Varies in size, form, and wall structure.

Parietal cytoplasm. Cytoplasm occurring next to the cell wall.

Passage cell. Cell in exodermis or endodermis of roots that retains thin walls when the associated cells develop thick secondary walls. Has Casparian strip in endodermis.

Pectic substances. A group of complex carbohydrates, derivatives of polygalacturonic acid. Occur in three general types, protopectin, pectin, and pectic acid. Main constituent of intercellular substance or middle lamella. Also present in cell walls.

Peltate hair. A trichome, consisting of a discoid plate of cells, borne on a stalk or attached directly to the foot.

Perforation plate. Part of the wall of a vessel member that is perforated.

Periblem. The meristem forming the cortex. One of the three histogens according to Hanstein.

Pericarp. Fruit wall developed from the ovary wall.

Periclinal. Parallel with the circumference.

Pericycle. Part of the ground tissue of the stele located between the phloem and endodermis. In seed plants regularly present in roots, absent in most stems.

Pericyclic fiber. A fiber located on the outer periphery of the vascular region commonly arising in the primary phloem (primary phloem fiber), sometimes outside of it (perivascular fiber).

Periderm. Secondary protective tissue derived from the phellogen (cork cambium) and replacing the epidermis, typically in stems and roots. Consists of *phellem* (cork), *phellogen,* and *phelloderm.*

Perimedullary region or zone. The peripheral region of the pith (medulla). Also called *medullary sheath.*

Perisperm. Storage tissue in seed similar to endosperm but derived from the nucellus.

Perivascular fiber. A fiber located on the outer periphery of the vascular region and originating outside the primary phloem as distinct from primary phloem fiber. Often called pericyclic fiber, as is also the primary phloem fiber.

Perivascular sclerenchyma. Sclerenchyma located on the outer periphery of the vascular system in seed plants and not originating in the primary phloem. Old term, *pericyclic sclerenchyma.*

Phellem (cork). Protective tissue composed of nonliving cells with suberized walls. Replaces the epidermis in one-year and older stems and roots of many plants and formed by the phellogen (cork cambium). Part of the periderm.

Phelloderm. A tissue formed by the phellogen in the opposite direction from the cork. Resembles cortical parenchyma. Part of the periderm.

Phellogen (cork cambium). A lateral meristem forming the periderm, a protective tissue common in stems and roots of dicotyledons and gymnosperms. Produces phellem (cork) toward the surface of the plant, phelloderm toward the inside.

Phelloid cell. A cell within the phellem, or cork, but distinct from the cork cell in having no suberin in its walls. May be a sclereid.

Phlobaphenes. Anhydrous derivatives of tannins; amorphous yellow, red, or brown substances; very conspicuous in sectioned material.

Phloem. The principal food-conducting tissue of the vascular plant, basically composed of sieve elements, parenchyma cells, fibers, and sclereids.

Phloem elements. Cells of the phloem tissue.

Phloem initial. A cambial cell on the phloem side of the cambial zone that divides periclinally one or more times and gives rise to cells that differentiate into phloem elements either with or without additional divisions in various planes. Sometimes called *phloem mother cell.*

Phloem mother cell. A cambial derivative that divides to produce certain elements of the phloem tissue, such as, sieve element and its companion cells or phloem parenchyma cells forming a parenchyma strand. Used also in a wider sense synonymously with *phloem initial.*

Phloem parenchyma. Parenchyma cells occurring in the phloem. In secondary phloem refers to axial parenchyma.

Phloem ray. That part of a vascular ray which is located in the secondary phloem.

Phloic. Pertaining to phloem.

Phloic procambium. That part of procambium which differentiates into primary phloem.

Photosynthetic cell. A chloroplast-containing parenchyma cell in which the process of photosynthesis occurs.

Phragmoplast. Fibrous structure that appears at telophase of a mitosis between the two nuclei and plays a role in the formation of the initial partition (*cell plate*) dividing the mother cell in two. Has the shape of a spindle connected to the nuclei at first. Spreads ring-like laterally later.

Phragmosome. Cytoplasmic plate formed across the cell in the plane of cell division before the *phragmoplast* appears in the same region.

Phyllotaxy (or phyllotaxis). The mode in which the leaves are arranged on the axis of a shoot.

Phylogeny. Evolutionary history of a species or larger plant group.

Pillar cell. One of several designations of the subepidermal sclereid in the seed coat of certain Leguminosae. See also *osteosclereid* and *hourglass cell.*

Pit. A recess or thin place in the cell wall. In primary wall designated also as *primordial pit* or *primary pit-field.* Usually a member of a *pit-pair.*

Pit aperture. Opening into the pit from the interior of the cell. If a pit canal is present in a bordered pit, two apertures are recognized, the *inner,* from the cell lumen into the canal, and the *outer,* from the canal into the pit cavity.

Pit canal. The passage from the cell lumen to the chamber of a bordered pit. (Simple pits in thick walls usually have canal-like cavities.)

Pit cavity. The entire space within a pit from pit membrane to the cell lumen or to the outer pit aperture if a pit canal is present.

Pith. Ground tissue in the center of a stem or root. Homology of pith in root and stem is controversial.

Pith ray. Interfascicular region in a stem. Medullary ray.

Pit membrane. The part of the intercellular layer and primary cell wall that limits a pit cavity externally.

Pit-pair. Two complementary pits of two adjacent cells. Essential components are two *pit cavities* and the *pit membrane.*

Plasma membrane. See *ectoplast.*

Plasmodesma (pl. plasmodesmata). A strand of cytoplasm passing through a pore in the cell wall and usually joining the protoplasts of two adjacent cells.

Plastid. A cytoplasmic body differentiated as a center of chemical and/or vital activity.

Plastochron (or plastochrone). The time interval between the inception of two successive repetitive events, as origin of leaf primordia, attainment of certain stage of development of a leaf, etc. Variable in length as measured in time units.

Plate collenchyma. See *lamellar collenchyma.*

Plate meristem. A meristematic tissue consisting of parallel layers of cells dividing only anticlinally with reference to the wide surface of the tissue. Characteristic of ground meristem of plant parts that assume a flat form, as a leaf.

Plerome. The meristem forming the core of the axis composed of the primary vascular tissues and associated ground tissue, such as pith. One of the three histogens according to Hanstein.

Plicate mesophyll cell. A mesophyll cell with folds or ridges of cell wall projecting into the lumen.

Polyarch. Primary xylem. Having many protoxylem strands, or protoxylem poles.

Polyderm. A special type of protective tissue in which layers of cells with endodermal characteristics alternate with layers of nonsuberized parenchyma cells.

Pore. In wood. A term of convenience for the cross section of a vessel.

Pore cluster. See *pore multiple.*

Pore multiple. In wood. A group of two or more pores (cross sections of vessels) crowded together and flattened along the surfaces of contact. *Radial pore multiple,* pores in radial file; *pore cluster,* irregular grouping.

Porous wood. Secondary xylem having vessels.

Precursory cell. A cell giving rise to others by division. Mother cell.

Primary body (of plant). The part of the plant, or entire plant if no secondary growth occurs, that arises from the apical meristems and their derivative meristematic tissues.

Primary cell wall. Version based on studies with the light microscope: cell wall formed chiefly while the cell is increasing in size. Version based on studies with the electron microscope: cell wall in which the cellulose microfibrils show various orientations— from random to more or less parallel—that may change considerably during the increase in size of the cell. The two versions do not necessarily coincide in delimiting primary from secondary wall.

Primary growth. The growth of the successively formed roots and vegetative and reproductive shoots, from the time of their initiation by the apical meristem until their expansion is completed. Has its inception in the apical meristems and continues in their derivative meristems, protoderm, ground meristem, and procambium and even in still older tissues.

Primary meristem. Often used for each of the three meristematic tissues derived from the apical meristem: protoderm, ground meristem, and procambium.

Primary phloem. Phloem tissue differentiating from procambium during primary growth and differentiation of a vascular plant. Commonly divided into the earlier *protophloem* and the later *metaphloem*. Not differentiated into axial and ray systems.

Primary phloem fiber. A fiber located on the outer periphery of the vascular region and originating in the primary phloem. Often called *pericyclic fiber*.

Primary pit-field. A thin area of the intercellular layer and primary cell wall within the limits of which one or more pits develop. **Syn.** *primordial pit.* (See also *pit.*)

Primary root. The taproot. Root developing in continuation of the radicle of the embryo.

Primary thickening meristem. A meristem derived from the apical meristem and responsible for the primary increase in thickness of the shoot axis. May appear as a distinct mantle-like zone. Often found in monocotyledons.

Primary vascular tissue. Vascular tissue (both xylem and phloem) differentiating from procambium during primary growth and differentiation of a vascular plant.

Primary xylem. Xylem tissue differentiating from procambium during primary growth and differentiation of a vascular plant. Commonly divided into the earlier *protoxylem* and the later *metaxylem.* Not differentiated into axial and ray systems.

Primordial pit. See *primary pit-field.*

Primordium (pl. primordia). An organ, a cell, or an organized series of cells in their earliest stage of differentiation, e.g., leaf primordium, sclereid primordium, vessel primordium.

Procambium. Primary meristem or meristematic tissue which differentiates into the primary vascular tissue. Also called *provascular tissue.*

Procumbent ray cell. In secondary vascular tissues. A ray cell elongated in radial direction, that is, a prostrate cell.

Proembryo. Embryo in early stages of development, often the stages before main body and suspensor become distinct.

Promeristem. In the apical meristem. The initiating cells and their most recent derivatives. The most distal part of the shoot or root.

Prophyll. One of the first leaves on a lateral shoot.

Proplastid. A plastid in its earliest stages of development; a primordial plastid.

Protective layer. In abscission zone. Layer of cells that, because of substances impregnating its walls, has a protective function in the scar formed by abscission of a leaf or other plant part.

Protoderm. Primary meristem or meristematic tissue giving rise to the epidermis. Epidermis in meristematic state. May or may not arise from independent initials.

Protophloem. The first-formed elements of the phloem in a plant organ. First part of the primary phloem.

Protophloem poles. Term of convenience for loci of phloem elements that are the first to mature in the vascular system of a plant organ. Applied to views in cross sections.

Protoplasm. Living substance. Inclusive term for all living contents of a cell or an entire organism.

Protoplast. The organized living unit of a single cell.

Protostele. The simplest type of stele, containing a solid column of vascular tissue.

Protoxylem. The first formed elements of the xylem in a plant organ. First part of the primary xylem.

Protoxylem lacuna. Space surrounded by parenchyma cells in the protoxylem of a vascular bundle. Appears in some plants after the extension and cessation of function of the tracheary elements of protoxylem.

Protoxylem poles. Term of convenience for loci of xylem elements that are the first to mature in the vascular system of a plant organ. Applied to views in cross sections.

Provascular tissue. See *procambium.*

Proximal. Nearest the place of origin or attachment.

Pulvinus. An enlargement of the petiole of a leaf or petiolule of a leaflet at its base. A structure that plays a role in the movements of a leaf or leaflet.

Quarter-sawed wood. Wood sawed along a radial plane so that the radial surface of the wood is exposed.

Radial parenchyma. See *ray parenchyma.*

Radial pore multiple. See *pore multiple.*

Radial section. A longitudinal section cut along a radius of a cylindrical body, such as that of stem or root.

Radial seriation. The occurrence in an orderly sequence in a radial direction.

Radicle. The embryonic root. Constitutes the basal continuation of the hypocotyl in an embryo.

Ramified. Branched.

Ramiform pit. Simple pit with coalescent canal-like pit cavity, as in a stone cell.

Raphe. A ridge along the body of the seed formed by the part of the funiculus which was adnate to the ovule (in an anatropous ovule).

Raphid. A needle-shaped crystal. Typically occurs with other crystals of the same type in a closely packed bundle.

Ray. A panel of tissue variable in height and width, formed by the ray initials of the vascular cambium and extending radially in the secondary xylem and secondary phloem.

Ray initial. A cambial ray cell in the vascular cambium that gives rise to ray cells of the secondary xylem and secondary phloem.

Ray parenchyma. In secondary vascular tissues. Parenchyma cells that are components of a vascular ray. As contrasted with axial parenchyma.

Ray system. In secondary vascular tissues. The total of all rays, as contrasted with the *axial system.* Syn. *radial system* and *horizontal system.*

Ray tracheid. In wood. A tracheid forming part of a ray. Found in certain conifers.

Reaction wood. Wood with more or less distinctive anatomical characteristics, formed in parts of leaning or crooked stems and in branches, and apparently tending to restore the original position. *Compression wood* in conifers, *tension wood* in dicotyledons.

Redifferentiation. A reversal in differentiation in a cell or tissue and subsequent differentiation into another type of cell or tissue.

Reproductive apical meristem. See *apical meristem.*

Residual meristem. Meaning a residuum of the least differentiated part of the apical meristem. A tissue that is relatively more highly meristematic than the associated differentiating tissues beneath the apical meristem. Gives rise to procambium and to some ground tissue.

Resin duct. A duct of schizogenous origin containing resin. Also called *resin canal.*

Reticulate cell-wall thickening. In tracheary elements. Secondary wall material deposited on the primary wall so as to give the appearance of a net.

Reticulate perforation plate. In vessel member. A type of multiperforate plate in which the perforations form a net-like pattern.

Reticulate sieve plate. A compound sieve plate with sieve areas arranged so as to form a net-like pattern.

Reticulate venation. In leaf blade. Veins form an anastomosing system, the whole resembling a net. *Netted venation.*

Reticulum. A net.

Rhexigenous. As applied to an intercellular space. Originating by tearing of cells.

Rhizodermis. The epidermis of the root. Term used to express the view that the epidermis of the root is not homologous with that of the shoot.

Rhytidome. The periderm and tissues isolated by it. Encloses masses of cortical and phloem tissue in the younger stems, secondary phloem in the older. A technical term for the outer bark. (The secondary phloem is the inner bark.)

Rib. Elongated protrusion.

Rib meristem. A meristematic tissue in which the cells divide at right angles to the longitudinal axis of the stem, root, or leaf and produce a complex of parallel files ("ribs") of cells. Characteristic of ground meristem of organs assuming a cylindrical form. Also called *file meristem.*

Ring bark. A type of rhytidome resulting from the formation of successive periderms approximately concentrically around the axis.

Ring-porous wood. Wood in which the pores (vessels) of the early wood are distinctly larger than those of the late wood and form a well-defined zone or ring in a cross section of wood.

Rootcap. A thimble-like mass of cells covering the apical meristem of the root.

Root hair. A type of trichome on root epidermis that is a simple extension of an epidermal cell. Concerned with absorption of soil solution.

Sapwood. Wood that in the living tree contains living cells and reserve materials.

Scalariform cell-wall thickening. In tracheary elements. Secondary wall material deposited on the primary wall so as to form a ladder-like pattern. Like a helix of low inclination with the coils interconnected at intervals.

Scalariform perforation plate. In vessel member. A type of multiperforate plate in which elongated perforations are arranged parallel. The remnants of the plate between the openings are called bars.

Scalariform pitting. In tracheary elements. Elongated pits arranged parallel so as to form a ladder-like (scalariform) pattern.

Scalariform-reticulate cell-wall thickening. In tracheary elements. Secondary wall material deposited on the primary wall in a pattern intermediate between those termed scalariform and reticulate.

Scalariform sieve plate. A compound sieve plate with elongated sieve areas arranged parallel to one another in a ladder-like (scalariform) pattern.

Scale. On epidermis. A scarious trichome. Usually disc-like and with a stalk attached to its lower surface (peltate).

Scale bark. A type of rhytidome in which the sequent periderms develop as restricted overlapping strata, each cutting out a scale-like mass of tissue.

Scar tissue, or cicatrice tissue. Composed of necrosed cells injured by the wounding and subjacent cells impregnated with protective substances.

Schizogenous. As applied to an intercellular space. Originating by separation of cell walls along the middle lamella.

Sclereid. A sclerenchyma cell, varied in form, but typically not much elongated, and having thick lignified secondary walls with many pits.

Sclerenchyma. A tissue composed of sclerenchyma cells. Also a collective term for sclerenchyma cells in the plant body or plant organ. Includes *fibers, fiber-sclereids,* and *sclereids.*

Sclerenchyma cell. Cell variable in form and size and having more or less thick, often lignified, secondary walls. Belongs to the category of supporting cells and may or may not be devoid of protoplast at maturity.

Sclerification. The act of becoming changed into sclerenchyma, that is, developing secondary walls, with or without subsequent lignification.

Sclerotic parenchyma cell. A parenchyma cell that through deposition of a thick secondary wall became changed into a sclereid.

Scutellum. The cotyledon in Gramineae. Considered to be one of two cotyledons if the epiblast is interpreted as a cotyledon also.

Secondary body. The part of the plant body that is added to the primary body by the activity of the lateral meristems, the vascular cambium and the phellogen. Consists of secondary vascular tissues and periderm.

Secondary cell wall. Version based on studies with the light microscope: cell wall deposited in some cells over the primary wall after the primary wall ceased to increase in surface. Version based on studies with the electron microscope: cell wall in which the cellulose microfibrils show a definite parallel orientation. The two versions do not necessarily coincide in delimiting secondary from primary wall.

Secondary growth. In gymnosperms, most dicotyledons, and some monocotyledons. A type of growth characterized by an increase in thickness of stem and root and resulting from formation of secondary vascular tissues by the vascular cambium. Commonly supplemented by activity of the cork cambium (phellogen) forming periderm.

Secondary phloem. Phloem tissue formed by the vascular cambium during secondary growth in a vascular plant. Differentiated into axial and ray systems.

Secondary phloem fiber. A fiber located in the axial system of secondary phloem.

Secondary root. See *branch root.*

Secondary thickening. Used widely for both deposition of secondary cell-wall material and secondary increase in thickness of stems and roots.

Secondary vascular tissues. Vascular tissues (both xylem and phloem) formed by the vascular cambium during secondary growth in a vascular plant. Differentiated into axial and ray systems.

Secondary xylem. Xylem tissue formed by the vascular cambium during secondary growth in a vascular plant. Differentiated into axial and ray systems.

Secretory cavity. Commonly refers to a space lysigenous in origin and containing a secretion derived from the cells that broke down in the formation of the cavity.

Secretory cell. A living cell specialized with regard to secretion or excretion of one or more usually organic substances.

Secretory duct. Commonly refers to a duct schizogenous in origin and containing a secretion derived from the cells (*epithelial cells*) lining the duct.

Secretory hair. See *glandular hair.*

Secretory structure. Any of a great variety of structures, simple or complex, external or internal, that produces a secretion.

Seed coat, or testa. The outer coat of the seed derived from the integument or integuments.

Seminal adventitious root. Root initiated in the embryo on the hypocotyl or higher on the axis.

Separation layer. See *abscission layer.*

Septate fiber. A fiber with thin transverse walls (septa), which are formed after the cell develops a secondary wall thickening.

Septum. A partition.

Sheath. A structure inclosing or encircling another. Applied to tubular or enrolled part of an organ, such as leaf sheath, and for a tissue layer surrounding a mass of another tissue.

358 ANATOMY OF SEED PLANTS

Sheathing base. Applied to the base of a leaf, either sessile or petiolate, when the leaf base encircles the stem.

Shell zone. In axillary bud primordia. A zone of parallel curving layers of cells, the entire complex shell-like in form. The result of orderly cell division along the proximal limits of the primordium.

Sieve area. A pit-like area in the wall of a sieve element with pores lined with callose and occupied by strands of protoplasmic material that interconnects the protoplasts of contiguous sieve elements.

Sieve cell. A type of sieve element that has relatively undifferentiated sieve areas, that is, sieve areas with narrow pores and thin connecting strands; and these sieve areas are rather uniform in structure on all walls; that is, there are no sieve plates. Typical of gymnosperms and lower vascular plants.

Sieve element. The cell in the phloem tissue concerned with mainly longitudinal conduction of food materials. Classified into *sieve cell* and *sieve-tube member*.

Sieve field. Old term for a relatively undifferentiated sieve area found on wall parts other than the sieve plates.

Sieve pitting. An arrangement of small pits in sieve-like clusters.

Sieve plate. The part of the wall of a sieve element bearing one or more highly differentiated sieve areas. Typical of angiosperms.

Sieve tube. A series of sieve elements (sieve-tube members) arranged end to end and interconnected through sieve plates.

Sieve-tube element. See *sieve-tube member*.

Sieve-tube member. One of the cellular components of a sieve tube. Also *sieve-tube element* and the obsolete *sieve-tube segment*.

Silica cell. Cell filled with silica, as in epidermis of grasses.

Simple laticifer. Laticifer that is a single cell. *Nonarticulated laticifer*.

Simple perforation plate. In vessel member. A perforation plate with a single perforation.

Simple pit. A pit in which the cavity becomes wider, remains of constant width, or only gradually becomes narrower during the growth in thickness of the secondary wall, that is, toward the lumen of the cell.

Simple pit-pair. An intercellular pairing of two simple pits.

Simple sieve plate. Sieve plate composed of one sieve area.

Siphonostele. A type of stele in which the vascular system appears in the form of a hollow cylinder; that is, pith is present.

Slime. In sieve element. A relatively viscous inclusion usually regarded as composed of protein.

Slime body. In sieve element. A body of apparently proteinaceous material. Frequently only in a young sieve element; disperses later in the vacuole as slime.

Slime plug. An accumulation of slime on a sieve area. Forms apparently in response to sectioning of the phloem.

Softwood. Wood produced by conifer trees.

Solenostele. A form of an amphiphloic siphonostele.

Solitary pore. In wood. A pore (cross section of a vessel) surrounded by cells other than vessel members.

Specialization. Change in structure of a cell, a tissue, a plant organ, or entire plant resulting in a restriction of functions, potentialities, or adaptability to varying conditions. May result in greater efficiency with regard to certain specific functions. Some specializations are irreversible, others reversible.

Spiral cell-wall thickening. See *helical cell-wall thickening*.

Spongy parenchyma. Leaf mesophyll parenchyma with conspicuous intercellular spaces. Cells vary in shape.

Spring wood. See *early wood.*

Square ray cell. In secondary vascular tissues, a ray cell approximately square as seen in radial section. (Considered to be of the same morphological type as the *upright ray cell.*)

Starch. An insoluble carbohydrate, the chief food-storage substance of plants, composed of anhydrous glucose residues of the formula $C_6H_{10}O_5$ into which it easily breaks down.

Starch sheath. Applied to the innermost layer (or layers) of the cortex when it is characterized by conspicuous and rather stable accumulation of starch.

Stele (column). Conceived by Van Tieghem as a morphologic unit of the plant body comprising the vascular system and the associated ground tissue (pericycle, interfascicular regions, and pith). The central cylinder of the axis (stem and root).

Stellate. Star shaped.

Stigmatoid tissue. A tissue of apparent cytologic and physiologic similarity to the tissue of the stigma, and which connects the stigma with the interior of the ovary. Other terms are *pollen conducting* or *pollen transmitting tissue.*

Stomatal crypt. A depression in the leaf, the epidermis of which bears stomata.

Stomium. A circular opening in the anther lobe through which the pollen is released. Its formation is a type of dehiscence.

Stone cell. See *brachysclereid.*

Storied cambium. Vascular cambium in which the fusiform initials and rays are arranged in horizontal series on tangential surfaces. *Stratified cambium.*

Storied cork. Protective tissue found in the monocotyledons. The suberized cells occur in radial files, each consisting of several cells all of which are derived from one cell.

Storied wood. Wood in which the axial cells and rays are arranged in horizontal series on tangential surfaces. *Stratified wood.* Rays alone may be storied. Cause of the ripple marks visible with the unaided eye.

Stratified cambium. See *storied cambium.*

Stratified wood. See *storied wood.*

Striate venation. See *parallel venation.*

Stoma. A pore in the epidermis and the two guard cells surrounding it. Sometimes applied to the pore only.

Stroma. In plastid. A supporting framework.

Styloid. An elongated columnar crystal with pointed or square ends.

Subapical initial. In leaf. A cell beneath the protoderm at the apex of a leaf primordium that by repeated divisions contributes cells to the interior tissue of the leaf.

Suberin. The same definition as for *cutin* with which it is closely related.

Suberization. Impregnation of the wall with suberin or deposition of suberin lamellae in the wall.

Submarginal initials. In leaf. Cells beneath the protoderm along the margins of a growing leaf lamina that contribute cells to the interior tissue of the leaf. Components of *marginal meristem* which is concerned with *marginal growth.*

Subsidiary cell. An epidermal cell associated with a stoma and at least morphologically distinguishable from the epidermal cells composing the groundmass of the tissue. Also called *accessory cell.*

Summer wood. See *late wood.*

Supernumerary cambium layer. Vascular cambium originating in phloem or pericycle outside the regularly formed vascular cambium. Characteristic of some plants with anomalous type of secondary growth.

Supporting cell. See *supporting tissue.*

Supporting tissue. Refers to tissue composed of cells with more or less thickened walls, primary (collenchyma) or secondary (sclerenchyma), that adds strength to the plant body. Also called *mechanical tissue.*

Symplastic growth. See *coordinated growth.*

Syndetocheilic. Stomatal type in gymnosperms. Subsidiary cells (or their precursors) are derived from the same protodermal cell as the guard-cell mother cell.

Tabular. Having the form of a tablet or slab.

Tangential. In the direction of the tangent; at right angles to the radius. May coincide with *periclinal.*

Tangential section. A longitudinal section cut at right angles to a radius. Applicable to cylindrical structures such as stem or root, but used also for leaf blades when the section is made parallel with the expanded surface. Substitute term for leaf, *paradermal.*

Tannin. General term for a heterogeneous group of phenol derivatives. Amorphous strongly astringent substance widely distributed in plants, and used in tanning, dyeing, and preparation of ink.

Tapetum. In anther, a layer of cells lining the locule and absorbed as the pollen grains mature.

Taproot. The first, or primary, root of a plant forming a direct continuation of the radicle of the embryo.

Taproot system. A root system based on the taproot, which may have branches of various orders.

Tasche. In German, pocket. Covering of primordium of lateral root derived from the endodermis, as distinguished from the rootcap, which is derived from the pericycle.

Tension wood. Reaction wood in dicotyledons, formed on the upper sides of branches and leaning or crooked stems and characterized by lack of lignification and often by high content of gelatinous fibers.

Tepal. In flower. A member of the kind of perianth that is not differentiated into calyx and corolla.

Terminal apotracheal parenchyma. See *boundary apotracheal parenchyma.*

Testa. The seed coat.

Tetrarch. Primary xylem of root. Having four protoxylem strands, or protoxylem poles.

Tissue. Material formed by the union of cells, which may be similar in character (simple tissue) or varied (complex tissue).

Tissue system. A tissue or tissues in a plant or plant organ structurally and functionally organized into a unit.

Tonoplast. The cytoplasmic membrane bordering the vacuole, as contrasted with the *ectoplast.*

Torus. In bordered pit. The central thickened part of the pit membrane, and consisting of middle lamella and two primary walls.

Trabecula. In a cell. A rod-like or spool-shaped part of a cell wall extending across the lumen of the cell.

Tracheary element. General term for a water-conducting cell, tracheid or vessel member.

Tracheid. A tracheary element of the xylem that has no perforations, as contrasted with a vessel member. May occur in primary and in secondary xylem. May have any kind of secondary wall thickening found in tracheary elements.

Transection. Transverse section.

Transfusion tissue. In gymnosperm leaves, a tissue surrounding or otherwise associated with the vascular bundle, and composed of nonliving tracheids and living parenchyma cells. See also *accessory transfusion tissue.*

Transfusion tracheid. A tracheid in *transfusion tissue.*

Transition region. In primary body. The region in which the contrasting structures of root and shoot are united, and which, therefore, shows transitional characteristics as compared with root and shoot.

Transition zone. In apical meristem. A zone of orderly dividing cells disposed about the inner periphery of the promeristem or, more specifically, of the central mother cells.

Transverse division (of cell). With reference to cell, division perpendicular to the longitudinal axis of the cell. With reference to plant part, division of the cell perpendicular to the long axis of the plant part.

Transverse section. A cross section. Section taken perpendicular to the longitudinal axis. Also called *transection.*

Traumatic resin duct. A resin duct developing in response to injury.

Triarch. Primary xylem of root. Having three protoxylem strands, or three protoxylem poles.

Trichoblast. Now used for epidermal cells of roots that give rise to root hairs.

Trichome. An outgrowth of the epidermis, variable in shape, size, and function. Various categories, including hairs, scales, and others.

Trichosclereid. A type of branched sclereid. With thin hair-like branches extending into intercellular spaces.

Trilacunar node. In stem. A node with three leaf gaps related to one leaf.

Tube cells. In the caryopsis (fruit) of grasses. Lignified cells elongated parallel with the long axis of the grain. Remaining cells of the inner epidermis of pericarp.

Tunica. Of apical meristem of shoot. Peripheral layer or layers of cells dividing only, or almost so, in the anticlinal plane, thus undergoing surface growth. Forms a mantle over the corpus.

Tunica-corpus concept. A concept of the organization of apical meristem of shoot according to which this meristem is differentiated into two regions distinguished by their method of growth: the peripheral, tunica, one or more layers of cells, shows surface growth (anticlinal divisions); the interior, corpus, shows volume growth (divisions in various planes).

Two-trace unilacunar condition. In stem. At a node, two leaf traces, related to one leaf, are associated with one gap.

Tylosis (pl. tyloses). In wood. Outgrowth from a ray or axial parenchyma cell through a pit cavity in a vessel wall, partially or completely blocking the lumen of the vessel.

Tylosoid. Resembling a tylosis. A proliferation of an epithelial cell into an intercellular canal, as a resin duct.

Undifferentiated. In ontogeny, still in a meristematic state or resembling meristematic structures. In a mature state, relatively unspecialized.

Unifacial leaf. A leaf having similar structure on both sides. Ontogenetically, a leaf developing from a growth center abaxial or adaxial to the initial apex of the primordium, and which thus, developmentally, consists of the abaxial or adaxial side of the leaf only.

Unilacunar node. A node with one leaf gap related to one leaf. If two or more leaves are attached at such a node, each would be associated with one gap.

Uniseriate. Consisting of one layer of cells.

Uniseriate ray. In secondary vascular tissues. A ray one cell wide.

Unit cup fruit. Fruit derived from a syncarpous (united carpels) epigynous (inferior ovary) flower. Fruit wall consists of pericarp and accessory parts (floral tube or receptacle).

Unit free fruit. Fruit derived from a syncarpous (united carpels) hypogynous flower (superior ovary). Fruit wall consists of pericarp only.

Upright ray cell. In secondary vascular tissues. A ray cell with its longest dimension oriented axially, that is, vertically in a stem.

Vacuole. Cavity within the cytoplasm filled with a watery fluid, the cell sap.

Vascular bundle. A strand-like part of the vascular system composed of xylem and phloem. Occurs in stem, leaf, and flower.

Vascular cambium. Lateral meristem that forms the secondary vascular tissues, secondary xylem and secondary phloem. Located between these two tissues and, by periclinal divisions, gives off cells in both directions.

Vascular cylinder. A term of convenience applied to the vascular tissues and associated ground tissue in stem or root. Refers to the same part of stem or root that is designated *stele* but without the theoretical implications of the stelar concept. Same as *central cylinder*.

Vascular meristem. General term applicable to the *vascular cambium* and *procambium*. See these terms.

Vascular ray. A ray of secondary xylem or secondary phloem. See *ray* and *ray system*.

Vascular system. The total of the vascular tissues in their specific arrangement in a plant or plant organ. In contrast to vascular cylinder, central cylinder, or stele, it does not include ground tissues.

Vascular tissue. A general term referring to either or both vascular tissues, xylem and phloem.

Vasicentric paratracheal parenchyma. In wood. Axial parenchyma cells forming complete sheaths around vessels. See also *paratracheal parenchyma*.

Vein. A strand of vascular tissue in a flat organ, as a leaf. Hence, leaf venation.

Vein rib. In leaf. A ridge of ground tissue occurring along a larger vein, usually on the lower side of the leaf.

Velamen. A multiple epidermis covering the aerial roots of some tropical epiphytic orchids and aroids. Occurs in some terrestrial roots also.

Venation. The arrangement of veins in the leaf blade.

Vertical parenchyma. See *axial parenchyma*.

Vessel. A tube-like series of vessel members the common walls of which have perforations.

Vessel element. See *vessel member*.

Vessel member. One of the cellular components of a vessel. Also *vessel element* and the obsolete *vessel segment*.

Wall. See *cell wall*.

Water vesicle. A type of trichome. An enlarged epidermal cell high in water content.

Wood. Secondary xylem.

Wound cork. See *wound periderm*.

Wound gum. Gum formed as a result of some injury. See also *gum*.

Wound periderm. Periderm formed in response to wounding or other injury.

Xeromorphic. Refers to structural features typical of plants (xerophytes) adapted to dry habitats.

Xylary procambium. The part of procambium that differentiates into primary xylem. Sometimes called xyloic procambium.

Xylem. The principal water-conducting tissue in vascular plants characterized by the presence of tracheary elements. The xylem may be also a supporting tissue, especially the secondary xylem (wood).

Xylem elements. Cells composing the xylem tissue.

Xylem fiber. A fiber of the xylem tissue. Two types are recognized in the secondary xylem, *fiber-tracheid* and *libriform fiber.*

Xylem initial. A cambial cell on the xylem side of the cambial zone that divides periclinally one or more times and gives rise to cells differentiating into xylem elements either with or without additional divisions in various planes. Sometimes called *xylem mother cell.*

Xylem mother cell. A cambial derivative that divides to produce certain elements of the xylem, such as axial parenchyma cells forming a parenchyma strand. Used also in a wider sense synonymously with *xylem initial.*

Xylem ray. That part of a vascular ray which is located in the secondary xylem.

Xylotomy. The anatomy of xylem or wood.

INDEX

(Numbers in bold face indicate illustrations located apart from the description of the subject in the illustration.)